Concentrating Solar Thermal Energy

SCIENCES

Energy, Field Directors – Alain Dollet and Pascal Brault

Renewable Energies, Subject Head – Abdelilah Slaoui

Concentrating Solar Thermal Energy

Fundamentals and Applications

Coordinated by
Gilles Flamant

iSTE

WILEY

First published 2022 in Great Britain and the United States by ISTE Ltd and John Wiley & Sons, Inc.

ISTE Ltd
27-37 St George's Road
London SW19 4EU
UK

www.iste.co.uk

John Wiley & Sons, Inc.
111 River Street
Hoboken, NJ 07030
USA

www.wiley.com

Library of Congress Control Number: 2022940279

British Library Cataloguing-in-Publication Data
A CIP record for this book is available from the British Library
ISBN 978-1-78945-079-8

ERC code:
PE8 Products and Processes Engineering
PE8_6 Energy processes engineering

Contents

Introduction

Gilles FLAMANT
PROMES-CNRS, Font-Romeu, France

Why a book on concentrated solar power?

Renewable energies play an important or even dominant role in the medium- and long-term energy strategy of most countries. In this context, any solution that may contribute to the emergence of a low-carbon world should be carefully considered. France, a country with a strong nuclear energy deployment, has also conducted groundbreaking research in the field of solar power, particularly concentrated solar power, which was not developed to the same level as nuclear power. Sixty-five years ago, at the end of World War II, Félix Trombe reversed the path of light on the parabolic spotlights of the anti-aircraft defense deployed by the German army around Paris and transformed them into solar concentrators. This is a magnificent symbol of revival, which in itself is sufficient to justify this book. However, it is obviously not the only reason. As a result of research and development, this technology reached the commercial stage for certain architectures of solar power plants, but concentrated solar power is still struggling to find its place next to wind power and photovoltaics. There are three main reasons for this delay, namely modularity, financing mechanisms and deployment area. Even though wind power and photovoltaics benefit from a positive scale effect, the latter is far more significant for concentrated solar power. Consequently, on the one hand, the manufacturing cost of small solar power plants (<10 MW) is not competitive unless cogeneration can be considered, and, on the other hand, building large power plants requires significant capital, therefore heavy financial expenses. The only modular systems adapted to small installation markets (for private consumers, for example) are the dish–Stirling systems whose operation and maintenance are more expensive

Concentrating Solar Thermal Energy,
coordinated by Gilles FLAMANT. © ISTE Ltd 2022.

than those of a PV system of equivalent power. This observation is related to a second point. The implementation of an attractive feed-in tariff for private consumers, then for ground power units, enabled the development of photovoltaics (accompanied by cost reduction) for a wide range of power and users. Equivalent mechanisms (negotiated feed-in tariff) were established for concentrated solar power in certain countries, but for a far narrower market. Indeed, the geographical area that is favorable to the deployment of this technology is smaller than that of PV. No one would consider installing solar thermal power plants in Germany, which was at the forefront of the introduction of photovoltaics on its territory! Finally, the multiplicity of possible configurations in which a solar power plant can be assembled from basic components is not only a strength, but also a weakness when considering the cost reduction associated with series production.

However, in this seemingly unfavorable context, concentrated solar power has its own place, in addition to wind power and photovoltaics. It offers solutions for low-cost bulk heat storage and can be deployed in fields of applications beyond electricity, such as industrial process heat and synthetic fuels (hydrogen, syngas, etc.). This book focuses on a presentation of various aspects of this solar technology, from components to the full system. It was written by researchers, professors, assistant professors (University of Perpignan Via Domitia – UPVD) and PhD alumni of PROMES (Processes, Materials and Solar Energy Laboratory) of CNRS, successors of Félix Trombe and Marc Foex, founders of the solar furnace of Odeillo-Font-Romeu. This book is structured into 11 chapters:

After a brief history, Chapter 1 presents the various existing technologies and their evolution perspectives in technical and economic terms.

Chapter 2 introduces the physical processes that condition the evolution of radiation intensity at the ground level, the measurement instruments (total and direct normal irradiation, DNI) and DNI prediction models.

Chapter 3 presents the physical laws of mirror-based concentration and the influence of associated defects on the performance of concentrators. It presents parabolic trough concentrators (linear concentration) and solar power towers (point concentration) as illustrations.

Echoing Chapter 3, Chapter 4 describes the receivers for parabolic trough and tower power plants. It details a physical model of a receiver for a tower power plant.

After a refresher of heat transfer physics, Chapter 5 presents fluid–wall heat transfer properties and coefficients of the main heat transfer fluids.

Chapter 6 is dedicated to the physical bases of transfers in turbulent flows and to the numerical developments required for the comprehension of complex interactions between the dynamic and thermal aspects of fluid flow.

Chapter 7 presents the spectral properties of bodies. Then, it describes materials adapted to various components of a solar power plant, and approaches the techniques for the evaluation of their service life.

Chapter 8 describes the storage technologies and the recent progress in thermocline storage, a key component of a modern solar thermal power plant. It also proposes a comparison of the costs of heat storage and electrochemical storage.

Chapter 9 analyzes and compares various concepts of hybridization between photovoltaic and heat converters, designed as combined solar energy conversion modes.

Chapter 10 describes the processes for the thermochemical production of hydrogen and synthetic gas from carbonaceous materials. This analysis is accompanied by technical and economic considerations.

Chapter 11 presents various cycles for thermochemical decomposition of H_2O and CO_2, as well as their development level. It analyzes the progress in terms of materials and processes.

1

Solar Power Plants: State of the Art

Gilles FLAMANT
PROMES-CNRS, Font-Romeu, France

1.1. Introduction

The concentration of the direct component of solar radiation allows the generation of heat at high temperature. This heat can be used directly in an industrial process or to generate mechanical energy or chemical energy carriers by thermochemical processes (Chapters 10 and 11). The focus is here on electric power generation. In this case, solar energy is transferred to a heat transfer fluid flowing through a solar receiver. Through a set of heat exchangers, this heat transfer fluid releases its heat to a working fluid that converts this thermal energy into mechanical energy through a thermodynamic cycle. Finally, a generator coupled to this power cycle produces electric power. The second part of this conversion chain is identical to that of a classical thermal power plant, but the source of primary energy is solar energy in the case of *concentrated solar power* (*CSP*) or *solar thermal electricity* (*STE*). These two terms will be used in this chapter.

By the end of 2019, the CSP installed in the world reached 6.3 GW for a generation of around TWh (IEA 2020a). The National Renewable Energy Laboratory (NREL) in the United States proposes a database containing the technical characteristics of various operating, under-construction or under-development solar power plants in the world (NREL 2020). Over 100 operating solar power plants with a power ranging between 1 and 280 MW are included in this database. With an installed power of 2.3 GW, Spain has the strongest development of this technology

Concentrating Solar Thermal Energy,
coordinated by Gilles FLAMANT.

to commercial stage. The list of countries with such power plants includes, in particular, the United States, Morocco, South Africa, Chile, some states in the Middle East (Dubai, Abu Dhabi, etc.), India and China. China has developed many demonstration projects based on various technologies since 2016.

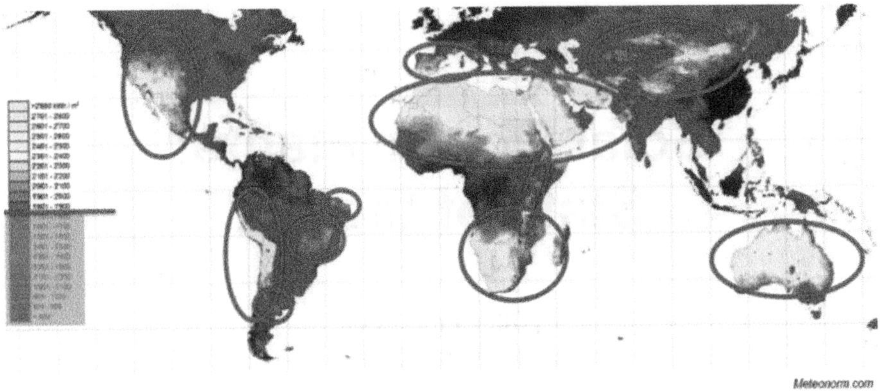

Figure 1.1. *Suitable geographical areas for concentrated solar power plant deployment (https://meteonorm.com/en/). For a color version of this figure, see www.iste.co.uk/flamant/energy.zip*

Since concentrating systems use only the direct component of solar radiation (DNI, Direct Normal Irradiation), the currently accepted minimal annual resource required for a solar power plant to become profitable is 1,800–2,000 kWh/m^2 per year. The suitable geographical areas are encircled in Figure 1.1 (see also Chapter 2). Europe, Spain, Portugal, Greece and some southern regions in Italy and in France meet this criterion. On the other hand, for the generation of industrial heat, this threshold may be significantly reduced, which explains the development of some applications in certain northern countries, such as Denmark (Aalborg 2020). Figure 1.2 illustrates this cogeneration project in Brønderslev (Denmark), which includes 16.6 MW$_{th}$ of parabolic trough concentrators.

The first section of this chapter presents a brief history of the development of solar power plants in the world. This is followed by a description of various concentrating technologies. Concentration makes it possible to reach high temperatures, while limiting radiation energy losses, which are proportional to the fourth power of temperature and to the receiving surface area. Section 1.4 presents the simple balance equations governing the evolution of the conversion efficiency of solar power plants. The following four sections are dedicated to the description of various solar power plant technologies. The implementation of linear concentrating systems leads to parabolic trough and linear Fresnel power plants, and the implementation of point concentrating systems allowed the development of tower power plants and dish–Stirling modules. Finally, the last section introduces considerations on the future deployment of solar thermal power plants, their integration in the grid, the evolution of generated electricity costs, the environmental impact of the technology and several recent developments that may be the future commercial CSP technologies.

Figure 1.2. *Solar field of heat generation in Brønderslev (Denmark). For a color version of this figure, see www.iste.co.uk/flamant/energy.zip*

1.2. History

The first prototypes of solar concentrators and their applications developed in the 18th and 19th centuries are reviewed in Chapter 3.

It can be considered that the first solar mechanical power demonstrator was developed at the beginning of the 20th century. In 1912, Frank Schuman (author of many patents) built at Maadi, in Egypt, a pumping system using steam generated by solar energy. This was a 60 m long parabolic trough concentrator with a 4 m aperture. The generated steam set in motion a 60 kW motor that pumped 1,400 m^3/h water used for irrigation. Because of the strong development of coal and oil industries, the efforts of the pioneers were unfortunately ignored. In France, it was not until the end of World War II that the interest in solar concentrated power was renewed (in 1946). Though his research focused on high-temperature chemistry, in

1949, Félix Trombe proposed in the *Science et Vie* magazine the construction of solar power plants in the desert. Baum et al. (1957) published the results of optical and thermal calculations for the evaluation of the performances of a steam-generating tower power plant. Finally, the works of Professor G. Francia in Genoa (Italy) in 1961 are also pioneering contributions (Royère 1981). His installation of about 100 kW_{th} demonstrated the capacity of concentrating systems to generate superheated steam. The development of solar power plants was motivated by the first oil crisis in 1973, following the Kippur war, which challenged the access to cheap oil as a basis for the economic growth of western countries. After this event, CNRS initiated an interdisciplinary research program on solar energy (PIRDES) in 1975 under the leadership of Robert Chabbal, founder of Engineering Sciences at CNRS and General Director of this research institution between 1976 and 1979 (Guthleben 2009). In this context, the French Solar Energy Authority (COMES) was established in 1978 (and dissolved in 1983) (Ferri 2001). At the international level, the International Energy Agency (IEA) initiated the Small Solar Power System (SSPS) Implementing Agreement (IA) in 1977, which is currently known as SolarPACES (SolarPACES 2020). This initiative materialized in the construction of several tower power plant demonstrators at the beginning of the 1980s in Spain, Italy, France, Japan, Russia and the United States (even if France was not part of the SSPS IA), implementing several architectures (heat transfer fluid, storage, thermodynamic cycle). This is the context in which the Thémis solar power plant (2.5 MW_e, 2 h storage), the first solar power plant using molten salt as heat transfer and storage fluid, was put into operation in 1983 and operated only for three years (Figure 1.3). In 1986, the Messmer plan (1974) for the deployment of nuclear power plants provided a national solution to the energy crisis and at the same time stifled the development of solar energy in France. Thémis was taken as a reference for the 10 MW_e solar power plant with molten salt Solar Two (after Solar One, 1982–1985), which operated in California at Barstow between 1997 and 2000.

IEA initiative allowed the construction of a demonstrator of a tower solar power plant and of a parabolic trough solar power plant on the reference site of SSPS program in Spain, near Almeria, which has since been known as Almeria Solar Platform (PSA). Similarly, the Thémis site became a pole of innovation of solar technologies and of experimentation at the prototype scale of new concepts of solar power plants.

By the end of the 1970s, the solar furnace at Odeillo (Figure 1.4) was the only high power installation in the world that had the capability to test prototypes of components, such as solar receivers (Royère 1981). The 5 MW_{th} experimental solar tower of Albuquerque (United States), operated by *Sandia National Laboratory* (SNL), dates from 1977. A first solar pilot power plant of 60 kW_e using oil as heat transfer fluid (see Chapter 5) was tested at the solar furnace in 1976–1977. This testing campaign was preceded by the tests, using the same installation, of a solar receiver prototype designed by the American company Martin Marietta that generated superheated steam at above 500°C.

Figure 1.3. *Thémis, operating solar power plant (Targasonne, 66). For a color version of this figure, see www.iste.co.uk/flamant/energy.zip*

The first commercial deployment of solar power plants was conducted by Luz company between 1985 and 1991. This company built and operated nine parabolic trough solar power plants in the Mojave Desert (California) for a total power of 354 MW_e (see section 1.5). These power plants, known as Solar Energy Generating Systems (SEGS), are still functioning in 2021, being operated by Kramer Junction Company (KJC). They served as industrial reference for the developments achieved after 2000. This industrial logic partly explains the choice of parabolic trough technology for the majority of Spanish solar power plants.

In the absence of fiscal incentives, no new commercial solar power plant was built for 25 years (1991–2006). Nevertheless, thanks to European and national funding, research continued despite this "solar eclipse" in Spain and Germany. In the United States, the national laboratories of the Department of Energy (DoE)

(NREL and Sandia) have also contributed to this research effort in collaboration with the industry. They facilitated the development of high-performance components and of innovative configurations of power plants.

Figure 1.4. *Solar furnace of CNRS at Odeillo-Font-Romeu. In the foreground, the dish–Stirling module being tested. For a color version of this figure, see www.iste.co.uk/flamant/energy.zip*

Starting in the mid-2000s, concerns related to climate change surpassed the issues associated with the price of oil and fossil fuel reserves. In 2006, the 64 MW_e Nevada Solar One (NSO) power plant, whose design is similar to those of SEGS power plants, was put into operation in the United States. At the same time, the proposal of preferential feed-in tariffs in certain European countries, such as Spain, led to a commercial revival of the sector. In 2007, the tower solar power plant PS10 with direct steam generation was put into operation by the company Abengoa near Seville. A first significant innovation was made with ANDASOL power plants (Guadix, Andalousia) starting in 2008: the integration of molten salt heat storage allowed electric power generation for 7.5 h at full capacity (50 MW_e). In 2011, the tower power plant Solar Tres (Gemasolar) of 19.9 MW_e with a 15 h storage at full capacity inaugurated the deployment of high capacity tower power plants with molten salt (150 MW_e for NOOR 3 in Morocco).

Due to the integration of heat storage, solar thermal power plants are highly appreciated by the power grid regulators for their dispatchability, which is the possibility to control generation depending on demand. Storage allows an increase in capacity factor. The capacity factor is the ratio of the electric power actually generated over a given period of time and the power generated by a power plant operating at nominal capacity during the same period (see section 1.9). Generally (except for peak operating power plants), the energy required to load the storage is provided by increasing the surface of the mirrors with respect to the surface required for nominal capacity generation. The criterion for qualifying this increase is known as solar multiple. The classical values of the solar multiple range between 1.5 and 2.5.

1.3. Various configurations of solar power plants

A logical way to present the configurations of solar power plants involves their classification according to two criteria, the concentration method and the mode of integration of the solar receiver with respect to the concentrator. Consequently, there are linear and point concentration technologies, and fix and mobile receivers. Figure 1.5 summarizes this classification.

In Figure 1.5, the concentration factor is denoted by C (see the definition in Chapter 5). Linear concentrators operate mobile optical systems along a single axis, generally with north–south orientation, and produce concentrations below 100. Point concentrators use mobile mirrors along two axes, in azimuth and elevation, in order to follow the Sun's path. Consequently, the concentrations obtained are much higher than those of linear concentrators, from several hundred to several thousand. The highest concentrations are obtained with paraboloid systems (of the order of 10,000). The elementary concentrator for tower power plants is referred to as heliostat (which gives a fixed image of the Sun).

Figure 1.6 shows images of experimental or commercial solar concentrators. The ranges of temperature shown in this figure are the maximal operating temperatures considered for these concentration systems. The common maximal operating temperatures are 400°C for linear concentrators, 565°C for tower power plants and of approximately 700°C for dish–Stirling modules. These latter devices are, by design, modular, as considering parabolic systems of over 500 kW_{th} at the industrial scale is unrealistic.

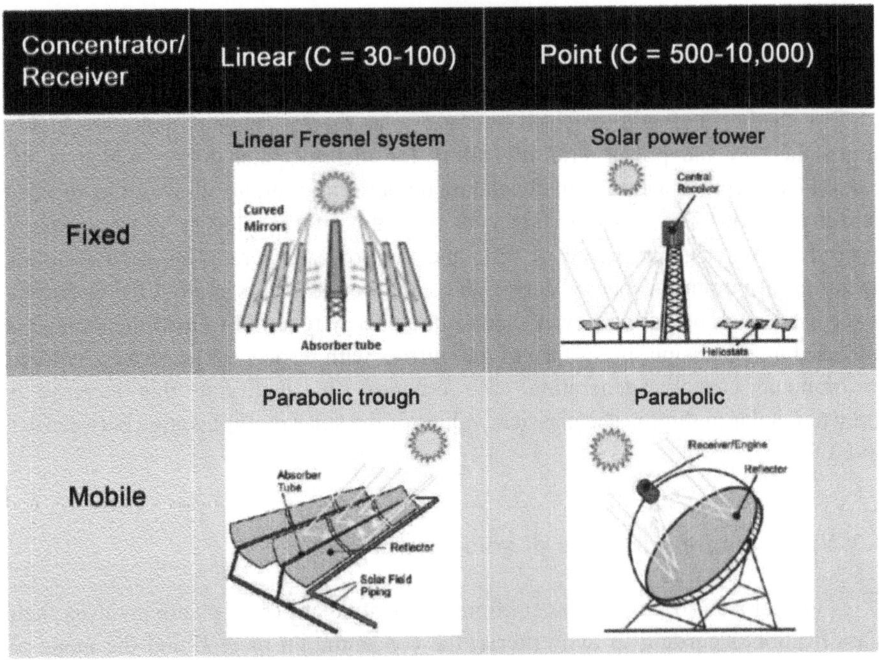

Figure 1.5. *Classification of solar power plants (Solar PACES 2020). For a color version of this figure, see www.iste.co.uk/flamant/energy.zip*

The previous figures show that all the concentration technologies developed at the industrial scale rely on principles of simple reflection. There are other principles allowing in particular the creation of a fixed focus. The double reflection solar furnace is a solution for obtaining very high concentrations (>10,000) and a horizontal incident beam. They are well adapted to research laboratories (see Chapter 3).

Beam-down devices are used to redirect the concentrated solar beam downward. They use an optical system with three reflections, as indicated in Figure 1.7. This optical system makes it possible to place the receiver on the ground. A compound parabolic concentrator (CPC) is generally placed at the entry of the solar receiver in order to increase the concentration affected by the first two reflections. The limits of this type of concentration optics were studied by Vant Hull (Vant Hull 2014).

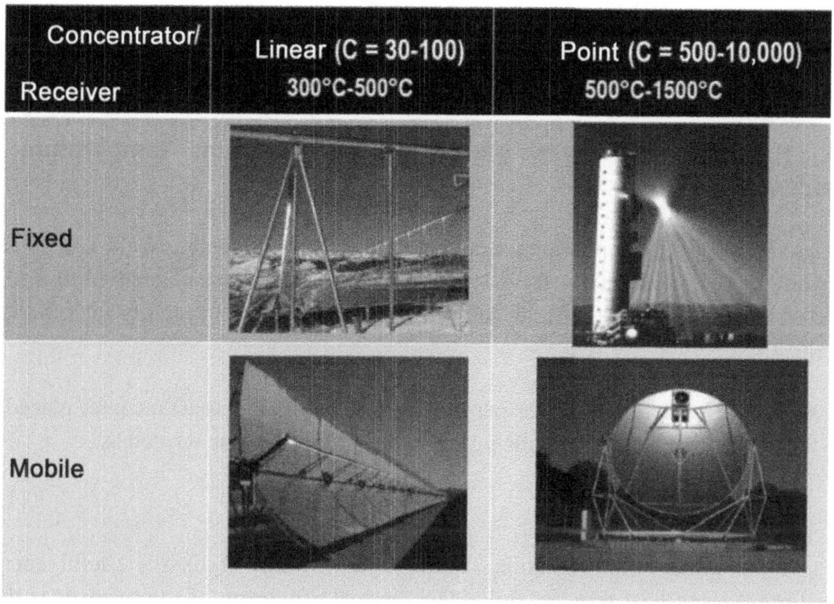

Figure 1.6. *Examples of various solar concentrators. For a color version of this figure, see www.iste.co.uk/flamant/energy.zip*

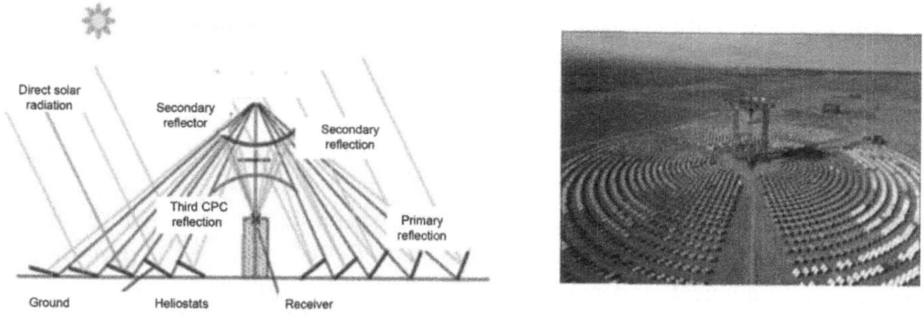

Figure 1.7. *On the left, the beam-down concentration optics principle adapted from Gordon and Feuermann 2019 and on the right the 17 MW_{th} Yumen Xinneng industrial installation (China). For a color version of this figure, see www.iste.co.uk/flamant/energy.zip*

The developments of solar power plants corresponding to these four concentration technologies are presented in sections 1.5–1.8. Chapter 4 presents the associated solar receivers. The next section is dedicated to thermodynamic considerations.

1.4. Paradigm of solar power plants, optimum temperature – concentration factor

As indicated in the previous section, the operating temperatures in CSP plants depend on the concentration factor. A simplified thermal balance applied to a solar receiver coupled with a thermodynamic engine can be used to readily interpret this result.

A thermal balance (in power per unit area) applied to a solar receiver placed at the focus of a concentrator producing a concentration C can be written as:

$$CI = Q_u + Losses \quad [1.1]$$

I is the direct solar radiation (DNI, in W/m^2) and Q_u is the useful energy transferred to the heat transfer fluid. The thermal efficiency of the receiver is then expressed by:

$$\eta_r = \frac{CI - Losses}{CI} \quad [1.2]$$

The heat losses are the sum of three contributions, the reflection of the incident solar radiation, the convection between the wall of the receiver and the ambient air and the infrared emission at the temperature of the receiver, T:

$$Losses = \rho CI + h_{conv}(T - T_{amb}) + \varepsilon\sigma(T^4 - T_{amb}^4) \quad [1.3]$$

where ρ, ε and σ are the reflectivity in the solar spectrum, the emissivity in the IR spectrum and the Stefan–Boltzmann constant ($5.67.10^{-8}$ $W.m^{-2}.K^{-4}$), respectively, and h_{conv} is the ambient air convection heat transfer coefficient.

In order to simplify the solution to this balance, the following hypotheses can be made: convection losses are significantly lower than radiative losses and are therefore neglected; the solar receiver is a gray body; therefore, α= ε=1-ρ, where α is the absorptivity in the solar spectrum. Under these conditions, the following can be written as:

$$\eta_r = \alpha - \frac{\varepsilon\sigma(T^4 - T_{amb}^4)}{CI} \quad [1.4]$$

Equation [1.4] shows the dominant influence of the absorptivity in the solar spectrum. For a black body, α= ε=1, equation [1.4] gets simplified as:

$$\eta_{r,BB} = 1 - \frac{\sigma(T^4 - T_{amb}^4)}{CI} \quad [1.5]$$

Equations [1.4] and [1.5] clearly show that the thermal efficiency of the receiver decreases with temperature.

When the receiver is coupled with a thermodynamic engine, the heat into the electricity conversion efficiency is limited by Carnot efficiency:

$$\eta_c = 1 - \frac{T_{amb}}{T} \quad [1.6]$$

This efficiency increases with temperature. As a last simplification, the temperature of the receiver surface and the operating fluid temperature are assumed identical, in order to have a clear picture of the main trends of evolution of the efficiency of the conversion system with temperature and concentration.

The simplified overall efficiency of the device for converting solar radiation into electric power η_{ov} is then expressed as follows:

$$\eta_{ov} = \eta_r . \eta_c = \left(\alpha - \frac{\epsilon\sigma(T^4 - T_{amb}^4)}{CI}\right).(1 - \frac{T_{amb}}{T}) \quad [1.7]$$

As the first term of equation [1.7] decreases and the second one increases with temperature, the overall efficiency exhibits an optimum that varies with the concentration C.

In order to illustrate this tendency, Figures 1.8 and 1.9 present the evolution of the overall efficiency of solar heat to electric power conversion for two reference cases: linear concentrator (low concentrations) and point concentrators (high concentrations). These examples show, on the one hand, the influence of the optical properties of the receiver (α and ε) and the hypotheses on the thermodynamic conversion efficiency on the other hand.

Figure 1.8 shows the critical importance of the optical properties of the absorber for the system efficiency (see Chapter 7). The following two situations are compared: a black body-like absorber (α = ε = 1) and an absorber coated with a selective material (α = 0.95 and ε=0.1). The comparison is striking in terms of the shift of maximum efficiency towards high temperatures and the increase of this efficiency.

Considering the example of a concentration of 50, the position of maximum efficiency shifts from 600 K to about 900 K, and the conversion efficiency increases from 42% to 58%. Specifically, the consequence of this result is that the parabolic trough solar power plants could not have been developed in the absence of high-performance selective coatings (see Chapter 7). Finally, the intersection of curves with the abscissa corresponds to the stagnation temperature for which no energy can be extracted from the system, as it is in radiative thermal equilibrium.

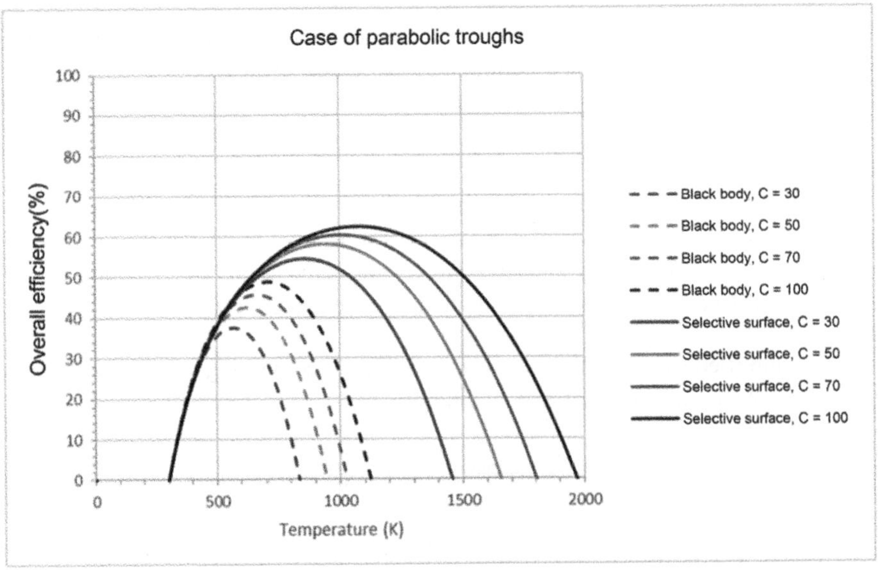

Figure 1.8. *Heat to electric power conversion efficiency for linear concentrators, selective surface: α = 0.95 and ε = 0.1; I = 900 W/m^2. For a color version of this figure, see www.iste.co.uk/flamant/energy.zip*

For point concentration systems, such as tower power plants, there is no selective coating that resists at air temperatures above 500°C. Figure 1.9 compares the conversion efficiency for two efficiency hypotheses, the Carnot efficiency (equation [1.6]) and a more realistic efficiency (referred to as the "corrected Carnot efficiency" in the figure), $\eta_{cc} = (1 - \sqrt{\frac{T_{amb}}{T}})$, (Askarov 2010).

For a concentration of 700, the efficiency is maximum around 1,000 K, which is close to the temperature of the receiver of tower power plants with molten salt. Higher-temperature (1,100–1,200 K) developments require an average concentration of 1,000 and more. The position of the maximum is practically not affected by the hypothesis on the Carnot efficiency. On the other hand, it decreases from 60% to 40%.

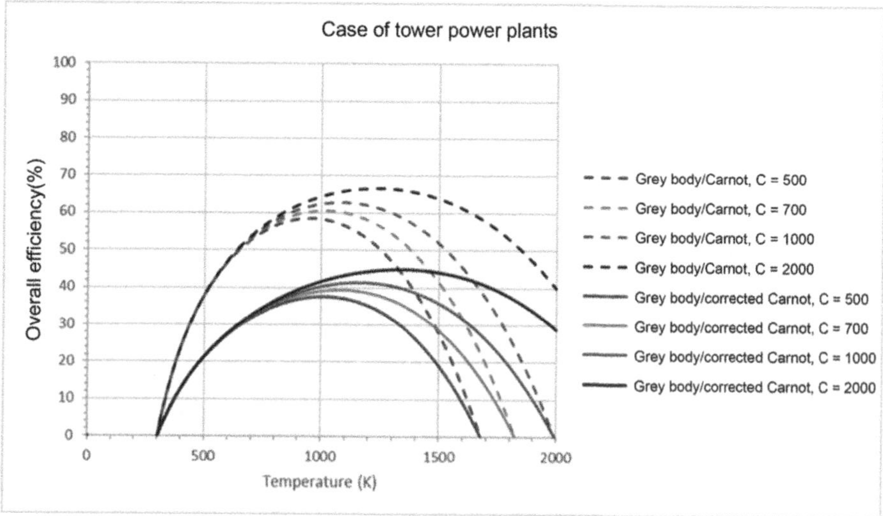

Figure 1.9. *Solar heat to electric power conversion efficiency for point concentrators, comparison of conversion efficiency for two hypotheses of converter efficiency formulation, gray body: $\alpha = \varepsilon = 0.9$; $I = 900\ W/m^2$. For a color version of this figure, see www.iste.co.uk/flamant/energy.zip*

Despite the hypotheses made for this calculation, it allows the estimation of the overall nominal efficiency of a tower power plant. If the heat losses of the conversion chain are considered negligible, this overall efficiency can be written as:

$$\eta_G = \eta_o.\eta_r.\eta_c \qquad [1.8]$$

η_o is the optical efficiency of the concentration system. For an average value of the optical efficiency of 65% and estimated losses of 10%, the overall nominal efficiency for the conditions illustrated in Figure 1.9 (C= 700 and corrected Carnot) is of 23%. This value is representative, but slightly overestimated with respect to reality; nominal efficiency of about 20–21%, in particular due to the uncertainty on the efficiency of the receiver and the heat losses in the conversion chain.

1.5. Parabolic trough solar power plants

As indicated in section 1.3, the parabolic trough solar power plants (PT) were the first to be developed commercially. The nine SEGS (354 MW_e) power plants built in California between 1985 and 1991 rely on an identical concept: power plants without storage with supplementary gas (hybridization) at a maximum share of 20% (solar fraction of 80%) with respect to the total thermal energy input. The principle is schematically represented in Figure 1.10.

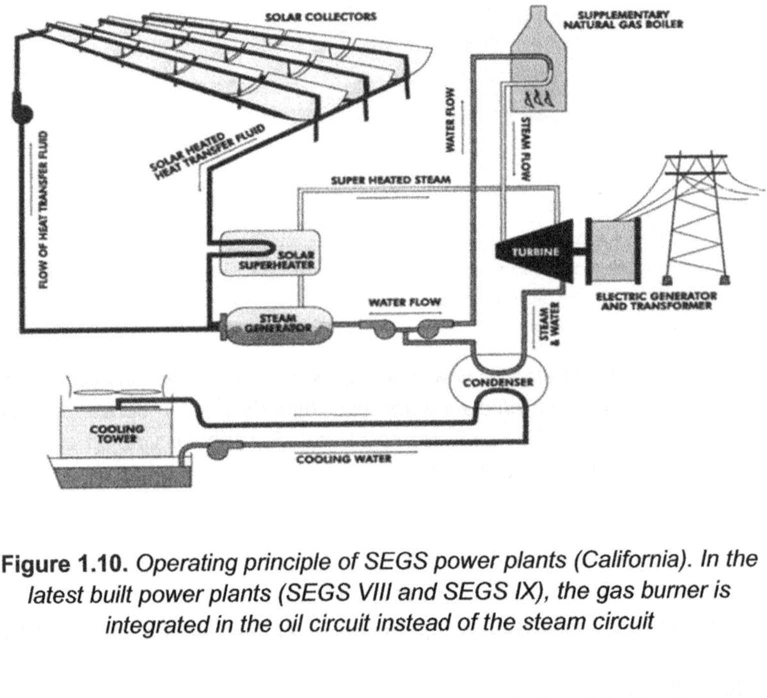

Figure 1.10. *Operating principle of SEGS power plants (California). In the latest built power plants (SEGS VIII and SEGS IX), the gas burner is integrated in the oil circuit instead of the steam circuit*

The heat transfer fluid is synthetic oil. The industrial development was conducted in stages concerning, on the one hand, the power and, on the other hand, the operating temperature. SEGS I has a power of 13.8 MW_e and operates with a fluid exiting the solar field at 307°C, while the power of SEGS VIII and SEGS IX is 80 MW_e operating at a temperature of 390°C, which is the nominal temperature of current power plants.

The solar power plant SHAM 1 (100 MW_e) inaugurated in 2013 at Abu Dhabi presents a first evolution compared to SEGS power plants. A double hybridization makes it possible, on the one hand, to compensate the absence of Sun and, on the other hand, to increase the temperature of the superheated steam in order to improve the conversion efficiency of the thermodynamic cycle. Despite this fossil fuel supplement, the solar fraction is still above 80% (82%) under nominal conditions (the solar fraction is the fraction of the total generated energy provided by the solar energy.). Figure 1.11 illustrates the design of the power plant.

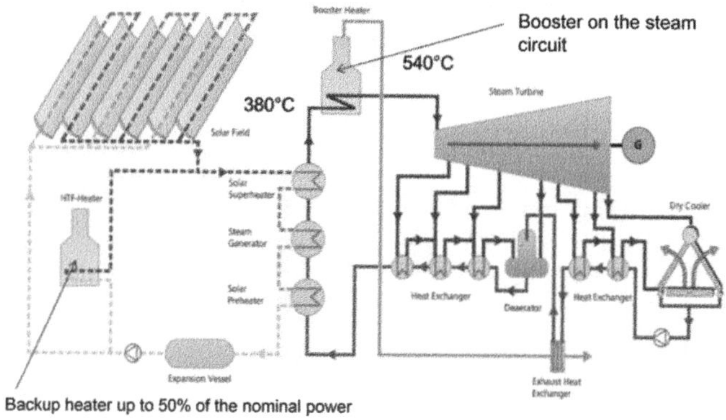

Figure 1.11. *Double hybridization architecture of SHAM 1 power plant. For a color version of this figure, see www.iste.co.uk/flamant/energy.zip*

The recent PT solar power plants integrate a molten salt storage. This innovation was introduced starting with 2008 on the ANDASOL commercial plant. The architecture of a PT solar power plant with molten salt storage is schematically represented in Figure 1.12.

As can be seen in Figure 1.12, three fluids are implemented: synthetic oil as heat transfer fluid in solar receivers, molten salt in storage tanks and steam as working fluid. The molten salt flows only between the hot and cold tanks. It exchanges heat with the oil coming from the solar field (charge) or going to the thermodynamic cycle (discharge). In this configuration, the temperature difference between the hot tank and the cold tank is 85°C, which involves the immobilization of a large quantity of salt per MWh stored by sensible heat. For example, 28,500 tons of solar salt (solar salt is a eutectic mixture of 60% sodium nitrate – $NaNO_3$ – and 40% potassium nitrate –KNO_3 – in weight) is necessary to store the energy required for a full capacity power generation for 7 h +30, ~1 GWh (Figure 1.13). This quantity is reduced by a factor of 3 in tower power plants that operate at higher temperature. The thermodynamic cycle is a Hirn cycle operating at 370°C and 100 bars with an efficiency of 37.4%.

The SOLANA solar power plant (Arizona) of 250 MW_e net is the parabolic trough solar power plant CP with the highest power operating since 2013, with 6 h storage at full capacity. The solar field has a surface area of 2,200,000 m^2, and the power block consists of two turbines of 140 MW_e. The storage system (~4 GWh thermal) involves six modules composed of a hot tank and a cold tank. The plant was dimensioned to generate 944 GWh/year.

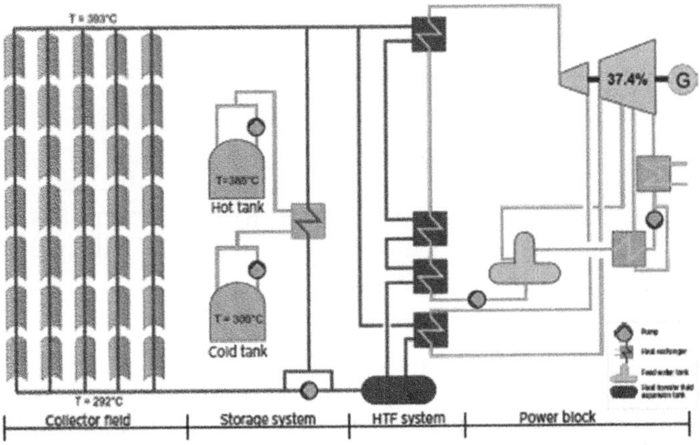

Figure 1.12. *The various sub-systems of a parabolic trough solar power plant with molten salt storage. Source DLR. For a color version of this figure, see www.iste.co.uk/flamant/energy.zip*

Figure 1.13. *Solar power plants ANDASOL 1.2 and 3 of 50 MWe (Spain). For a color version of this figure, see www.iste.co.uk/flamant/energy.zip*

A characteristic of parabolic trough solar power plant CP is the hydraulic circuit of the heat transfer fluid, as it comprises kilometers of solar receivers and pipes for the supply of receivers with the heat transfer fluid and the extraction of collected heat (Figure 1.14).

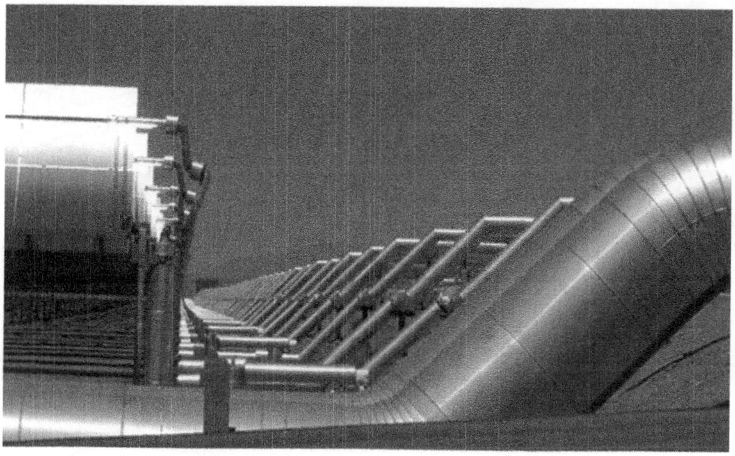

Figure 1.14. *A part of the hydraulic circuit of NOOR 1 power plant (Morocco). For a color version of this figure, see www.iste.co.uk/flamant/energy.zip*

Overall, oil flow is designed to maximize the collected thermal power and to minimize the capacity losses. The architecture of parabolic trough power plants is similar for all the installations; the solar field is square, having at its center the heat storage and power block. The heat collection system is modular. Three levels can be identified: elementary collector, solar collecting element (SCE) and solar collector assembly (SCA). The SCE is composed of series manufactured elementary collectors, generally 4 m long. The collector assembly comprises N (8–12)-aligned modules supported by pylons. The collector assembly is a basic unit, as a central pylon equipped with a motor tracks the Sun. Depending on the manufacturer, the length of a collector assembly ranges between 100 and 190 m. For example, a collector assembly composed of 12 modules comprising 3 elementary collectors has a total length of 144 m (Figure 1.15).

The solar field is built by arranging the collector assemblies into lines of length ranging between 300 and 400 m. The resulting stacks are paired to make a loop linked to the hydraulic circuit. Hence, the hydraulic circuit is designed to optimize the architecture of connections between the solar energy collection loops and the main pipes for the supply of cold heat transfer fluid and the extraction of hot fluid. H or I-shaped architectures are possible (Figure 1.16).

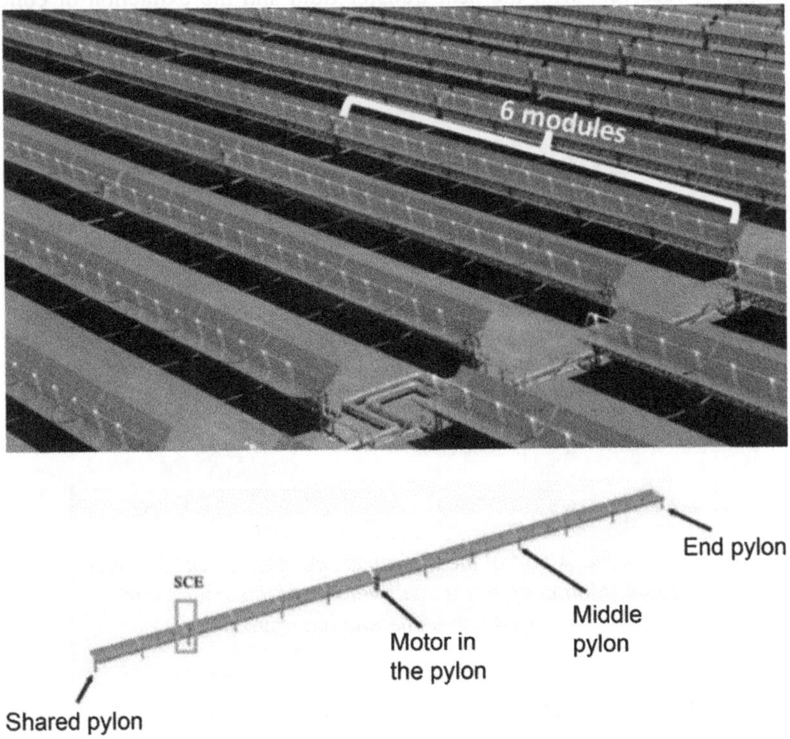

Figure 1.15. *Example of solar collector assembly (SCA) in the parabolic trough solar power plant. For a color version of this figure, see www.iste.co.uk/flamant/energy.zip*

The specificity of the parabolic trough solar power plants is the overall movement of the concentrator and of the solar receiver related to the optical principle of this concentration mode. The incident radiation must remain parallel to the symmetry plane of the concentrator. Therefore, the solar receiver must turn with the concentrator. From a technological standpoint, this specificity involves the existence of mobile connections at the end of the stack of collectors (Figure 1.17). These connections are the cause of possible leaks of heat transfer fluid. Consequently, they require vigilance within the maintenance of these solar power plants. The characteristics of solar receivers for PT are presented in Chapter 4.

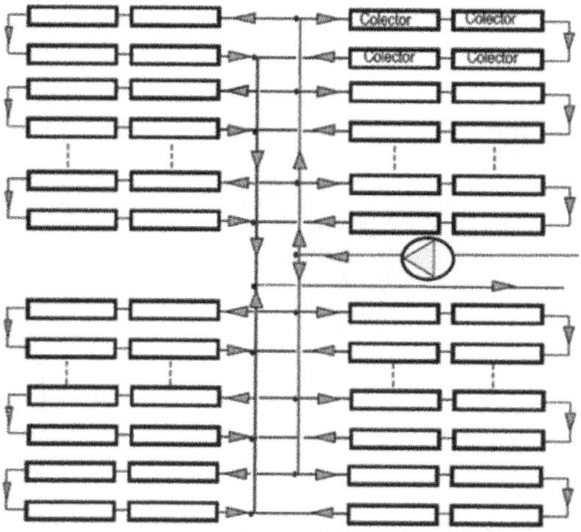

Figure 1.16. *I-shaped architecture of the hydraulic circuit of a PT solar power plant. For a color version of this figure, see www.iste.co.uk/flamant/energy.zip*

Figure 1.17. *Mobile connections at the end of stacks of collectors (NOOR 1 power plant, Morocco). For a color version of this figure, see www.iste.co.uk/flamant/energy.zip*

Finally, the technology of elementary collectors evolves, in particular towards an increase in the aperture of concentrators (Figure 1.18). The aperture of classical PT concentrators is 5.77 m, but some solar power plants (NOOR 2, for example) are equipped with 6.78 m aperture concentrators, and a version with 7.51 m aperture was also manufactured (Estela 2020).

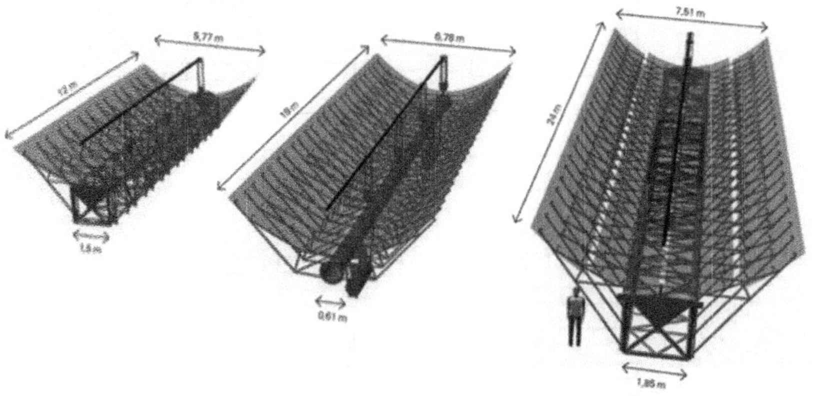

Figure 1.18. *Evolution of the size of parabolic-trough concentrators. For a color version of this figure, see www.iste.co.uk/flamant/energy.zip*

1.6. Solar power plants with linear Fresnel concentrators

In France, the first linear Fresnel concentrator (LFC) was developed in 1964 by G. Francia at Marseille (Francia 1968). The concentration principle is the main but not the only difference between solar power plants with LFC and with parabolic-trough concentrators. As schematically represented in Figure 1.19, the concentration is achieved on a fixed focal area by concave long mirrors moving along the axis of the receiver.

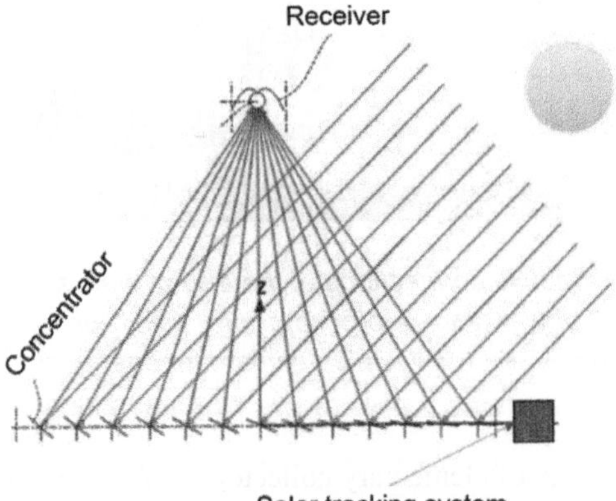

Figure 1.19. *Principle of linear Fresnel concentrators (Ko 2019). For a color version of this figure, see www.iste.co.uk/flamant/energy.zip*

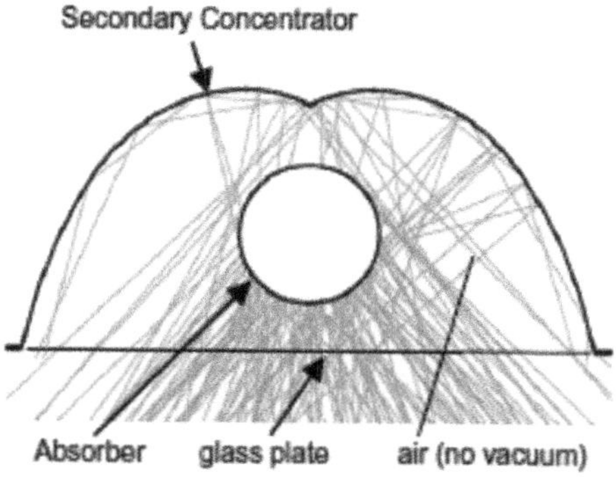

Figure 1.20. *Cross-section of a receiver for an LFC solar power plant, example (Morin et al. 2012). For a color version of this figure, see www.iste.co.uk/flamant/energy.zip*

The resulting concentration is below that of PT concentrators (10–30), and the receiver is composed of one or several absorber tubes inserted in a cavity generally having a secondary concentrator (Figure 1.20).

The optical principle of LFC results in a lower optical efficiency compared to that of PT concentrators. A comparison of the two systems on the same site was published by Morin et al. (2012). The yearly optical efficiency of PT concentrators is 57% and 41% for LFC, in this example. Even though the absolute values may vary depending on the technological options, the gap between the two values remains substantially constant at about 30%. Finally, LFCs have the advantage of simple and compact design, without hot mobile part (fixed receiver), but at the cost of low optical efficiency, especially in the morning and in the evening (cosine effect; see Chapter 5). LFC solar power plants generally operate in the direct steam generation mode. The steam generated in the receivers is directly sent to the turbine. A "steam storage" (pressurized water actually) can be added to the system.

At the industrial level, the 30 MW_e PE2 power plant installed at Murcia (Spain) in 2012 is the first commercial power plant put into operation by NOVATEC Solar (later FRENEL, Frenel 2020), generating steam at 270°C. The Dhursar power plant (Rajasthan, India) put into operation in 2018 is the largest existing unit with 100 MW_e power (EurObserv'ER 2019).

The only French CSP plant uses LFC technology, being a 9 MW_e "eLlo" power plant with 4 h storage, which is operational since 2018 in Cerdagne (Eastern Pyrenees) several kilometers away from the solar furnace and Thémis solar tower. It was built by SUNCNIM Company. The principle is schematically presented in

Figure 1.21. Steam is generated in the receivers at 286°C and 70 bars. The steam is either directly sent to the turbine or stored in nine pressurized tanks.

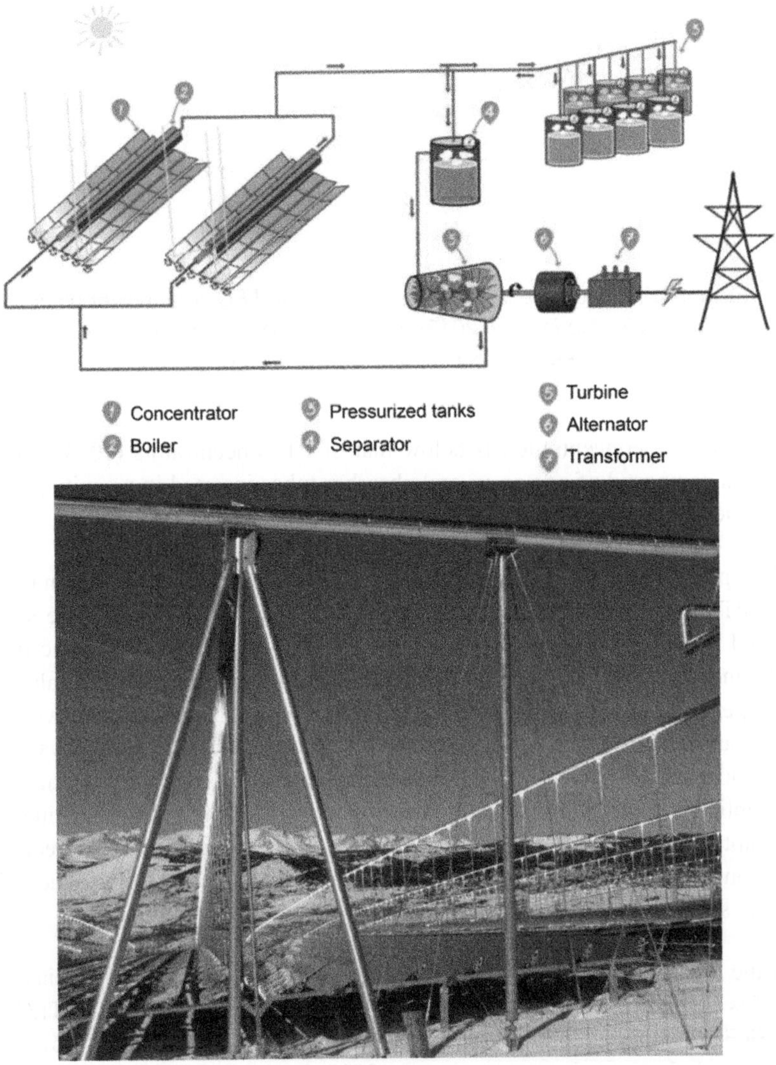

Figure 1.21. *Operating principle of eLlo LFC power plant and image of the collectors. For a color version of this figure, see www.iste.co.uk/flamant/energy.zip*

In terms of performances, the technology is disadvantaged by the temperature level (<300°C) which does not allow a high conversion efficiency, and by the direct generation of steam, because the low-cost bulk storage of heat is difficult. A technological breakthrough is under development. It involves changing the heat transfer fluid and using solar salt (see Chapter 5) at 500°C, instead of steam. This innovation may provide a solution to the previously mentioned limitations. A 2 MW_{th} demonstrator is being tested (Frenel 2020).

1.7. Tower power plants

The first commercial tower power plant PS10 (11 MWe) was put into operation in 2007 in Sanlûcar near Seville (Spain). Similar to its big sister PS20, it generates saturated steam at 250°C and 45 bars. This very conservative choice of steam characteristics made by the construction company Abengoa is explained by the lack of industrial reference at that time, which curbed any innovation due to financing constraints (performance guarantee). Further tower power plants with direct steam generation were subsequently built, particularly the three tower power plants of Ivanpah project totaling up to 392 MW_e (built by BrightSource Energy, Primm, Nevada, USA), as illustrated in Figure 1.22.

Figure 1.22. *Ivanpah solar power plant, three 130 MW_e towers (source: BrightSource Energy). For a color version of this figure, see www.iste.co.uk/flamant/energy.zip*

Due to the evolution of long-term storage needs, tower power plants with molten salt (solar salt) presented an advantage, namely the integration of high-performance bulk thermal storage.

After the demonstration projects of Thémis (France) and Solar Two (USA), the first commercial power plant of 19.9 MW_e, Solar Tres (Gemasolar), was put into operation in 2011 near Écija, in Andalusia (Spain) (Figure 1.23).

Figure 1.23. *Gemasolar solar power plant, Spain, source SENER (Sener 2020). For a color version of this figure, see www.iste.co.uk/flamant/energy.zip*

The 15 h heat storage at full capacity involves 8,500 tons of solar salt (670 MWh). This storage allows continuous power generation during summer months. The field of heliostats is composed of 2,650 heliostats for a total glass surface of 310,000 m^2. The thermal power of the receiver is 120 MW_{th} (Sener 2020).

The architecture of solar tower power plants with molten salt is illustrated in Figure 1.24. The transfer and storage fluid is the same molten salt, and the operating fluid is superheated steam at 540°C and 100–150 bar. The efficiency of the corresponding thermodynamic cycle is 42.8%. Hot salt is stored at 565°C (maximum) and cooled to 290°C in the exchangers, which means a temperature difference of 275°C. This difference is the triple of that implemented in parabolic trough power plants. The specific storage capacity per unit mass follows the same ratio.

After this first successful commercial experiment, which could be achieved thanks to the preferential feed-in tariff implemented in Spain and with European support, the field developed towards building higher capacity solar power plants, in order to benefit from scale effect. The failure of Crescent Dunes project (110 MW_e) conducted by the US company SolarReserve showed the importance of a gradual increase in power compatible with the technological difficulties of this type of power plants.

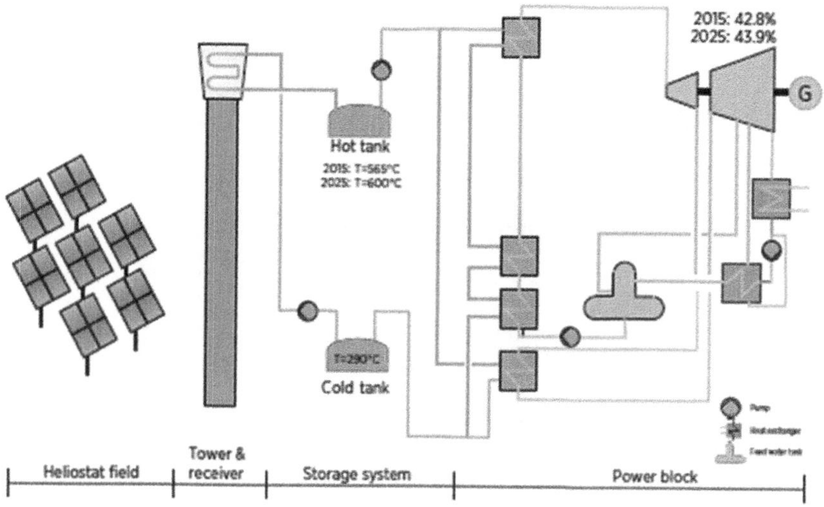

Figure 1.24. *Architecture of tower power plants with molten salt. Source: DLR. For a color version of this figure, see www.iste.co.uk/flamant/energy.zip*

The incident heat flux densities on the solar receiver are an order of magnitude higher than those received by the tubes of parabolic trough power plants. With an average value of solar radiation of 500 kW/m^2 and a maximum of 1 MW/m^2 on the receiver tubes, the control of the heliostat field becomes essential to prevent exceeding the threshold of constraints acceptable by the materials. Furthermore, the management of salt in order to avoid its solidification is also a technological difficulty. The NOOR 3 solar power plant of 150 MW_e and 7.5 h storage at full capacity, put into operation by the end of 2018 on the Ouarzazate site (Morocco), is presently the most powerful solar tower power plant with molten salt. It was built by the ACWA Power/SENER/MASEN consortium. The solar field is composed of 7,400 heliostats of 178.5 m^2. China has recently put into service several solar power plants with molten salt, particularly Delingha power plant (50 MW_e, 7 h storage at full capacity) built by Zhejiang Supcon Solar Technology starting in early 2019 (Figure 1.25).

To facilitate the understanding of the scale change stakes in engineering terms, Figure 1.26 represents the evolution of the thermal power of the solar receiver from the Thémis demonstrator (10 MW_{th}) to the NOOR 3 solar power plant (600 MW_{th}). The power of the solar receiver and of the solar power plant was multiplied by 60 and the storage duration by 4. The characteristics of the solar receivers for solar tower power plants are presented in Chapter 4.

Figure 1.25. *50 MW_e solar power plant in Delingha (China). Source: HELIOSCSP/Supcon. For a color version of this figure, see www.iste.co.uk/flamant/energy.zip*

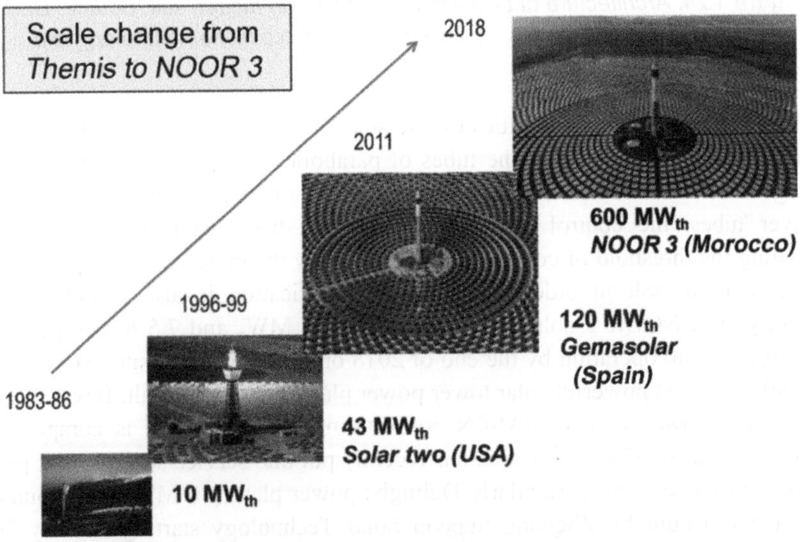

Figure 1.26. *Evolution of the thermal power of the solar receiver from Thémis to NOOR 3. For a color version of this figure, see www.iste.co.uk/flamant/energy.zip*

Except for the power and the heat transfer fluid, the solar tower power plants differ by the shape of the heliostat field, the unit surface of the heliostats and the height of the tower.

The location of the heliostat field is either south or north of the tower or circular around the tower (Figure 1.27). The solar fields extending towards the north, east and west with respect to the base of the tower (e.g. PS10 and PS20) are used for power plants of average power. They have good overall optical efficiency (see Chapter 3), and they are preferable for solar power plant implementation at high latitudes. Circular fields are chosen for high power units and geographical locations near the tropics. Nevertheless, the optical efficiency differs between the northern and southern parts (and depends on the hemisphere).

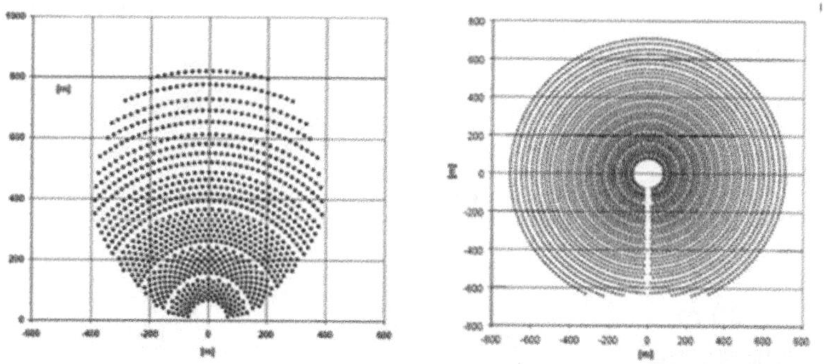

Figure 1.27. *Configuration of the heliostat fields of solar tower power plants. For a color version of this figure, see www.iste.co.uk/flamant/energy.zip*

Concerning the unit surface of the heliostats, there is no consensus among manufacturers. It varies from 15 m^2 (Brightsource Energy) to 178 m^2 (SENER) with an intermediate scale of 47.5 m^2 (SBP). Figure 1.28 illustrates this variety of size (and shape).

Figure 1.28. *Heliostat field of 15 m^2, 47.5 m^2 and 178 m^2 (from left to right). For a color version of this figure, see www.iste.co.uk/flamant/energy.zip*

The arguments in favor of small heliostats are optical quality, ease of adjustment, chain production capacity and the ease of installation. The arguments in favor of large heliostats are the limited number of units for a given power, the smaller number of sun-tracking motors and the simpler control.

The height of the tower increases with the power of the installation, as the larger the number of heliostats, the further away from the tower are the successive rows. Hence, the distance between the last rows of heliostats and the tower ranges between 2 and 3 km for high power solar power plants. In order to limit the losses caused by shading and blocking (see Chapter 3), the position of the solar receiver must be lifted. Figure 1.29 shows the evolution of the height of the towers depending on power.

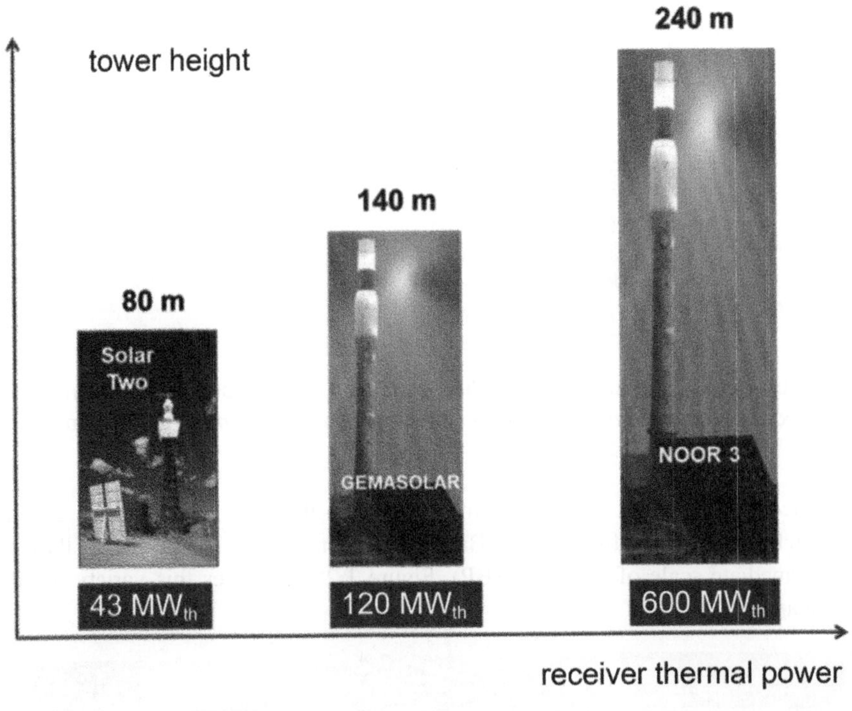

Figure 1.29. *Evolution of the height of solar power plant with the power of the receiver. For a color version of this figure, see www.iste.co.uk/flamant/energy.zip*

The international advances in this field are discussed in section 1.9.

1.8. Dish–Stirling modules

As indicated in Figures 1.5 and 1.6, the dish–Stirling systems are modular by design, since several MW_{th} dishes tracking the Sun are not realistic. The currently used motors are Stirling motors, but micro-turbines can also be used (Brayton Energy 2020). These concentrators have an excellent optical efficiency (of about 85–90%) as they are always in the Sun's axis in the tracking position, and the solar energy into electricity conversion efficiency may exceed 30%. Nevertheless, their development was stopped by the absence of a heat storage solution. Being in direct competition with the less expensive photovoltaic systems, this technology has not yet entered the market.

The developments of dish–Stirling systems started at the beginning of the 1980s with, for example, the 25 kW_e Vanguard module of Advanco Corporation. The power of most of the tested systems ranges between 10 and 25 kW_e as it corresponds to industrially available Stirling motors. It is important to mention two long-term test campaigns in the United States and in Europe: on the one hand, the 25 kW_e Suncatchers developed by SES company and tested by Sandia laboratory in the United States and, on the other hand, the 10 kW_e Eurodish developed by SBP and tested by various European laboratories, including PROMES in France (Figure 1.30).

Figure 1.30. *On the left, the (25 kWe) Suncatcher, on the rights, the (10 kWe) Eurodish module installed at PROMES, Font-Romeu. For a color version of this figure, see www.iste.co.uk/flamant/energy.zip*

The tests of Suncatcher lasted 10 years, from 2003 to 2013. The Stirling motor operated with hydrogen as heat transfer fluid at 720°C and 160 bars. The overall measured conversion efficiency reached 30%. Eurodish stems from the Distal II system tested in Almeria (PSA, Spain) in 1997. A campaign focused on the inter-comparison of the performances of the device in several countries (Spain, Italy, Germany, France) was organized between 2003 and 2005. The Stirling motor (Solo 161) also operated with hydrogen at 650°C and 140 bars. The efficiency measured by PROMES laboratory reached 21.6% (11.1 kW_e with a DNI of 974 W/m^2) (Nepveu et al. 2009).

Attempts to scale change were done. For example, SBP (Germany) developed a parabolic concentrator of 500 m^2 named HelioFocus (Keck et al. 2015). The concentrator is illustrated in Figure 1.31.

Figure 1.31. *HelioFocus parabolic concentrator of 500 m^2. Source SBP. For a color version of this figure, see www.iste.co.uk/flamant/energy.zip*

The company Cleanergy developed a 10 kW_e commercial module named Sunbox, 10 units of which were installed and tested in Dubai in 2015–2016 (Nilsson et al. 2017). Low power (13 kW_e) solar Stirling technology is currently promoted by the Swedish company Azelio that proposes a system composed of a 13 h latent heat storage with AlSi (Tf = 577°C) as eutectic and a receiver using sodium as heat transfer fluid. But this technology uses a mini-tower concentration system instead of a paraboloid system.

1.9. Perspectives: deployment, capacity factor, costs, environmental impact and new concepts

1.9.1. *Commercial deployment*

1.9.1.1. *Installed electrical power*

In its latest report (IEA 2020a), the International Energy Agency expects an installed CSP electrical power between 9 and 13 GW by 2025 with a minimum generation (for 9 GW) of 26 TWh (Figure 1.32).

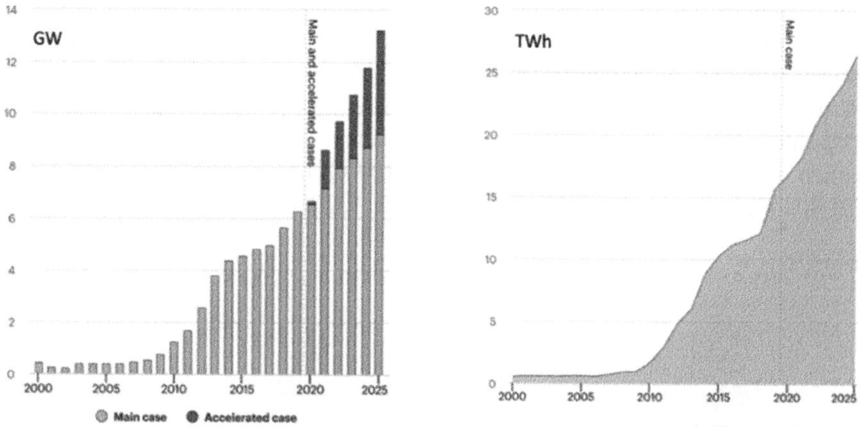

Figure 1.32. *Evolution of the installed power and of electricity generation of solar thermal power plants (IEA 2020a). For a color version of this figure, see www.iste.co.uk/flamant/energy.zip*

The heat storage (see Chapter 8) makes it possible to adapt the production of solar thermal power plants in order to meet the demand of the local grids and in addition to other sources of renewable energy. One of the relevant criteria to evaluate this positioning is the capacity factor. There are base load power plants (capacity factor of 65–70%), half-base load power plants (capacity factor of 45–50%) or peakers (capacity factor of about 20%). For the first two options, the power plant integrates a storage of 14–15 h and 7–8 h, respectively. The peaker mode operation corresponds to energy generation in phase with the consumption peaks, generally in the evening and in the morning in Europe. Under these conditions, the power plant does not generate during the day but stores all this collected energy in its heat storage system in order to release it when the solar resource is weak or zero. Therefore, there is a full gap between resource availability and power plant production.

According to this scenario, solar thermal power plants are the ideal complement to photovoltaic power plants without storage. Indeed, heat storage is less expensive than electrochemical storage in batteries (Feldman et al. 2016 and Chapter 8). Since the energy price is significantly higher than the average price during these periods, better use compensates for the short generation time.

1.9.1.2. *Industrial heat*

Industrial heat generation is another domain in which CSP is able to provide relevant solutions to energy decarbonation in the industry. Globally, in 2014, the industry accounted for 32% of the worldwide energy consumption and heat represented 74% of the industrial energy consumption (Figure 1.33). Half of this fraction corresponded to low-temperature (< 100°C) heat and medium-temperature (150-400°C) heat, as indicated in Figure 1.33 (IEA, 2020b).

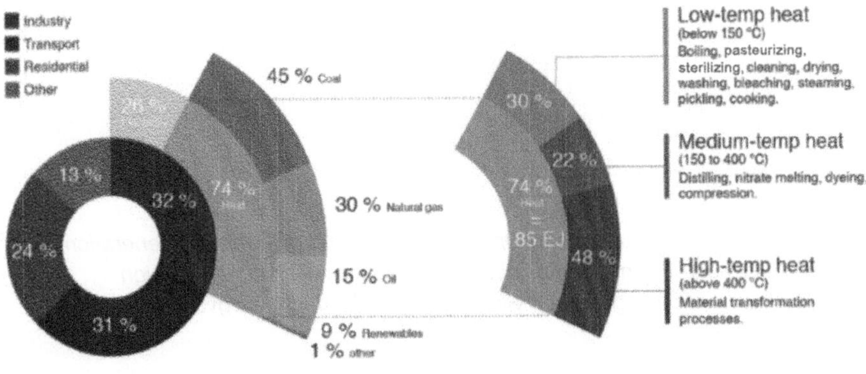

Figure 1.33. *Energy consumption of the industry (360 EJ en 2014, 1 EJ = 1018 Joule) and part of the heat at various temperature levels in this consumption (IEA 2020b). For a color version of this figure, see www.iste.co.uk/flamant/energy.zip*

Solar concentrators are needed to generate heat at medium (150–400°C) and high (>400°C) temperatures. At medium temperatures, heat is essentially consumed as vapor, particularly in food, textile, chemical and paper-making industries. Figure 1.34 presents a diagram of a hybrid (solar-fuel) installation adapted to this type of applications.

The high-temperature heat-intensive industries are the chemical, mineral and metallurgical (ferrous and nonferrous metals) industries. Unlike the medium-temperature applications, for which there are industrially available technologies, the development of high-temperature CSP applications requires significant research efforts. In this field, solar calcination at 850–900°C was recently demonstrated at pilot scale (Esence et al. 2020).

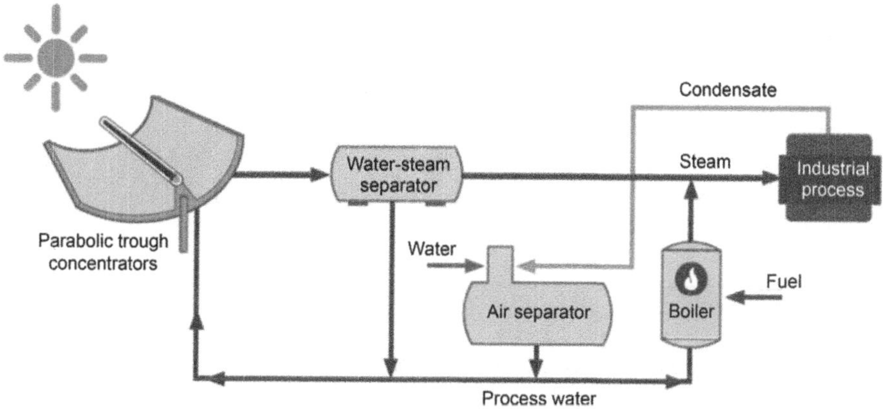

Figure 1.34. *Block diagram of a solar installation for steam generation. Parabolic trough concentrators can be replaced by linear Fresnel concentrators. For a color version of this figure, see www.iste.co.uk/flamant/energy.zip*

1.9.2. *Capacity factor*

IRENA (International Renewable Energy Agency) studied the capacity factor of solar power plants (Irena 2019). Figure 1.35 illustrates these results.

In Figure 1.35, the capacity factor corresponding to power plants without storage varies between 22 and 33%; with 4–8 h storage, it ranges between 35 and 45% (essentially depending on the solar resource). The power plants with more than 8 h storage have a capacity factor generally within 50–65%. The configuration corresponding to peak generation is not represented in this figure, as there is yet no power plant operating in this mode. Finally, this figure shows an ascending tendency of the capacity factor in time, as all the new solar thermal power plants integrate a storage unit.

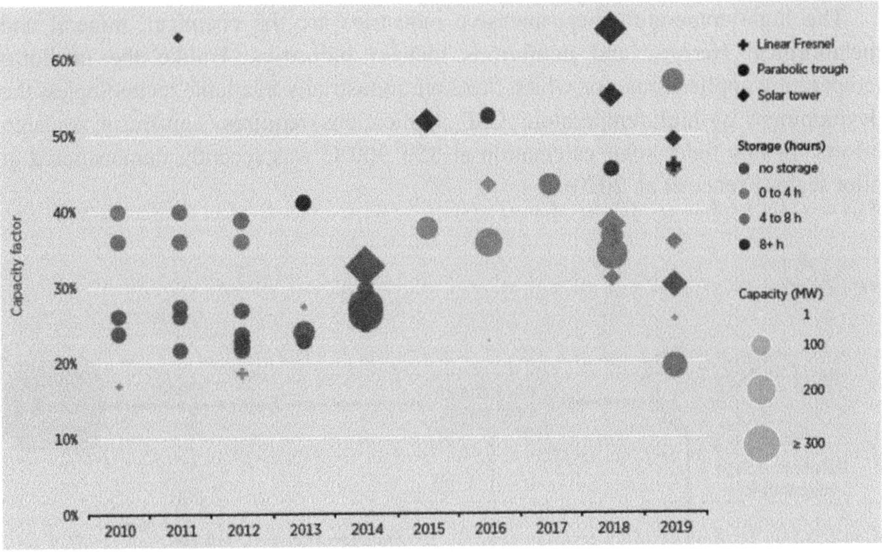

Figure 1.35. *Capacity factor of operating solar power plants depending on their storage capacity. Source: IRENA Renewable cost database (IRENA 2019). For a color version of this figure, see www.iste.co.uk/flamant/energy.zip*

1.9.3. *Cost of electricity*

The evolution of the cost of electricity generated by the solar thermal power plants was studied by IRENA (IRENA 2019). In this case, the commonly used indicator is the levelized cost of electricity (LCOE). The LCOE of the electricity generated by the solar thermal power plants dropped by 47% between 2010 and 2019 (Figure 1.36). The learning curve (reduction of LCOE associated with doubling of the generation capacity) of the CSP is very close to that of photovoltaic power, with a value of 30% and 35%, respectively. Figure 1.36 shows the costs of about 70 USD/MWh in 2021 (with storage). These figures correspond to the feed-in tariff accepted by the power plant project developer when the PPA contract was signed. This significant drop was observed in particular for the Dubai project of 700 MW_e (3×200 MW_e parabolic trough solar power plants with 10 h storage and 1×100 MW_e tower solar power plant with 15 h storage). The mechanisms that allowed this offer were analyzed (Lilliestam and Pitz-Paal 2018). This study shows that an exceptional duration of 35 years of the feed-in agreement and advantageous financing conditions explain this figure. Indeed, building a solar thermal power plant is very costly in terms of capital expenditures (CAPEX), which explains the sensitivity of the production cost to financing conditions.

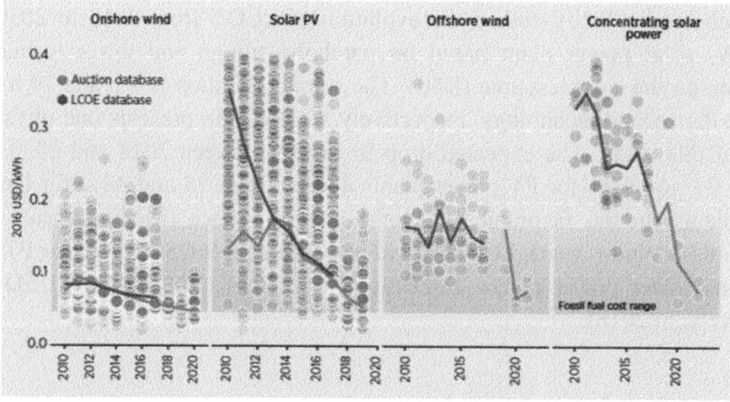

Figure 1.36. *Evolution of the levelized cost of electricity and medium value for onshore wind, photovoltaic, offshore wind and concentrated solar power. Each point corresponds to a project. Source: IRENA Renewable cost database and auction database (Irena 2019). For a color version of this figure, see www.iste.co.uk/flamant/energy.zip*

In the United States, the actual cost of the electricity generated by a tower power plant with molten salt is 100 USD/MWh (Murphy et al. 2019). Figure 1.37 compares the costs of STE in 2010 and 2017, and the goals displayed by the US Department of Energy (DOE). The goal for 2030 is 50 USD/MWh for the baseload power plants with 12 h storage and 100 USD/MWh for peaker power plants.

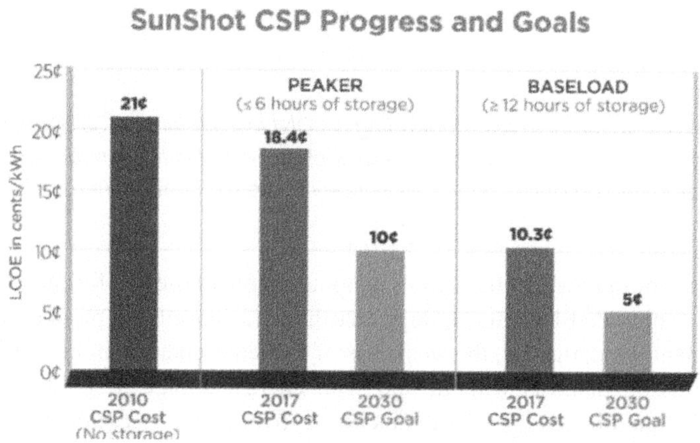

Figure 1.37. *Evolution of the levelized cost of solar electricity generated by thermal power plants in the United States and national goals for 2030 (Murphy et al. 2019). For a color version of this figure, see www.iste.co.uk/flamant/energy.zip*

Dersch et al. (2020) studied the evolution of LCOE from 2018 to 2030 for a 150 MW_e solar power plant based on parabolic trough and tower technologies depending on the solar resource (DNI). The storage duration is 7 h and 10 h for the first and the second technology, respectively. Figure 1.38 presents one of the main results of this study. The expected drop in LCOE between 2018 and 2030 ranges between 31 and 44% for PT power plants and between 35 and 48% for the tower power plants. In the favorable sites of Southern France (Mediterranean region, 1,900–2,000 kWh/m^2.year), LCOE would be 130 and 109 USD/MWh for PT power plants and tower power plants, respectively, by 2030 (193 and 170 USD/MWh, respectively, in 2018).

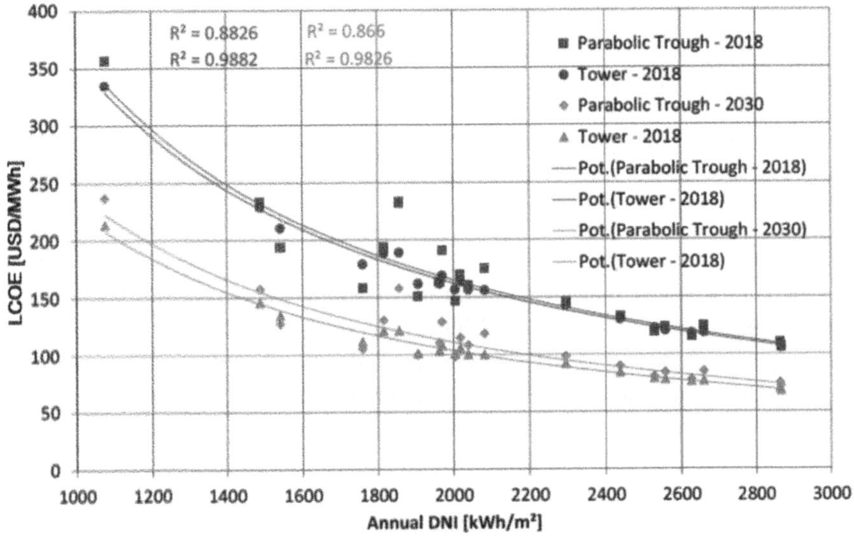

Figure 1.38. *Evolution of LCOE depending on DNI in 2018 and 2030 (Dersch et al. 2020). The points correspond to the results of calculations for various sites. For a color version of this figure, see www.iste.co.uk/flamant/energy.zip*

It is important to note that the LCOE is not the proper criterion for the evaluation of the service provided by a technology that has the capacity to produce in phase with the demand. In particular, the weakness of the representation in Figure 1.36 lies in the comparison of costs without and with energy storage. It is clear that the analysis lacks one dimension. The distinction between cost goals for baseload and peaker power plants (Figure 1.37) is a significant first step. On the other hand, some countries such as South Africa had implemented a feed-in tariff differentiating between off-peak production and peak production capacity.

1.9.4. *Environmental impact*

The analysis of the life cycle of parabolic trough and tower solar power plants was made by Burkhardt et al. (2012). The emissions amount to 26–38 g CO_2-eq/kWh, 10 times less than for a natural gas power plant. The energy payback time is about one year. Furthermore, no rare material was used, except silver (for the mirror), which can be easily recycled. A critical point concerns water consumption if wet cooling is used, which requires 2 to 3 liters of water per generated kWh. Dry cooling can be used to avoid this consumption, which lowers production by 4–5% (efficiency reduction due to cold source temperature increase). In this case, the only remaining water consumption is that required for mirror washing, estimated at 0.25 l/kWh. Finally, the occupied ground area is about 2 ha per MW_e without storage; the area increases depending on storage duration (see solar multiple).

1.9.5. *Technological evolutions and new generations (GEN3)*

The complementarity of CSP and photovoltaic solar power is increasingly obvious due to the growing need for storage capacity along with the increasing role of intermittent energies in the energy mix. A new solar power plant architecture was retained for the 800 MW_e NOOR Midelt project (Morocco), which associates parabolic trough concentrators and photovoltaic panels according to the diagram in Figure 1.39. Other hybridization modes are presented in Chapter 9.

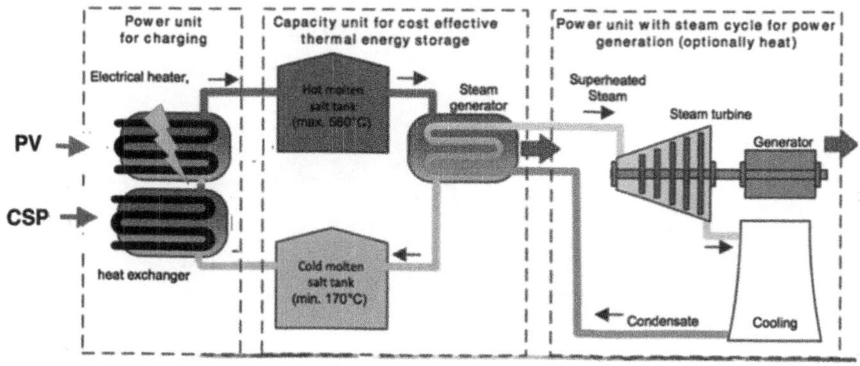

Figure 1.39. *Architecture of NOOR Midelt CSP-PV hybrid power plant, Morocco (HELIOSCSP 2020). For a color version of this figure, see www.iste.co.uk/flamant/energy.zip*

Another technological evolution involves the replacement of the heat transfer oil in PT power plants or the steam in LFC power plants with solar salt in order to increase the operating temperature and the specific storage capacity. The technological problem is related to controlling a liquid that solidifies at about 220°C in kilometers of linear receivers. A 50 MW_e LFC power plant using molten salt as heat transfer and storage fluid was put into operation in June 2020 in China (Dunhuang Dacheng). In Europe, the Partanna project in Sicily includes a 4.26 MW_e LFC power plant with molten salt (Falchetta et al. 2020).

After the first attempts to use sodium as heat transfer fluid on the site of Almeria at the end of the 1970s, the Australian company VAST Solar challenged this technology in 2009 (Vast Solar 2020). It is important to recall that due to the high thermal conductivity of liquid sodium, the fluid–wall heat transfer coefficient is one to two orders of magnitude greater than for any other single-phase fluid (see Chapter 5). This exceptional characteristic makes it possible to design very compact solar receivers exposed to the solar flow of the order of 2-3 MW/m^2. Jemalong CSP Pilot Plant of 1.1 MW_e (6 MW_{th}) has been operating since 2018 (Figure 1.40). The pilot is composed of five modules, and the total solar field includes 3,500 heliostats of 3.6 m^2. The height of the towers is 27 m, and the surface area of each receiver is 2.25 m^2 (Wood and Drewes 2019).

Figure 1.40. *Jemalong CSP Pilot Plant, pilot solar power plant using sodium as heat transfer fluid (Vast Solar 2020). For a color version of this figure, see www.iste.co.uk/flamant/energy.zip*

The technological breakthroughs are related to the new generation of solar power plant (GEN3) whose goal is to exceed the limit of 565°C imposed by the solar salt in order to integrate high-efficiency thermodynamic cycles (~50%) such as the combined cycles and sCO_2 cycles (supercritical CO_2). The options considered for heat transfer fluids are high-pressure gases, high-temperature salts and particles (Mehos et al. 2017). The use of particles makes it possible to employ the same medium as heat transfer fluid and storage material (which is not the case for pressurized gases) while avoiding solidification problems (case of chloride and carbonate salts). This concept was first proposed in 1980 (Flamant 1980).

The principle of a solar power plant with solid heat transfer medium is schematically represented in Figure 1.41 for a combined cycle (the same principle can be adopted with a sCO_2 or sH_2O cycle). The particles must reach approximately 750°C for sCO_2 cycles and 850°C for combined cycles. These two thermodynamic cycles enable us to reach conversion efficiencies of 50% and more, or a 17% increase at least with respect to the state of the art. As shown in Figure 1.41, hybridization is possible, but not necessary if the required temperatures are obtained at the output of the solar receiver.

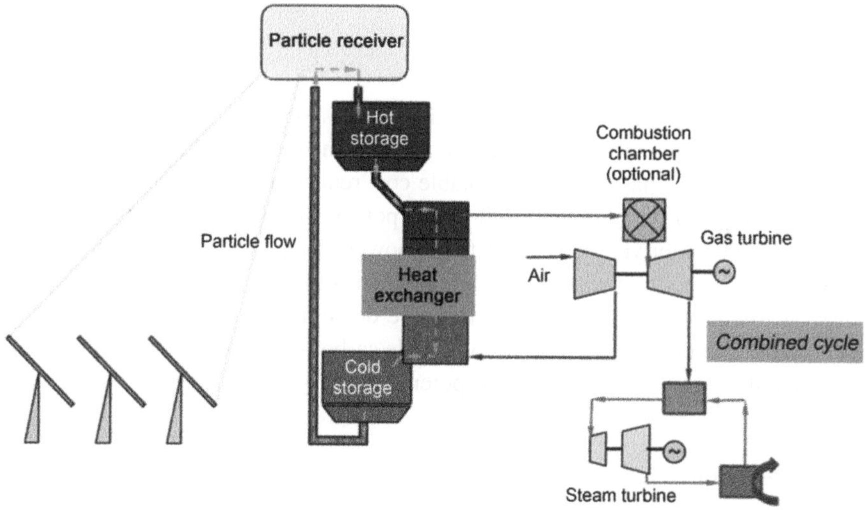

Figure 1.41. *Block diagram of a solar power plant with solid heat transfer medium associated with a combined cycle. For a color version of this figure, see www.iste.co.uk/flamant/energy.zip*

Several concepts of solar receivers are being developed internationally, in particular the particle curtain (Ho et al. 2017), the centrifugal receiver (Ebert et al. 2019) and the fluidized bed (Flamant and Hemati 2010). Prototypes at the MW_{th} scale are being tested (2021+). The proposed solutions also differ in terms of the size of the particles employed, from 300 to 700 μm for the first two technologies and 50 μm for the technology of dense fluidized beds flowing through tubes subjected to concentrated radiation. A temperature of at least 750°C was reached by all the receivers in the pilot test phase (Benoit et al. 2015). The concept of solid heat transfer medium imposes the vertical conveying of the particles in the cold storage tank to the receiver and the integration of particle to working fluid heat exchanger, which is an important component of the power plant. The use of fine particles allows the use of multistage fluidized bed heat exchangers. As an example, the performances of a solar power plant of about 20 MW_e integrating a tubular receiver with fluidized bed, a storage and a cycle combined with an existing gas turbine are attractive with a nominal solar energy into an electricity conversion efficiency of 21–24.7% (Behar et al. 2020).

Cost analyses show that the concept of solar receivers with solid heat transfer medium would contribute to reaching the DOE goal of 50 USD/MWh on favorable sites (strong annual DNI) with long-term storage (Buck and Giuliano 2019).

1.10. Conclusion

The development of CSP was subjected to many shocks for half a century, which slowed down its commercial deployment. Though the cumulated installed power is low (~6 GW_e), there has been a remarkable cost reduction during the last 10 years. The technology of parabolic trough solar power plants has reached industrial maturity, which is not yet the case for tower power plants, whose scaling up is more complex. The capacity of solar thermal power plants to bulk store the collected solar energy at low cost is an asset for their future deployment. Finally, the thermal or thermochemical applications of CSP to provide carbon-free heat to industry are only in an early stage and their development potential is very high.

1.11. References

Aalborg (2020). Available at: https://www.aalborgcsp.com/projects/project-overview/ [Accessed 27 December 2020].

Askarov, D. (2010). Endoreversible thermodynamics: Curzon-ahlborn engine. *Coursework for Physics*, 240, Stanford University. Available at: http://large.stanford.edu/courses/2010/ph240/askarov2/ [Accessed 2 January 2021].

Baum, V.A., Aparasi, R.R., Garf, B.A. (1957). High-power solar installations. *Solar Energy*, 1(1), 6–12.

Behar, O., Grange, B., Flamant, G. (2020). Design and performance evaluation of a modular solar power plant using a gas turbine combined cycle with fluidized particle-in-tube receiver and direct thermal storage. *Energy Conversion & Management*, 220, 11310.

Benoit, H., Perez Lopez, I., Gauthier, D., Sans, J.-L., Flamant, G. (2015). On-sun demonstration of a 750°C heat transfer fluid for concentrating solar systems: Dense particle suspension in tube. *Solar Energy*, 118, 622–633.

Brayton Energy (2020). Available at: https://www.braytonenergy.net/ [Accessed 2 January 2021].

Buck, R. and Giuliano, S. (2019). Solar tower system temperature range optimization for reduced LCOE. *AIP Conference Proceedings*, 2126, 030010.

Burkhart, J.J., Heath, G., Cohen, E. (2012). Life cycle greenhouse gas emissions of trough and tower concentrating solar power electricity generation. *Journal of Industrial Ecology*, 16, 93–109.

Dersch, J., Dieckmann, S., Hennecke, K., Pitz-Paal, R., Taylor, M., Ralon, P. (2020). LCOE reduction potential of parabolic trough and solar tower technology in G20 countries until 2030. *AIP Conference Proceedings*, 2303, 120002.

Ebert, M., Amsbeck, L., Rheinländer, J., Schlögl-Knother, B., Schmitz, S., Sibum, M., Uhlig, R., Buck, R. (2019). Operational experience of a centrifugal particle receiver prototype. *AIP Conference Proceedings*, 2126, 030018.

Esence, T., Guillot, E., Tessonneaud, M., Sans, J.-L., Flamant, G. (2020). Solar calcination at pilot scale in a continuous flow multistage horizontal fluidized bed. *Solar Energy*, 207, 367–378.

ESTELA (2020). Available at: https://www.estelasolar.org/flabeg-fes-ultimate-trough-will-be-integrated-in-the-first-commercial-solar-power-plant-in-the-kingdom-of-saudi-arabia/ [Accessed 22 December 2020].

EurObserv'ER (2019). Baromètres. Solaire thermique et thermodynamique. Available at: https://www.eurobserv-er.org/.

Falchetta, M., Mazzei, D., Russo, V., Campanella, V.A., Floridia, V., Schiavo, B., Venezia, L., Brunatto, C., Orlando, R. (2020). The partanna project: A first of a kind plant based on molten salts in LFR collectors. *AIP Conference Proceedings*, 2303, 040001.

Feldman, D., Margolis, R., Denholm, P., Steckeli, J. (2016). Exploring the potential competitiveness of utility-scale photovoltaics plus batteries with concentrating solar power, 2015–2030. National Renewable Energy Laboratory Technical Report, NREL/TP-6A20-66592.

Ferri, L. (2001). Le commissariat à l'énergie solaire, COMES. *Archives Nationales. Répertoire* (20010454/1-20010454/42). Available at: www.siv.archives-nationales.culture.gouv.fr/siv/IR/FRAN_IR_019327 [Accessed 25 December 2020].

Flamant, G. (1980). Thermochimie solaire à hautes températures, résultats expérimentaux, quelques perspectives d'applications. *Revue Phys. Appliquée*, 15, 503–511.

Flamant, G. and Hemati, M. (2010). Dispositif collecteur d'énergie solaire [Device for collecting solar energy]. French patent FR, 1058565, PCT extension WO2012052661, 2012.

Francia, G. (1968). Pilot plants of solar steam generating stations. *Solar Energy*, 12, 51–64.

FRENEL (2020). Available at: https://www.frenell.de/company/ [Accessed 21 December 2020].

Gordon, J.M. and Feuermann, D. (2019). Aplanatic beam-down solar towers. *Proceedings Volume 11120, Nonimaging Optics: Efficient Design for Illumination and Solar Concentration XVI*; 111200E. Available at: https://doi.org/10.1117/12.2525231.

Guthleben, D. (2009). Histoire de l'énergie solaire au CNRS. 40 ans du four solaire d'Odeillo, 25 September 2009. Available at: www.promes.cnrs.fr/40ans/downloads/Denis_Guthleben_Odeillo_250909.pdf [Accessed 20 December 2020].

HELIOSCSP (2020). Available at: http://helioscsp.com/category/nacional/ [Accessed 2 January 2021].

Ho, C., Christian, J., Romano, D., Yellowhair, J., Sigel, N., Savoldi, L., Zanino, R. (2017). Characterization of particle flow in a free-falling solar particle receiver. *ASME Journal of Solar Energy Engineering*, 139, 021011-1/9.

IEA (2020a). Renewables 2020. Analysis and forecast to 2025, Fuel report – November 2020. Available at: https://www.iea.org/reports/renewables-2020 [Accessed 28 December 2020].

IEA (2020b). Solar heat integrations in industrial processes. Available at: https://task49.iea-shc.org/publications [Accessed 29 December 2020].

IRENA (2019). Renewable power generation costs in 2019. International Renewable Energy Agency, Abu Dhabi. Available at: www.irena.org.

Keck, T., Balz, M., Blumenthal, Y. (2015). Large is beautiful – Progress of HelioFocus 500 m² dish. *Energy Procedia*, 69, 1597–1602.

Ko, G.K. (2019). Etude et modélisation dynamique d'un concentrateur à miroir de Fresnel linéaire. Thesis, Université de Perpignan. Available at: https://tel.archives-ouvertes.fr/tel-02368068/document.

Lilliestam, J. and Pitz-Paal, R. (2018). Concentrating solar power for less than USD 0.07 per kWh: Finally the breakthrough? *Renewable Energy Focus*, 26, 17–21. Available at: https://www.nrel.gov/docs/fy19osti/71912.pdf.

Mehos, M., Turchi, G., Vidal, J., Wagner, M., Ma, Z., Ha, C., Kolb, W., Andraka, C., Kruizenga, A. (2017). Concentrating solar power – Gen3 demonstration roadmap. *National Renewable Energy Laboratory Technical Report 2017*, NREL/TP-5500-67464.

Morin, G., Dersch, J., Platzer, W., Eck, M., Häberle, A. (2012). Comparison of linear Fresnel and parabolic trough collector power plants. *Solar Energy*, 86, 1–12. 1850, 050004. Available at: https://doi.org/10.1063/1.4984408.

Murphy, C., Sun, Y., Cole, W., Maclaurin, G., Turchi, G., Mehos, M. (2019). The potential role of concentrating solar power within the context of DOE's 2030 solar cost targets. National Renewable Energy Laboratory Technical Report, NREL/TP-6A20-71912.

Nepveu, F., Ferrière, A., Bataille, F. (2009). Thermal model of a dish/Stirling system. *Solar Energy*, 83, 81–89.

Nilsson, M., Jamot, J., Malm, T. (2017). Operational data and thermodynamic modeling of a Stirling-dish demonstration installation in desert conditions. *AIP Conference Proceedings*, 1850, 050004.

NREL (2020). Available at: https://solarpaces.nrel.gov/ [Accessed 28 December 2020].

Royère, C. (1981). Les applications du four solaire de 1000 kW du CNRS à Odeillo. *Entropie*, 97, 147–160.

SENER (2020). Available at: https://www.energy.sener/projects/gemasolar [Accessed 20 December 2020].

SolarPACES (2020). Available at: https://www.solarpaces.org/ [Accessed 28 December 2020].

Vant-Hull, L. (2014). Issues with beam-down concepts. *Energy Procedia*, 49, 257–264.

Vast Solar (2020). Available at: https://vastsolar.com/ [Accessed 30 December 2020].

Wood, G. and Drewes, K. (2019). Vast solar: Improving performance and reducing cost and risk using high temperature modular arrays and sodium heat transfer fluid. SolarPACES2019. Available at: https://www.solarpaces.org/wp-content/uploads/Vast-Solar-improving-performance-and-reducing-cost-and-risk-using-high-temperature-modular-arrays-and-sodium-heat-transfer-fluid-1.pdf.

2

Solar Resource Management, Assessment and Forecasting

Stéphane THIL and Stéphane GRIEU
PROMES-CNRS, UPVD,
Perpignan, France

This chapter presents solar resource-related notions essential for its measurement, assessment or forecasting. The first part introduces the basics of solar resource modeling and measurement. The second part presents the main approaches to solar resource forecasting.

2.1. Measurement and assessment of the solar resource

Solar radiation reaching the ground depends on the position of the Sun, extra-terrestrial radiation and the atmospheric specificities of the site being studied. A general presentation of these concepts is proposed. The components of solar radiation as well as the associated measurement instruments are also detailed.

2.1.1. *Earth–Sun pair*

Knowledge of the celestial mechanics governing the Earth–Sun pair is an essential prerequisite for assessing the various components of solar radiation. It can be used to find out the zenith angle and the azimuth angle of the Sun at each moment from any observation point on the globe. These angles are then used for

Concentrating Solar Thermal Energy,
coordinated by Gilles FLAMANT.

modeling or measuring the direct and indirect components of the terrestrial solar radiation (see sections 2.1.3 and 2.1.4).

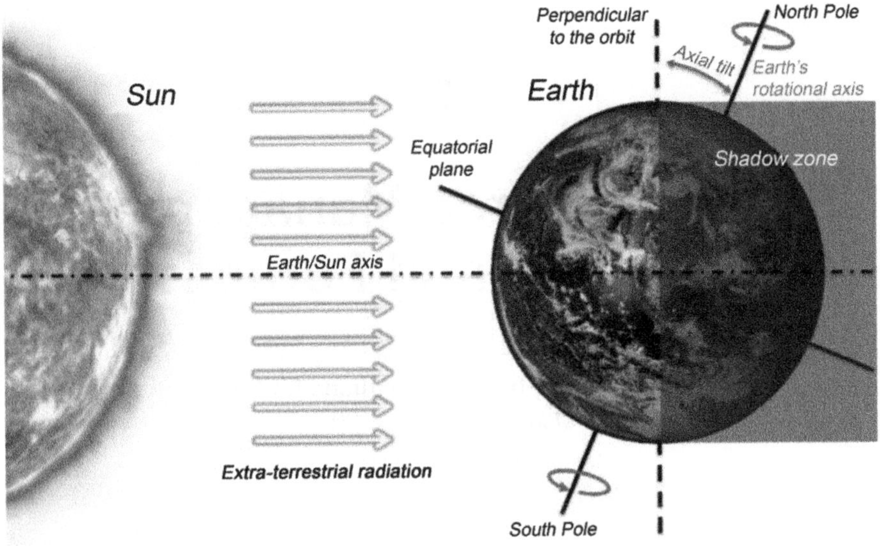

Figure 2.1. *Simplified view of the Earth/Sun system. For a color version of this figure, see www.iste.co.uk/flamant/energy.zip*

Calculating from the Earth, the position of the Sun at a given moment and at a given place is not a trivial problem. Indeed, this position is influenced by many phenomena, such as the interactions between planets, which influence both the rotation of the Earth about its axis and its revolution about the Sun (Figure 2.1). Nevertheless, there are now many references in the scientific literature that accurately present in detail the calculation of this position (Bretagnon and Francou 1988; Meeus 1999). There is also a variety of algorithms for calculating this position, each of them having a calculation time, precision and period of validity meeting a specific need. An overview of the development of these algorithms and their specificities is presented here.

In what follows the solar/zenith angle (SZA) is the angle between the Sun and the zenith, and the solar/azimuth angle (SAA) is the angle between the Sun and the azimuth (Figure 2.2). It should be noted that the azimuth angle is measured with respect to the north and in the clockwise direction (e.g. east corresponds to an azimuth of 90°).

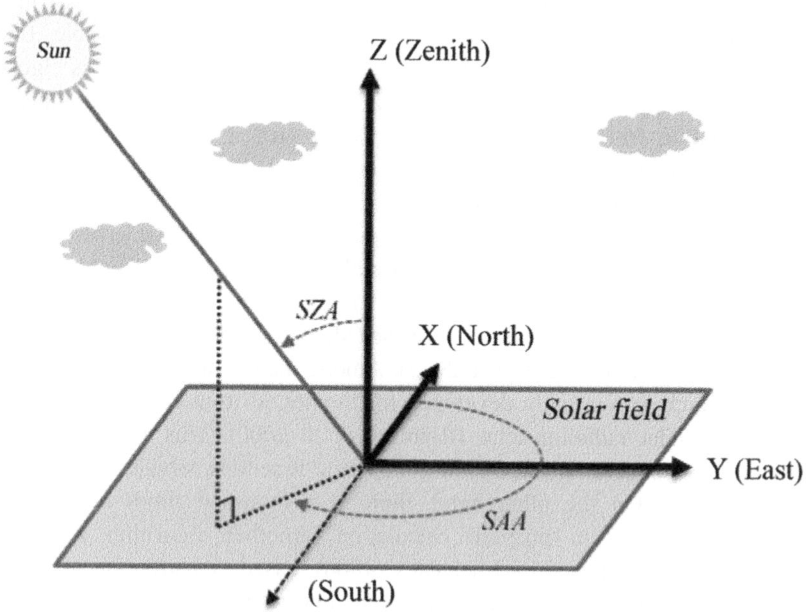

Figure 2.2. *Solar/zenith angle (SZA) and solar/azimuth angle (SAA). For a color version of this figure, see www.iste.co.uk/flamant/energy.zip*

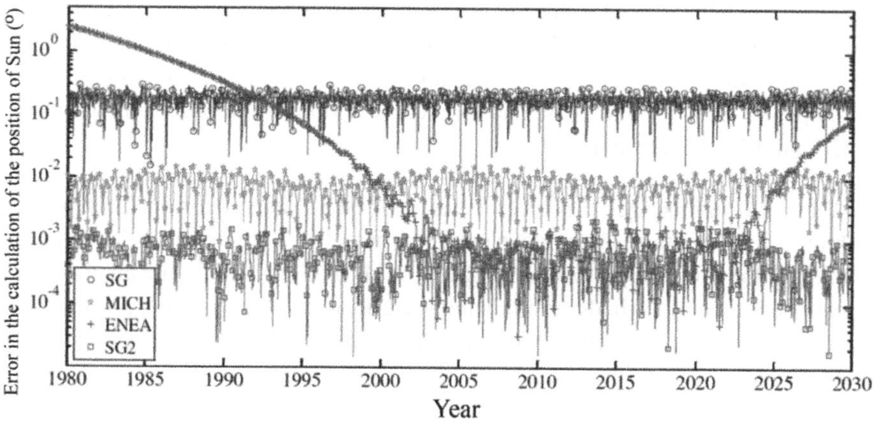

Figure 2.3. *Error in the calculation of the position of the Sun, with respect to SPA, for SG, MICH, ENEA and SG2 algorithms, throughout 1980–2030 (Blanc and Wald 2012). For a color version of this figure, see www.iste.co.uk/flamant/energy.zip*

The commonly used algorithms for calculating the position of the Sun include the sun position algorithm (SPA) developed by National Renewable Energy Laboratory (NREL) in the United States (Reda and Andreas 2004). This algorithm can be used for estimating SZA and SAA over a very extended period of time, from the year −2000 to +6000, with an uncertainty of about 1″ (≈0.0003°). It is thus used as a reference by most of the other algorithms estimating the position of the Sun. Nevertheless, with nearly 2,300 operations and over 300 calls of trigonometric functions, this algorithm is relatively computationally expensive. This constraint may prove impossible to meet for certain real-time applications (Blanc and Wald 2012). This explains the development of alternatives that are less computationally resource-intensive but inevitably less precise (Michalsky 1988; Blanco-Muriel et al. 2001; Blanc and Wald 2012; Grena 2012). Among them is the *solar geometry* (SG) algorithm, which was initially developed for the period from 1980 to 2030, within the European solar radiation atlas (Rigollier et al. 2000). This algorithm requires approximately 60 operations and 25 trigonometric functions, which makes it clearly faster than SPA. On the other hand, there is a thousand times higher (≈0.3°) uncertainty on the Sun position calculation. Another algorithm with higher performances was proposed by Michalsky (1988), based on astronomical almanacs from 1985. This algorithm (MICH) reduces the uncertainty to less than 0.02° over a period from 1950 to 2050 and for a calculation time similar to SG. An additional order of magnitude was reached for the ENEA algorithm developed by Grena (2008). Nevertheless, the latter is optimized for a shorter period, from 2003 to 2023. Its uncertainty increases significantly beyond this period (Figure 2.3). In order to extend the validity period while maintaining an uncertainty similar to that of the ENEA algorithm, an improved version (SG2) of the SG algorithm was developed (Blanc and Wald 2012). SG2 thus makes it possible to retain an error close to 0.002° over a period from 1980 to 2030, and for a computational time equivalent to the SG, MICH and ENEA algorithms.

Table 2.1 summarizes the specificities of these various algorithms. In parallel with the SG2 algorithm, Grena developed a series of five algorithms calculating the position of the Sun over a period from 2010 to 2110 (Grena 2012), the precision and computation time of each algorithm being specific to a given type of application; the computation time and uncertainty of the one with the highest precision are similar to those of the SG2 algorithm.

The algorithms presented here show that a compromise between precision and computation time must be reached. This choice depends on the intended application.

Characteristics	SPA	SG	MICH	ENEA	SG2
Validity (years)	[−2000; 6000]	[1980; 2030]	[1950; 2050]	[2003; 2023]	[1980; 2030]
Uncertainty (°)	0.0003	0.31	0.0128	0.0029	0.0025
Additions	1,000	25	20	40	20
Multiplications	1,300	35	35	40	30
Trigonometric functions	300	25	25	25	23
Computation time	t_{SPA}	$\approx t_{SPA}/20$	$\approx t_{SPA}/20$	$\approx t_{SPA}/20$	$< t_{SPA}/20$

Table 2.1. *Specificities of SPA, SG, MICH, ENEA and SG2 algorithms*

EXAMPLE.– As it will be seen further on, sky imagers can be used for solar resource measurement (see section 2.1.4) or for its short-term forecasting (see section 2.2.3.2). The position of the Sun is required for the calibration of the extrinsic parameters of such a camera. Once calibrated, it is then possible to know the position of the Sun on the image at any moment, even when it is blurred (an image taken by a sky imager, as well as the Sun's path, can be seen in Figure 2.4). This position is also used by other algorithms, notably for finding out the angle subtended by the Sun and each pixel of the image. High precision is therefore generally expected here.

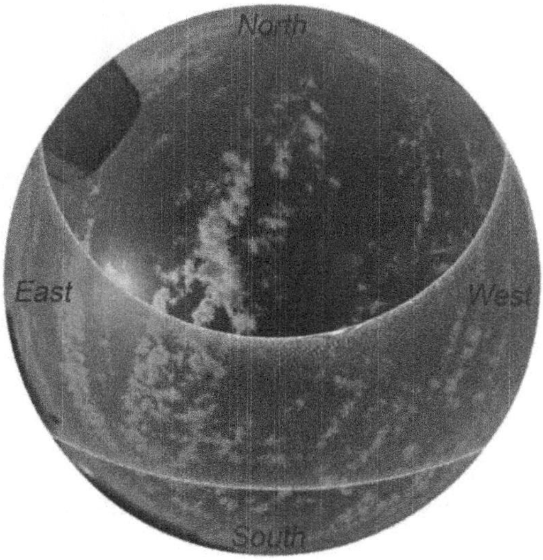

Figure 2.4. *Sun's path in 2015 above the PROMES-CNRS laboratory, calculated using the SG2 algorithm. For a color version of this figure, see www.iste.co.uk/flamant/energy.zip*

2.1.2. *Extra-terrestrial solar irradiance*

Besides the extra-terrestrial solar radiation, that is, the electromagnetic radiation emitted by the Sun and getting under normal incidence to the outer layer of the atmosphere (Figure 2.5), there is an energy irradiance[1]. Since the Earth–Sun distance is not constant, this irradiance is not constant either. However, given the principle of energy conservation, the following can be written:

$$I_1 \frac{S_{\text{Earth}}}{D_1^2} = I_2 \frac{S_{\text{Earth}}}{D_2^2} \Rightarrow I_1 = \left(\frac{D_1}{D_2}\right)^2 I_2 \qquad [1.1]$$

where I_1 and I_2 are the irradiances received by the Earth at distances D_1 and D_2, respectively, and S_{Earth} is the surface of projection of the Earth on a plane that is normal to the radiation and passes through the center of the Earth. It is therefore sufficient to know a particular value of the irradiance in order to deduce its evolution during the year. By convention, this value, called "solar constant" and denoted by I_{UA} in this chapter, is chosen at mid-distance between the Earth and the Sun ($D_{\text{UA}} = 1\ \text{UA} \approx 1.5 \times 10^8$ km).

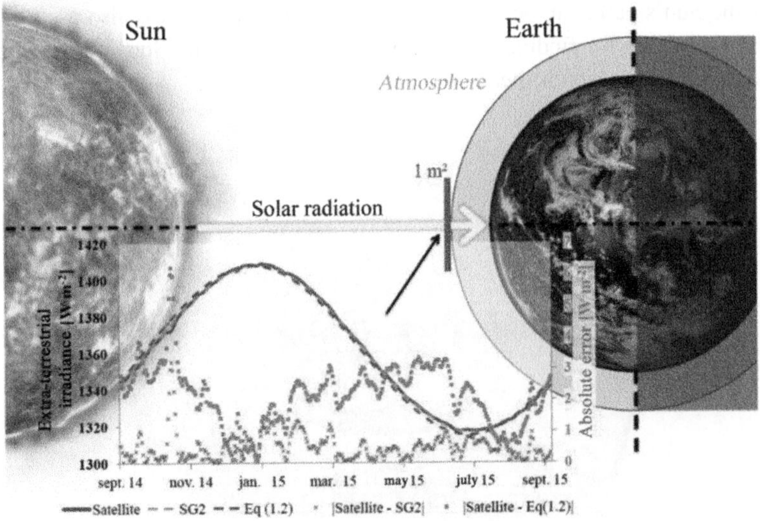

Figure 2.5. *Annual variation of the extra-terrestrial irradiance. For a color version of this figure, see www.iste.co.uk/flamant/energy.zip*

1 It is important to note that energy irradiance is defined as the density of energy flux per unit area incident at a point of an actual or imaginary surface, and is expressed in W m^{-2}. As there is no possible confusion here with *luminous* illumination (in lm·m^{-2}), we will simply speak of illuminance in the following.

Assessing the solar constant was the objective of many satellite missions, given its important role in the development of climate models (Lean and Rind 2009; Kopp 2014). The first measurements started at the end of the 1970s during the launch of Nimbus 7 satellite dedicated to monitoring extra-terrestrial radiation (Kyle et al. 1993). Until the early 1990s, the solar constant was assessed at about (1,366 ± 7) W.m^{-2} (Stoffel et al. 2010). Then, in the 1990s, new measurements yielded a more precise value (1,365.4 ± 1.3) W.m^{-2}. However, in 2011, a new solar constant was proposed after the discovery of a systematic bias in the previous measurements and thanks to the data provided by a new and more accurate instrument: I_{UA} = (1,360.8 ± 0.5) W.m^{-2} (Kopp and Lean 2011). This latter value proved to be particularly precise, since after having studied a large database with data covering 42 years, a recent study has slightly modified it, yielding I_{UA} = (1,361.1 ± 0.5) W.m^{-2} (Gueymard 2018). It is important to note that the solar activity (dark spots and bright spots) generates variations of the solar constant in time. These variations lead to fluctuations of I_{UA} by ± 0.2% during solar cycles.

When the Earth/Sun distance is unknown, it is also possible to assess the extraterrestrial irradiance using a simplified formula, such as:

$$I_0 = \left(\frac{D_0}{D_{\mathrm{UA}}}\right)^2 I_{\mathrm{UA}} \approx \left(1 + 0.034\cos\left(\frac{360n}{365}\right)\right) I_{\mathrm{UA}} \quad [2.2]$$

where I_0 is the incident extra-terrestrial solar irradiance at the Earth/Sun distance D_0 and n is the number of the day in the year (example: n = 32 on February 1). It is important to note that the irradiance variation due to the elliptical orbit of the Earth is approximately ± 3 % (Figure 2.5). Moreover, there is a good correlation between the assessment of this irradiance and the satellite measurements: the mean absolute error is below 0.5 W.m^{-2} using the SG2 formula and below 2.5 W.m^{-2} using equation 1.2, for the period from September 2014 to September 2015 (Figure 2.5).

2.1.3. *Solar irradiance interaction with the atmosphere*

Extra-terrestrial solar irradiance is partially absorbed and dispersed along its path through the atmosphere. This process is at the origin of the direct and indirect (diffuse) components of the irradiance received by an observer at the ground level. Focusing on the evolution of its direct component, the following general equation can be written (Bouguer–Beer–Lambert law, this equation being valid only for monochromatic quantities):

$$I = I_0 e^{-m\tau} \quad [2.3]$$

where I_0 is the incident extra-terrestrial solar irradiance at the Earth/Sun distance D_0, m is the relative optical air mass and τ is an attenuation coefficient taking into account the composition of the atmosphere.

2.1.3.1. Relative optical air mass

The relative optical air mass corresponds to the ratio of the distance traveled by light through the atmosphere and the distance traveled by this beam through a standard atmosphere, at the sea level and with the Sun at the zenith. The position of the Sun is therefore necessary to calculate this air mass. Assuming a flat atmosphere in which light propagation is rectilinear, the relative optical air mass is calculated as follows (Figure 2.6):

$$m = \frac{1}{\cos \mathrm{SZA}} \qquad [2.4]$$

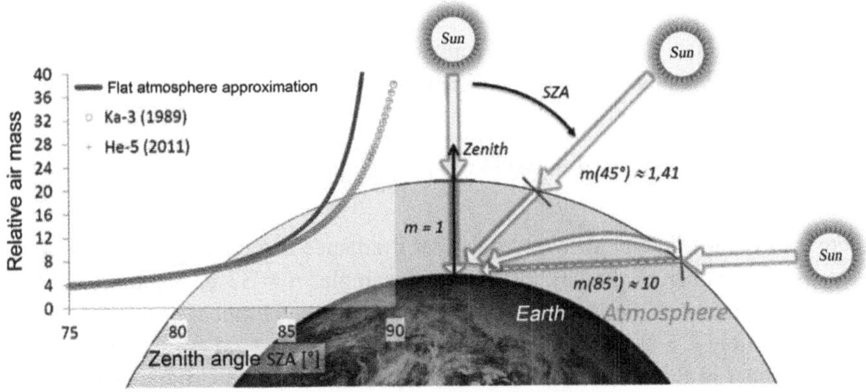

Figure 2.6. *Schematic representation of the relative optical air mass. For a color version of this figure, see www.iste.co.uk/flamant/energy.zip*

However, the Earth's curvature and the variable densities of the atmosphere lengthen the mean path of solar radiation, particularly for wide zenith angles (Figure 2.6). On the other hand, the site altitude shortens this path. In order to take such phenomena into account, more complex formulas were developed. In 2011, Rapp-Arrarás and Domingo-Santos conducted a comparative study of 26 formulations of m, at the central wavelength of the solar spectrum (727.4 nm) and for three different atmospheric profiles, with an observer located at the sea level (Rapp-Arrarás and Domingo-Santos 2011). The 26 equations were grouped into five categories, depending on the number of adjustable parameters in the equation; the equations having a larger number of parameters being generally more precise. The maximum error between the empirical air mass and the reference one is obtained when the Sun is at the horizon, therefore when the effects of optical distortions are the most notable. For each category, Rapp-Arrarás and Domingo-Santos (2011) have then identified the empirical formula leading to best results.

Category	Name of the model	Max. absolute error	Max. relative error
1	Dr-1	1.70	4.45%
2	Ra-2	2.91×10^{-1}	0.76%
3	He-3	3.28×10^{-2}	0.086%
4	He-4	2.49×10^{-3}	0.0065%
5	He-5	3.24×10^{-4}	0.0008%
3	Ka-3	6.90×10^{-2}	0.18%

Table 2.2. *Synthesis of the most successful formulas (Rapp-Arrarás and Domingo-Santos 2011)*

Table 2.2 synthesizes the best results obtained for each category and for the US Standard Atmosphere (USSA-76) (Champion et al. 1985; Kivalov 2007). Generally speaking, it can be noted that each time a new parameter is added to the equation, the precision of m calculation gains one order of magnitude. Moreover, the best equation for categories 3–5 relies on the same type of mathematical expression. This expression was initially introduced for modeling the delay of radio signals traveling through the atmosphere (Mendes et al. 2002; Boehm et al. 2006). The formula for the He-5 model is:

$$m = \frac{1 + a_1/\left(1 + a_2/\left(1 + a_3/\left(1 + a_4/(1 + a_5)\right)\right)\right)}{\cos \mathrm{SZA} + a_1/\left(\cos \mathrm{SZA} + a_2/\left(\cos \mathrm{SZA} + a_3/\left(\cos \mathrm{SZA} + a_4/(\cos \mathrm{SZA} + a_5)\right)\right)\right)} \qquad [2.5]$$

As a comparison, the result obtained with the Ka-3 formula (Kasten and Young 1989) was also added to this table. This equation is very frequently cited and currently used by the scientific community; it can, however, be noted that it does not yield better results, under the test conditions. Its formulation is:

$$m = \frac{1}{\cos \mathrm{SZA} + \dfrac{a_1}{(a_2 - \mathrm{SZA})^{a_3}}} \qquad [2.6]$$

The relative optical air mass calculated with equations [2.4]–[2.6] can be seen in the diagram in Figure 2.6. It can be noted that the difference between these three formulas is small for solar heights above 10° (SZA < 80°). On the other hand, as the Sun is approaching the horizon, it can be noted that the optical air mass for equations [2.5] and [2.6] increases at a lower rate than the one calculated using equation [2.4]. This makes it possible to reach a non-divergent value when the Sun is at the horizon ($m(90°) \approx 38$). Nevertheless, the difference between equations [2.5] and [2.6] is not perceptible on this diagram: He-5 and Ka-3 appear to overlap. This result is not surprising; though He-5 is better than Ka-3, their absolute errors are very small (see Table 2.2). Generally speaking, most of the formulations tested by Rapp-Arrarás and Domingo-Santos may be suitable for the majority of applications.

NOTE.– The precision of these formulas may be misleading. Indeed, variations of over 10% can be observed between the formula and the reference when the structure and the properties of the layers of the atmosphere are significantly different from the standard conditions. A more robust formulation could be used in order to minimize the error in case of "atypical" conditions. The challenge is therefore to determine the evolution of He-5 parameters depending on the specificities of the place of study (altitude, humidity, pressure, temperature) and wavelength. According to Rapp-Arrarás and Domingo-Santos, such a formulation may lead to a precise evaluation of the optical air mass, clearly surpassing the precision of equation Ka-3.

2.1.3.2. *Attenuation coefficient*

As equation [2.3] clearly shows, the attenuation coefficient τ also modifies the extra-terrestrial radiation. Until present, for the sake of simplification, the spectral specificities of radiation have been ignored. This is no longer possible when the focus is on its interaction with the atmosphere.

The Sun can be assimilated to a black body whose surface temperature is around 5,770 K. Therefore, the radiation emitted by the Sun has a spectral distribution that can be modeled by Planck's law. Within this distribution, 10% of the energy is in the UV region, the rest being more or less equally distributed between the visible and the infrared, with a maximum for the wavelength λ = 500 nm (Figure 2.7). When traveling through the atmosphere, the extra-terrestrial radiation is partly absorbed or diffused, reducing the direct radiation received on the ground. This phenomenon corresponds to the transmission coefficient $e^{-m\tau}$ of equation [2.3].

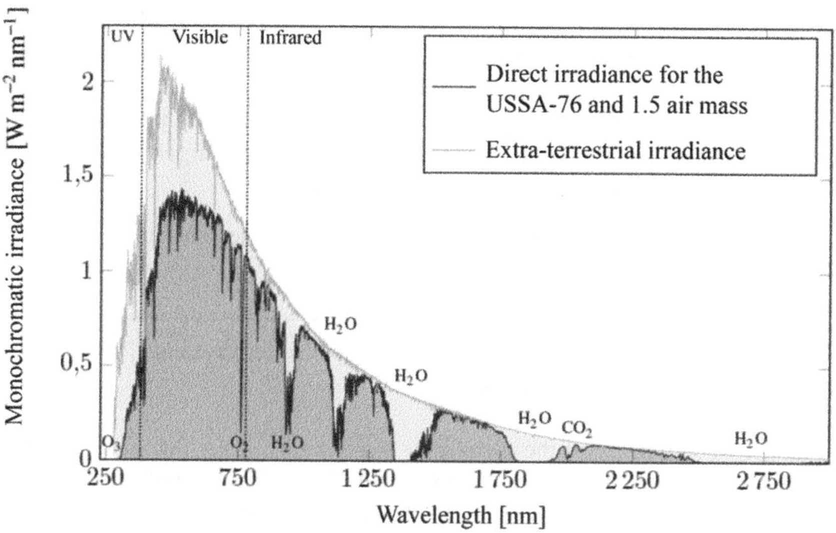

Figure 2.7. *Spectral distribution of the extra-terrestrial irradiance and of the irradiance at the ground for a USSA-76 and a 1.56 air mass (NREL data, https://www.nrel.gov/grid/solar-resource/spectra-am1.5.html). For a color version of this figure, see www.iste.co.uk/flamant/energy.zip*

The absorption of solar radiation by the atmosphere is selective. Four components are mainly responsible for this absorption: ozone, oxygen, carbon dioxide and water vapor (Figure 2.7). The ozone absorbs a part of UV; oxygen absorbs certain narrow bands of the visible spectrum; carbon dioxide absorbs essentially in the far-infrared range; finally, water vapor absorbs many spectral bands, mainly in the near-infrared range.

If radiation is not absorbed, it can be deflected. Similar to absorption, scattering depends on the wavelength. But in this case, the density and the size of atmospheric particles have also an influence. Two types of scattering are predominant: Rayleigh scattering and Mie scattering. Rayleigh scattering occurs when the size of the particles is significantly below the incident wavelength ($<\lambda/10$). This happens for gaseous molecules (O_2, N_2, CO_2) and fine dust particles. This scattering is all the stronger as the wavelength is shorter, which explains why the sky is blue. Mie scattering occurs when the size of the particles is comparable to the wavelength of the incident radiation. This is the case for water vapor and aerosols (dusts, fumes, pollens). However, the latter is less sensitive to the wavelength: Mie scattering is approximately proportional to $1/\lambda$ (Ångström law), while Rayleigh scattering is proportional to $1/\lambda^4$. Consequently, Mie scattering generates a paler color. Moreover, this scattering is stronger in the direction opposite to that of the incident wave. This phenomenon is all the stronger as the particle is larger. Based on these various characteristics, the following can be deduced:

– Mie scattering is predominant at low altitude (where aerosols are concentrated) and in the Sun's corona (given its angular distribution);

– Rayleigh scattering is predominant in the rest of the sky, except when the atmosphere is overloaded with aerosols, water vapor or ice crystals. This is the case in highly polluted areas and near deserts.

Finally, when the size of the particles is significantly larger than the wavelength of the incident radiation, a non-selective scattering, which is independent of the wavelength, occurs. This is, for example, the case of clouds or fog, which appear to be white.

Clouds have a significantly stronger influence on direct irradiance than any other atmospheric phenomenon: there can be total attenuation when a thick cloud overshadows the Sun. Moreover, this attenuation does not depend on the incident wavelength, unlike other phenomena. This is why in the modeling of solar irradiance, the "clear-sky" situation is generally considered separately from the "cloudy sky" situation. Equation [2.3] can be rewritten as follows:

$$I = I_0 e^{-m(\tau_{\text{cs}}+\tau_{\text{nuage}})} = e^{-m\tau_{\text{nuage}}} \cdot I_0 e^{-m\tau_{\text{cs}}} = k_c \cdot I_{\text{cs}} \qquad [2.7]$$

where k_c is the clear-sky index and I_{cs} is the irradiance directly received under clear-sky conditions, depending on the attenuation coefficient under clear-sky conditions τ_{cs}. This coefficient takes into account the absorption and selective scattering phenomena presented in this section. The clear-sky index indicates the presence or absence of a cloud in front of the Sun. This coefficient varies between 0 and 1: $k_c = 0$ signifies that a thick cloud overshadows the Sun, while $k_c = 1$ signifies the absence of clouds in front of the Sun. Finally, intermediate cases appear when thin clouds, such as cirrus, overshadow the Sun (Figure 2.8). In what follows, "clear-sky conditions" are applicable when $k_c = 1$.

Equation [2.7], which dissociates the attenuation of direct solar irradiance produced by the atmosphere from its attenuation produced by the clouds, is fundamental and is a starting point for many DNI forecasting models.

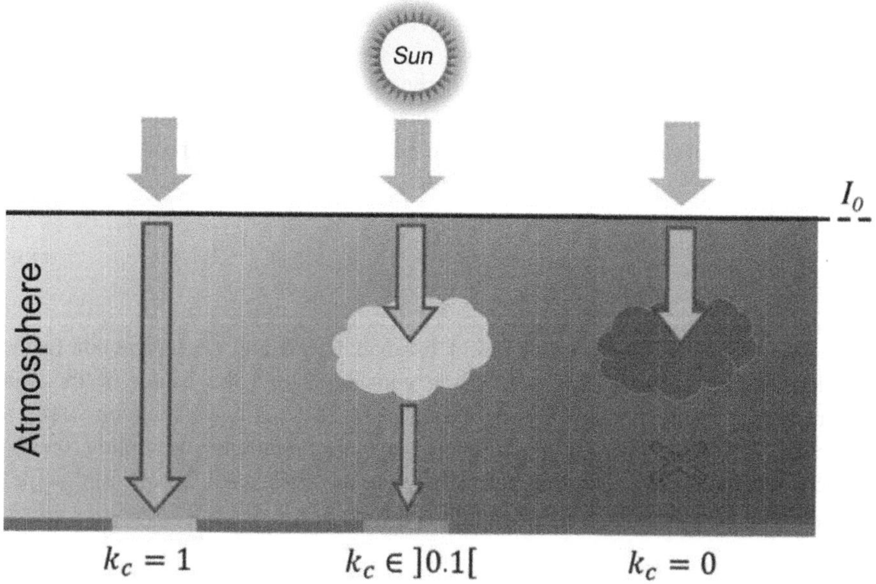

Figure 2.8. *Schematic representation of the clear-sky index k_c. For a color version of this figure, see www.iste.co.uk/flamant/energy.zip*

NOTE.– It is important to note that if, under clear-sky conditions, there is no cloud in front of the Sun (from the observer's point of view), this does not mean that there is no cloud in the sky (still from the observer's point of view). The latter case is rather referred to as "cloudless sky".

2.1.4. *Components of solar irradiance and associated instruments*

As noted in the previous section, a fraction of the extra-terrestrial irradiance was absorbed while traveling through the atmosphere. Another fraction of this radiation is scattered, leading to an indirect radiation for the observer. This section offers a detailed presentation of the direct and indirect components of the irradiance, as well as the associated measurement instruments.

2.1.4.1. *Direct and circumsolar irradiance*

The direct component of solar irradiance is referred to as DNI, which stands for *direct normal irradiance*. DNI is more precisely defined by ISO-9488 (ISO 1999) as "the quotient of the radiant flux on a given plane receiver surface received from a small solid angle centered on the Sun's disk to the area of that surface." If the plane is perpendicular to the axis of the solid angle, direct normal solar irradiance (DNI) is received. Let us denote by I the radiance expressed in W m^{-2} sr^{-1}, emanating from the solar disk and from its near neighborhood. Under clear-sky conditions, it is reasonable to assume that radiance has a radial symmetry in the vicinity of the Sun. The ISO-9488 definition of DNI can then be formulated as follows (Blanc et al. 2014):

$$I(\xi_{\max}) = 2\pi \int_0^{\xi_{\max}} I(\xi) \sin \xi \, d\xi \qquad [2.8]$$

where I corresponds to the measured or estimated DNI and ξ corresponds to the angular distance between the point being considered and the center of the Sun (Figure 2.9). Even though this definition can be readily understood from a theoretical point of view, it is the object of interpretations, depending on the respective field of application, due to the vagueness of the term "small solid angle" related to the parameter $\xi_{\max}$ of equation [2.8].

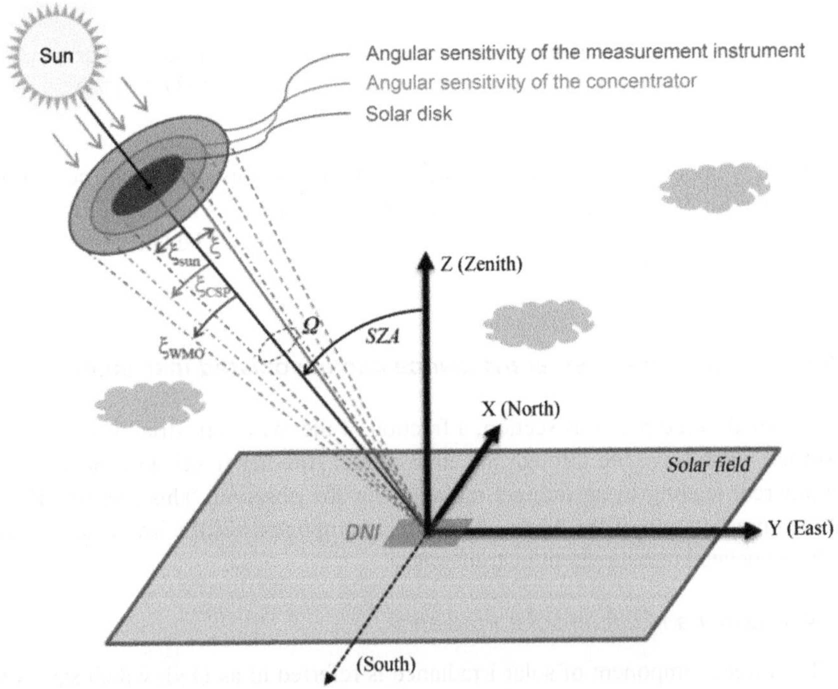

Figure 2.9. *Schematic representation relative to DNI definition. For a color version of this figure, see www.iste.co.uk/flamant/energy.zip*

In the context of the modeling of radiative exchanges in the atmosphere, the strict definition of DNI refers to photons that did not interact with the atmosphere along their path to the observer (Berk et al. 1998; Gueymard 2001). Then, the angle ξ_{max} of equation [2.8] corresponds to the apparent radius of the solar disk. From the Earth, this apparent radius is $\xi_{\text{Sun}} = 0.2666° \pm 1.7\%$ (Figure 2.9), the variation of ξ_{Sun} being due to the variations of the Earth–Sun distance (see section 2.1.2). It can be deduced that DNI of the solar disk, denoted by I^{Sun}, is then defined by $I^{\text{Sun}} = I(\xi_{\text{max}})$. It is important to note that DNI of the solar disk is sometimes distinguished from the strict DNI initially mentioned. Indeed, while strict DNI considers only the photons that did not interact with the atmosphere, I^{Sun} also takes into account the diffuse irradiance of the solar disk. Nevertheless, this subtle difference is generally considered negligible (Blanc et al. 2014).

At this stage, it is important to mention that DNI is the only component of the solar irradiance used by concentrating systems, hence by "concentrated solar power" (CSP) plants. It is for this reason that our focus is on the interaction between direct solar irradiance and the atmosphere in section 2.1.3. In practice, the angular sensitivity of solar concentrators is, however, higher than the apparent radius of the Sun (Rabl and Bendt 1982; Buie and Monger 2004). In other terms, irradiance concentrated on the receiver is composed of the irradiance from the solar disk and also from a part of the solar corona, which is referred to as circumsolar irradiance (Figure 1.9). Similarly to DNI, circumsolar irradiance can be formulated as follows (Blanc et al. 2014):

$$\mathring{I}(\xi_{\text{min}}, \xi_{\text{max}}) = 2\pi \int_{\xi_{\text{min}}}^{\xi_{\text{max}}} I(\xi) \sin \xi \, \mathrm{d}\xi \qquad [2.9]$$

where $\mathring{I}$ corresponds to the irradiance of the solar aureole between the angles ξ_{min} and ξ_{max}. Since this aureole influences the optical performances of concentrators, it is preferable to include its contribution in the assessment of the availability of the solar resource. This is why it makes sense to include the solar corona in the definition of DNI, in other terms, to choose ξ_{max} above ξ_{Sun}. It can be noted that equation [2.8] can be rewritten as:

$$I(\xi_{\text{max}}) = I^{\text{Sun}} + \mathring{I}(\xi_{\text{Sun}}, \xi_{\text{max}}) \qquad [2.10]$$

The angular sensitivity of solar concentrators is, however, not constant. Indeed, it varies depending on the type of technology employed (Rabl 1976) and on the position of the Sun (Blanc et al. 2014). For a CSP plant, ξ_{CSP} normally ranges between 0.7° and 2.3°. There may be an infinity of possible definitions for DNI, depending on the concentrator employed. On the other hand, it is recommended to conduct all the DNI measurements with the same aperture angle, to allow for a comparison between measurements conducted with various instruments.

In the field of metrology of solar radiation, the World Meteorological Organization (WMO) recommends the measurement of DNI with half-aperture angle $\xi_{WMO} = 2.5°$ (WMO 2010). It can be noted that ξ_{WMO} is nearly 10 times larger than ξ_{Sun}. On the other hand, this angle is slightly larger than ξ_{CSP}, which makes it possible to include all the range of angular sensitivities covered by CSP technologies (Figure 2.9). Nevertheless, an overestimation of DNI is likely, since the measurement systematically records a surplus of irradiance from the solar corona. Moreover, this overestimation is not constant. Indeed, as already noted, the solar corona was dominated by Mie scattering, which is essentially a function of the distribution and size of the aerosols. Thin clouds also influence this corona. Since these phenomena are variable (Gueymard 2012), the distribution of irradiance inside the solar corona is also variable. This is why it is difficult to know the DNI received by the concentrator, relying only on the DNI measured by the instrument. But this is highly desirable for an optimized management of the power plant. Given this observation, new measurement instruments are developed in order to be able to assess the DNI distribution in the solar corona (see section 2.1.4.1.2).

Before attempting to correct this DNI overestimation, it must be quantified in order to eliminate the possibility of a false problem. In order to quantify the fraction of DNI in the solar corona, the *circumsolar ratio* (CSR) is generally used, which is defined as follows (Blanc et al. 2014):

$$\mathrm{CSR}(\xi_{max}) = \frac{\mathring{I}(\xi_{Sun}, \xi_{max})}{I(\xi_{max})} \qquad [2.11]$$

Numerical simulations (Gueymard 2010; Blanc et al. 2014) and measurement campaigns (Wilbert et al. 2013; Wilbert 2014) tried to estimate CSR. Simulations have also shown that the contribution of circumsolar irradiance for a half-angle of aperture ξ_{WMO} rarely exceeds the 5% of DNI under clear-sky conditions. Nevertheless, when a thin cloud overshadows the Sun, CSR may exceed 20%. For a DNI close to 600 W·m^{-2}, a CSR over 20% was measured in 7% of cases (Blanc et al. 2014). This means that for a solar concentrator with an angular sensitivity nearing ξ_{Sun}, less than 80% of the measured DNI is effectively collected. The "thin cloud" case is a real challenge for the management of CSP plants. This is why the measurement of DNI distribution in the solar corona is fully justified.

2.1.4.1.1. DNI measurement instruments

It is important to recall that CSP plants use only the direct component of the solar irradiance. Moreover, since solar energy is transformed into thermal energy, the spectral specificities of irradiance are less important; what counts is the DNI integrated over the whole spectrum.

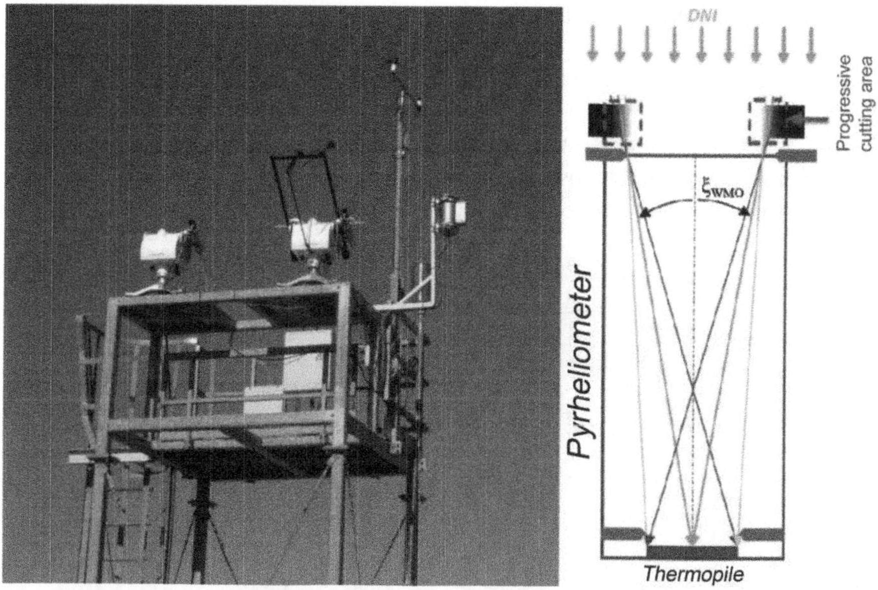

Figure 2.10. *Left: image of two pyrheliometers on solar trackers. Right: diagram of a pyrheliometer. For a color version of this figure, see www.iste.co.uk/flamant/energy.zip*

Pyrheliometer. The most commonly used instrument for the measurement of DNI is the pyrheliometer. It takes the form of a tube whose dimensions define the aperture angle and inside which a thermopile is placed (Figure 2.10). In order to measure the direct solar flux, the pyrheliometer must be aligned with the axis of the Sun; a solar tracker is therefore needed for the automated DNI acquisition. The uncertainty on the measurement strongly depends on the calibration, as well as on the maintenance of the instrument (Geuder and Quaschning 2006; Pape et al. 2009). For a properly calibrated and daily maintained pyrheliometer this uncertainty is around ±2% (WMO 2010). While, in the past, many aperture angles were tested (Blanc et al. 2014), all the modern pyrheliometers presently follow the WMO recommendations, namely $\xi_{WMO} = 2.5°$ (Sengupta et al. 2015). Hence, having $\xi_{Sun} \ll \xi_{WMO}$ makes it possible to reduce the errors related to the imperfect alignment of the pyrheliometer with the Sun. As a last important point, the measurement of DNI is conducted throughout the surface of the thermopile; it is not a punctual measurement. Consequently, the aperture angle of the pyrheliometer, seen from the thermopile, varies depending on the considered point of measurement: the solar flux does not stop short at the cut-off angle ξ_{WMO} (Figure 2.10, right). This progressive cut-off area, known as shadow function, affects the measurement of DNI and therefore must be taken into account.

Alternatives to pyrheliometer. The cost of a pyrheliometer is about 2.5 k€ HT (CHP1, Kipp & Zonen). That of a solar tracker is 10 k€ HT (SOLYS 2, Kipp & Zonen). Moreover, in order to optimally operate, such a system requires daily maintenance. Due to these cost and maintenance constraints, less costly and more robust instruments were developed.

Figure 2.11. *Two RSI installed at the PROMES-CNRS laboratory, Perpignan. For a color version of this figure, see www.iste.co.uk/flamant/energy.zip*

The rotating shadowband irradiometer (RSI) is currently without any doubt the most fashionable instrument. This instrument features a horizontally assembled photodiode used for the measurement of global irradiance (Figure 2.11). A rotating band periodically overshadows the photodiode in order to measure the diffuse radiation, which is the global radiation without its direct component. Based on measurements of global and diffuse radiation and knowledge of the Sun's position, DNI can then be estimated (see section 2.1.4.3). This instrument has the advantage of providing several measurements of irradiance, unlike the pyrheliometer. The fact of having a simple rotating band instead of a solar tracker on two axes also reduces the complexity of the system while improving its robustness. Moreover, studies have shown that RSI was far less sensitive to soiling than the pyrheliometer (Geuder and Quaschning 2006; Geuder et al. 2014). This reduced sensitivity is notably explained by the geometry of the photodiode: the scattering of solar radiation by the dust is less constraining. Finally, the RSI costs about 7 k€ HT (Twin-RSI, CSP Services), which is approximately 60% less expensive than a pyrheliometer with tracker. Besides these various advantages, a precise spectral calibration of the photodiode is required. This calibration is complex, generating a higher uncertainty on the DNI than for a pyrheliometer: this is typically about ± 5%. A further drawback: the definition of an aperture angle is problematic, since DNI is measured indirectly. It is, however, estimated that $\xi_{\mathrm{RSR}} \approx 2.9°$ (Wilbert 2014).

It is important to note that a camera with a wide angle of view without a shading device (see section 2.2.3.2) can potentially be used as a DNI measurement instrument (Chauvin et al. 2019a, 2019b). Indeed, the camera is a device measuring the light intensity of the scene it observes. The Sun being in the field of view of the system, it is possible to measure its intensity and therefore deduce from it the DNI. There are many advantages: no mobile element, an ideal shadow function, a reasonable cost. Nevertheless, the problem of pixel saturation in the circumsolar area, as well as the reflections occurring in the optical system, can prove to be a stumbling block for reaching a reliable measurement of DNI.

2.1.4.1.2. Measurement instruments for circumsolar irradiance

In recent decades, there were many attempts to measure circumsolar irradiance. The first notable experiment was conducted by *Lawrence Berkeley National Laboratory* (LBNL). The latter recorded nearly 180,000 solar profiles at 11 different sites in the United States between 1976 and 1981 (Grether et al. 1980). The instrument employed was a telescope supported by a solar tracker and equipped with a circular aperture such that $\xi_{\mathrm{LBNL}} = 3.2°$. A reduced and filtered database was extracted from these measurements and used as a starting point for the development of models of circumsolar irradiance (Buie et al. 2003). Later on, the DLR (German aerospace center) acquired 2,300 solar profiles by means of a CCD camera on three different sites in Europe (Neumann and von der Au 1997). The cameras were accompanied by a pyrheliometer allowing their radiometric calibration.

Following these first attempts, an instrument dedicated to the measurement of the solar profile was developed: *Sun & Aureole measurement* (SAM). This instrument, accompanied by a solar spectrometer, is actually the system with the best performances in the measurement of the solar profile (Wilbert et al. 2013). This profile is measured at the wavelength $\lambda_{\mathrm{SAM}} = 670$ nm and then estimated over the whole spectrum using a radiative transfer model (Gueymard 1993, 2001), even though multispectral measurements are also possible. It is important to note that this instrument has a half-aperture angle $\xi_{\mathrm{SAM}} \approx 7°$.

Three SAM dedicated to solar applications are nowadays operating in the world. One of them is installed on the roof of the PROMES-CNRS laboratory at Odeillo. An example of measurement of the solar profile conducted with this instrument is shown in Figure 2.12.

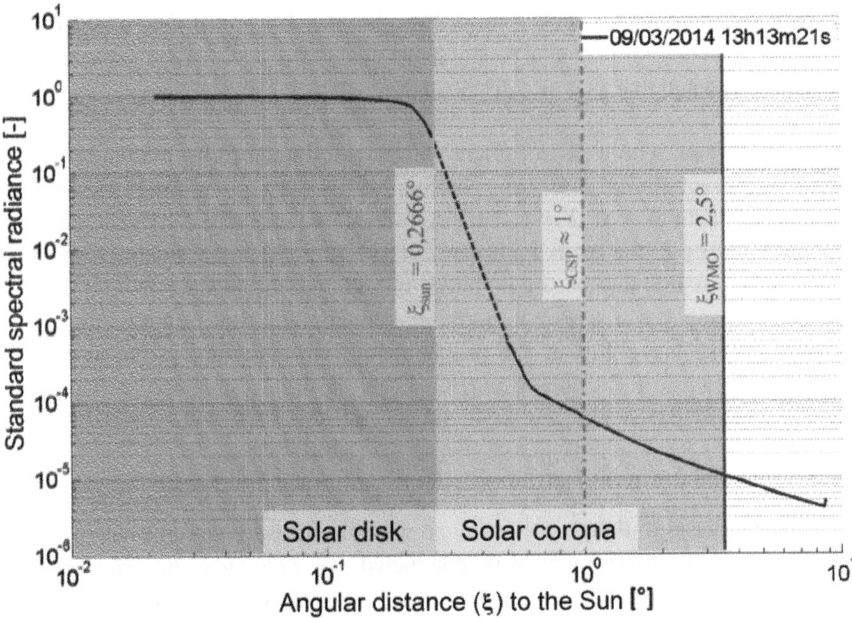

Figure 2.12. *Example of the measurement of the solar profile using SAM (Odeillo). For a color version of this figure, see www.iste.co.uk/flamant/energy.zip*

The drawback of SAM is its high cost: the full system costs approximately €80k. This is why less expensive alternatives were developed. One of them involves the use of a pair of pyrheliometers (Figure 2.13b) (Wilbert et al. 2012) with different aperture angles ($\xi_{P1} \in [1.25° ; 2.5°]$ and $\xi_{P2} = 2.5°$). The comparison between the two measurements makes it then possible to calculate $\mathring{I}(\xi_{P1}, \xi_{P2})$. On the other hand, the method is less precise than SAM because only two points of the solar profile are evaluated. Moreover, this instrument uses a photodiode that is not sensitive to the wavelengths of circumsolar irradiance, entailing a risk of underestimating the latter. Figure 2.13 presents an image of SAM and of the pair of pyrheliometers.

a) SAM at PROMES-CNRS laboratory (Odeillo) – Visidyne

b) BPI CSR 460 sensor (Wilbert 2014) – Black Photon Instruments

Figure 2.13. *Systems for the measurement of circumsolar irradiance. For a color version of this figure, see www.iste.co.uk/flamant/energy.zip*

Here too, a sky imager may be compared with these various systems (Chauvin et al. 2019a, 2019b). Indeed, it may be possible to deduce the circumsolar irradiance from the examination of the evolution of the light gradient near the Sun. It is important to note, however, that this type of measurement is also plagued by the previously mentioned saturation and reflection problems.

2.1.4.2. *Diffuse irradiance*

Diffuse irradiance corresponds to the irradiance that interacted with the atmosphere before reaching the observer. It is generally measured on a horizontal surface, and in this case, it is referred to as DHI, which stands for *diffuse horizontal irradiance*. Theoretically speaking, DHI includes the part of diffuse irradiance located in the solar corona. In practice, DHI does not contain this part, as it is included in the measurement of DNI (see section 2.1.4.1). The measurement of DHI therefore integrates only the solar irradiance over the totality of the celestial vault, for angles above ξ_{WMO} (Sengupta et al. 2015).

The most commonly used instrument for the measurement of DHI is certainly the irradiometer. Similarly to the pyrheliometer, the irradiometer uses a thermopile in order to convert the solar flux into a thermal flux and then into an electric signal. In the case of DHI measurement, the irradiometer is generally installed on a solar tracker equipped with a shading ball masking the Sun. The uncertainty for a conveniently calibrated and regularly maintained irradiometer is of about 3.2% (Sengupta et al. 2015).

Similar to the estimation of DNI, a camera with a wide angle of view is potentially capable of providing a measurement of DHI. The advantages are similar to those mentioned for the measurement of DNI. Moreover, the measurement of DHI is facilitated here by the absence of the Sun in the measurement area: the problem of pixel saturation is not important.

2.1.4.3. *Global irradiance*

The global irradiance corresponds to the sum of direct and indirect irradiances reaching the ground. If it is measured horizontally, it is referred to as GHI, which stands for *global horizontal irradiance*. The global horizontal irradiance can be expressed as a function of DHI and DNI, as follows:

$$\mathrm{GHI} = \mathrm{DHI} + \mathrm{DNI} \cdot \cos(\mathrm{SZA}) \qquad [2.12]$$

If DNI, DHI and the position of the Sun are known, then GHI can be obtained. However, in practice, GHI is measured using an irradiometer for which the Sun is apparent. A solar tracker is therefore not needed for the measurement. Consequently, the measurement of GHI is the less complex of all. This is why equation [2.12] is generally reversed in order to obtain DNI from GHI, DHI and the position of the Sun. It is the basic principle of RSI presented in section 2.1.4.1.1.

Similarly to the DHI and DNI, a camera with a wide angle of view is potentially capable of providing this type of measurement (Chauvin et al. 2019a, 2019b).

2.2. Forecasting of direct normal irradiance

DNI forecasting is essential if its availability and variability are to be anticipated. To this date, the main studies conducted focus, however, on GHI forecasting, in order to forecast the power generation of photovoltaic solar power plants (Chow et al. 2011; Stefferud et al. 2012; Fu and Cheng 2013; Bernecker et al. 2014; West et al. 2014; Yang et al. 2014; Schmidt et al. 2015; Voyant et al. 2017; Tolba et al. 2020). These works are, however, not directly transposable to DNI, even though, for the forecasting by sky imager, the techniques for the detection of clouds and their motion are identical (Law et al. 2014). Indeed, DNI variability is far more important than that of GHI, due to the greater influence of clouds on direct irradiance. On the other hand, for applications such as CSP, it is more important to forecast ramps and other sudden changes that may occur, which often requires a higher demand in terms of precision, forecasting horizon and space–time resolution.

2.2.1. *Definitions and needs of an operator of CSP plant*

As far as forecasting is concerned, it is important to start by defining the space–time characteristics; all the more so as depending on these characteristics, some approaches are more appropriate than others (see section 2.2.2).

The first notion is certainly the forecasting time horizon: it corresponds to the duration between the present moment and the moment for which the forecasting is made. This horizon is linked to the needs of the operator; in the energy sector, the "very short term" refers to an intra-hour horizon[2] and the "short term" to an intra-day horizon. Beyond this, there is daily, seasonal or annual forecasting. The temporal resolution step corresponds to the sampling period of the forecasting. For example, a forecasting can be made up to 1 h (horizon) with a time step of 5, 10 or 20 min (the resolution step).

The space horizon and the spatial resolution step are related to the surface on which the forecasting is made and to its sampling. Even though the huge majority of the existing works until present deal with *point forecasting*, it may be important to focus on the spatial validity of this forecasting: one forecasting may be precise at one point of a CSP plant without being so throughout the solar field. For example, a forecasting can be made over a 10 km^2 surface (horizon) cut into 1 km^2 elements (resolution step).

Finally, in the case of a real-time forecasting, a duration between two updates of the forecasting can be defined. This duration must be shorter than the chosen time horizon; on the other hand, it is independent of the temporal resolution. It is indeed possible to refresh the forecasting as soon as a new measurement is obtained, all the more so as a high refreshing speed makes it possible to reduce the forecasting error. The refreshing speed of the forecasting is, however, limited by the running time of the forecasting model; generally speaking, the objective is to get as close as possible to the running time of the forecasting model, in order to minimize the forecasting errors while operating in real time.

The diversity of forecasting horizons results from the diversity of needs of an operator of the CSP plant. Figure 2.14 provides a non-exhaustive summary of the needs of an operator of a solar power plant in terms of forecasting horizon, for various objectives of power plant control.

2 Some use the term *nowcasting*, an oxymoron that will hopefully fall into oblivion.

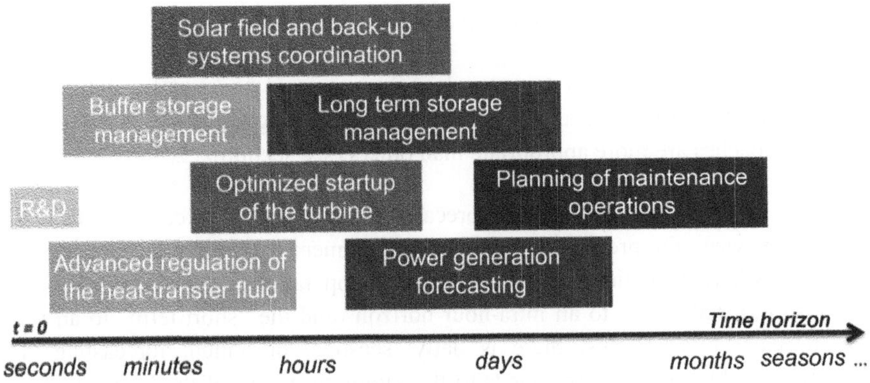

Figure 2.14. *Needs of an operator in terms of forecasting horizon, depending on the intended application*

EXAMPLE 1.– For an optimized startup of a turbine, DNI forecasting over a time horizon equivalent to the turbine characteristic time (up to several hours) is useful, with a sufficiently short resolution step in order to best apprehend the rise in temperature of the equipment and to anticipate the potential need for an electric backup. Referring to the spatial aspect of the forecasting, an averaged value over the entire power plant may be sufficient.

EXAMPLE 2.– At a finer scale, the regulation of the heat-transfer fluid flow rate in real time would allow the optimization of the performances of a parabolic trough power plant. If the average time of passage of the heat-transfer fluid through a receiver is of about 20 min, a forecasting over a time horizon nearing 1 h, with a temporal resolution step of about 5 min, seems relevant. The space horizon is the characteristic length of the power plant, but the spatial resolution step must be of the order of magnitude of the parabolic trough modules.

2.2.2. *The main DNI forecasting techniques*

Depending on the considered forecasting horizon, various approaches are possible.

At an intra-hour scale, the cameras with a wide angle of view are probably the most adequate systems for DNI forecasting. Thanks to their high spatial–temporal resolution (Figure 2.15), they allow the fine observation of the local motion of clouds and also the anticipation of the power variations related to rapid DNI variations. Further details on this approach are provided in section 2.2.3.2.

At an intra-day scale, the satellite imagery is adapted to DNI forecasting (Figure 2.15): it brings indeed a more global view of cloud motion. Moreover, these images are collected at various wavelengths, allowing a spectral measurement of the interactions between the solar radiation and the atmosphere. These radiance measurements, combined with radiative transfer equations applied to various layers of the atmosphere, make it possible to assess the solar resource throughout the globe. Another alternative allowing the assessment of the solar resource involves the calibration of the pixels of the satellite images with irradiance measurements at the ground level. Examples related to the resource assessment by means of satellite images can be found in Hammer et al. (2003), Rigollier et al. (2004), Aguiar et al. (2015) and Alonso-Montesinos and Battles (2015). Parallel to the resource assessment, an assessment of cloud motion allows the anticipation of the irradiance variations in the coming hours (Hamill and Nehrkorn 1993). In order to make the forecasting, it is generally assumed that the optical depth of clouds and their velocity are constant, and that clouds do not undergo rigid deformations.

Beyond the intra-day scale, the solar resource forecasting is generally made by means of numerical weather forecasting (NWP) models (Figure 2.15). These models forecast over several days the evolution of weather parameters (temperatures, winds, pressures, etc.) using a fine modeling of the atmosphere. The initial state of the atmosphere is therefore essential for this type of approach; the closer this initial state is to reality, the better the forecasting. These NWP models rely on thousands of code lines that required decades of research. Their complexity is therefore limiting their refresh rate. The best known NWP models include the *Weather and Research Forecasting*, the *Global Forecast System*, the *European Centre for Medium-Range Weather Forecast* or the *Mesoscale Model MM5* (Inman et al. 2013; Law et al. 2014; Lauret et al. 2016; Lopes et al. 2018). Once the evolution of the atmosphere is known, a model of radiative transfer can be used to deduce the evolution of the solar resource. However, the small spatial resolution step of NWP models involves a coarse precision on a local evaluation of DNI variability. The major interest of NWP models for the operator is then to optimize its electric power offer for the coming days.

The three above-mentioned approaches share a common point, namely that they use weather information[3] at least in part. It is, however, also possible to resort to

3 This is an umbrella term for data from sky image or satellite image, to weather data requiring cutting-edge instrumentation, such as that required by NWP models.

purely statistical forecasting models, relying on the analysis of a database and machine learning. Though these models may have difficulties in successfully forecasting the occurrence of particular atmospheric phenomena (such as the passage of a cloud in front of the Sun, for example), they are used to make predictions over various horizons and may certainly prove useful for DNI forecasting. In particular, a model taking advantage of both the weather information and the capacity of generalization and inference characteristic of artificial intelligence techniques may make it possible to do without the fine modeling of very complex weather phenomena (see the discussion in section 2.2.3.2.2).

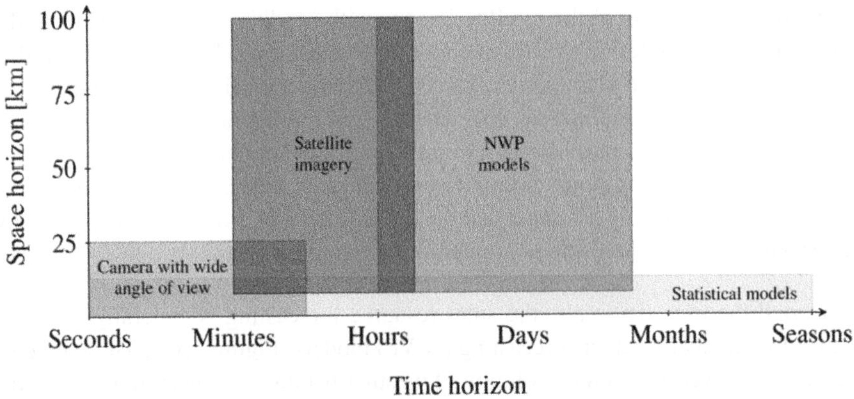

Figure 2.15. *Approaches for solar resource forecasting, depending on the spatiotemporal horizon (adapted from Ramirez and Vindel 2017). For a color version of this figure, see www.iste.co.uk/flamant/energy.zip*

Figure 2.15 summarizes the type of approach depending on the expected spatiotemporal horizons. A closer look at the needs of the operator (Figure 2.14) shows that they overlap the various existing forecasting solutions (Figure 2.15). This is why the optimal forecasting solution for the operator would probably be a combination of forecasting by sky imager, by satellite and by NWP models (see, for example, Inman et al. 2013 and Law et al. 2014).

The results on the spatiotemporal forecasting of DNI are, to this day, rare and result from exploratory work (see the end of section 2.2.3.2.2). In what follows, the focus is on forecasting (in time) localized in space.

2.2.3. *Intra-hour DNI forecasting models*

This part gives several examples of intra-hour DNI forecasting models. As already noted, DNI could be decomposed into a product of irradiance under clear-sky conditions I_{cs} and the clear-sky index k_c (equation [2.7]). DNI forecasting amounts to making a forecasting of these two components:

$$\hat{I}(t+\Delta t) = \hat{k}_c(t+\Delta t) \cdot \hat{I}_{cs}(t+\Delta t) \quad [2.13]$$

where Δt is the forecasting horizon. The major part of the existing models relies on this principle: its major interest is to separate the deterministic component of the irradiance (clear-sky irradiance I_{cs}) from its stochastic component (clear-sky index k_c)[4].

A "clear-sky model" is used for I_{cs} forecasting. Most of the existing clear-sky models originate in the works of Linke (1922), who proposed to express DNI under clear-sky conditions as a function of the optical depth of a clean and dry atmosphere δ_{cda} and a turbidity coefficient T_L, known as Linke turbidity factor:

$$\hat{I}_{cs} = I_0 e^{-\delta_{\text{cda}} \cdot T_L \cdot m} \quad [2.14]$$

The coefficient T_L corresponds to the number of clean and dry atmospheres leading to the noted attenuation. Depending on the manner in which δ_{cda} is calculated, various expressions of the Linke turbidity factor T_L are obtained, leading in the end to various clear-sky models. The best-known formulations of T_L include the works of Kasten (1980) and those of Ineichen (Ineichen and Perez 2002). Further details can be found in the state of the art on clear-sky models given in Chauvin et al. (2018).

2.2.3.1. *Classical persistence models*

2.2.3.1.1. DNI persistence

DNI persistence is undoubtedly the simplest way to achieve forecasting: this approach involves assuming a DNI invariance over the considered horizon. This model, often used as a reference model, can be expressed as follows:

$$\hat{I}_1(t+\Delta t) = I(t) \quad [2.15]$$

4 This deterministic/stochastic distinction should obviously not be taken literally: I_{cs} is actually not *fully deterministic*, while k_c is not *fully stochastic*.

where I corresponds to the measured DNI, $\hat{I}_1$ to DNI prediction and Δt to the forecasting horizon. This model does not require any particular knowledge on the atmosphere and only requires real-time access to DNI measurement. Though simple, it proves extremely efficient for short horizons, or when signal variability is low over the studied forecasting horizon.

2.2.3.1.2. Clear-sky index persistence

DNI persistence has its limits when signal variability is high over the studied forecasting horizon. One part of this variability results from the apparent motion of the Sun in the sky; this motion being precisely known, it is possible to correct its effect.

Based on equation [2.13], a persistence of the clear-sky index, $\hat{k}_c(t+\Delta t) = \hat{k}_c(t)$, yields:

$$\hat{I}_2(t+\Delta t) = \hat{k}_c(t) \cdot \hat{I}_{cs}(t+\Delta t) \qquad [2.16]$$

and using equation 2.7, the following can be written as:

$$\hat{I}_2(t+\Delta t) = \frac{I(t)}{\hat{I}_{cs}(t)} \cdot \hat{I}_{cs}(t+\Delta t) \qquad [2.17]$$

This clear-sky index persistence model[5], whose performance is generally higher than that of DNI persistence, requires real-time access to DNI measurement, and also to a "clear-sky model" allowing the prediction of clear-sky irradiance I_{CS}. It is important to note, however, that attention should be paid to the chosen clear-sky model, as the errors committed on I_{CS} forecasting may rapidly become detrimental (Chauvin et al. 2018).

Forecasting examples using DNI and k_c persistence are given in Figure 2.16, highlighting the superiority of k_c persistence and the inadequacy of these techniques for forecasting ramps or sudden variations of short duration, events that are generally highly important for a CSP plant.

5. It is known as "smart persistence" or "scaled persistence".

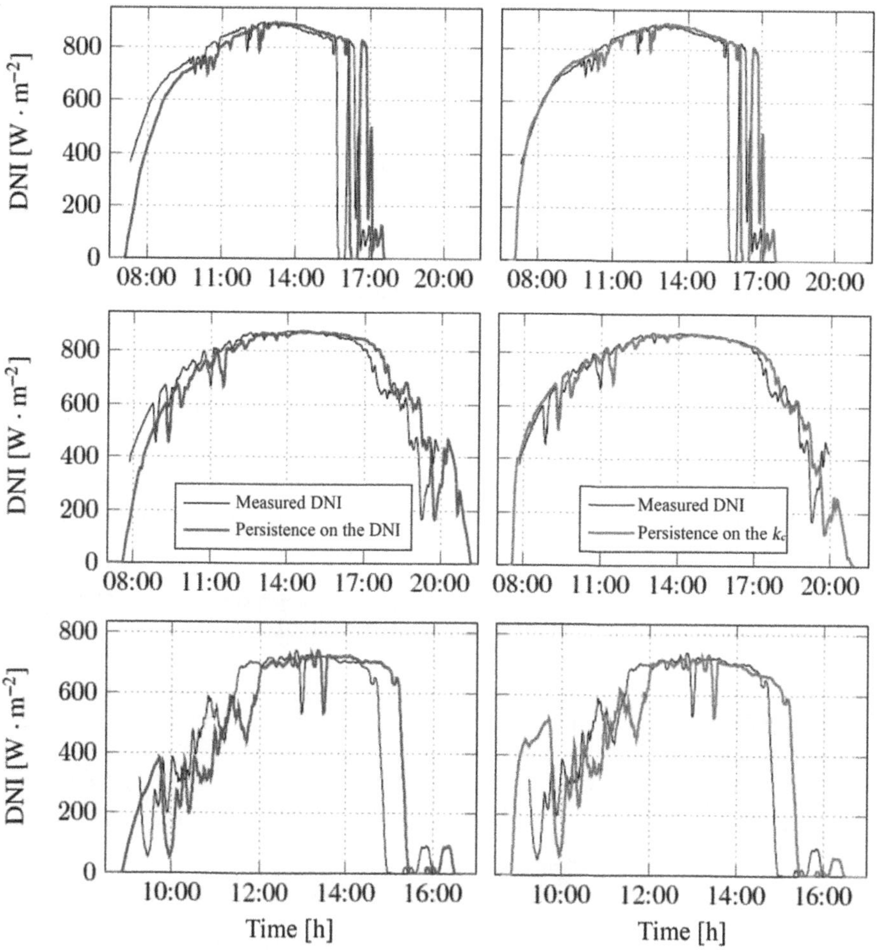

Figure 2.16. *Forecasting examples at* $\Delta t = 30\,min$ *using DNI persistence (on the left) and clear-sky index* k_c *persistence (on the right), over three different days (sampling of measurements: 1 min). For a color version of this figure, see www.iste.co.uk/flamant/energy.zip*

2.2.3.2. *Models relying on a sky imager*

2.2.3.2.1. PROMES sky imager

There are several different systems (commercial solutions, systems developed in the laboratory) that can be used to obtain sky images. They are mainly distinguished by:

– the type of capture: either indirect capture (e.g. when the camera is directed towards the nadir, where a hemispheric mirror is located), or a direct capture (the camera is directed towards the zenith);

– the presence of a Sun overshadowing system: this prevents the saturation phenomena, but it also obscures a part of the sky, which may prove detrimental for the subsequent processes;

– more generally, by the type and characteristics of the material employed: CCD or CMOS sensor, resolution, high dynamic range imagery (HDR), etc.

The systems that do not partly cover the sky (either due to indirect capture or to an overshadowing system) are mostly able to provide essential information concerning the solar resource, particularly if DNI is the focus. However, when Sun appears in the image, saturation phenomena are observed, and it is recommended to minimize them (e.g. by generating HDR images), to avoid losing as little information as possible. A review of the existing systems can be found in Chauvin et al. (2015) and the references it cites.

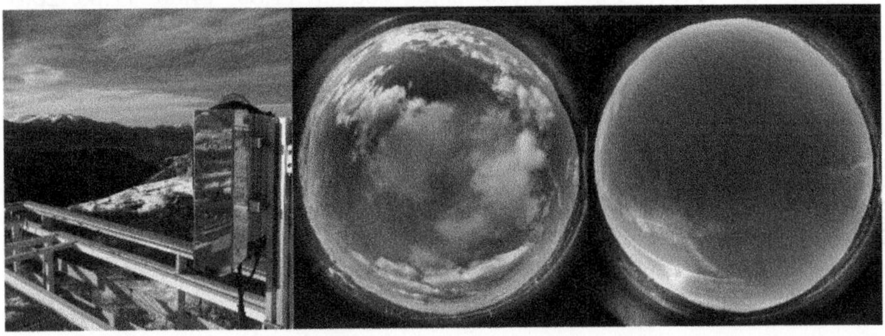

Figure 2.17. *The PROMES sky imager installed at Odeillo and several HDR images acquired (after tone correspondence). For a color version of this figure, see www.iste.co.uk/flamant/energy.zip*

In order to compensate for the limitations of the existing systems and have full setting control, PROMES-CNRS laboratory chose to develop its own physical solution, PSI (*PROMES sky imager*). The current generation, PSI4, makes it possible to produce an HDR image of 1.5 MP (1300 × 1216) every 10 s (see Figure 2.17).

2.2.3.2.2. Forecasting using a sky imager

As already noted (equation [2.13]), DNI forecasting amounts to clear-sky index k_c forecasting and DNI forecasting under clear-sky conditions I_{cs}. The idea of forecasting using a sky imager is therefore simple: it involves the use of a clear-sky model for I_{cs} forecasting and the sky images for k_c forecasting. Indeed, based on a sequence of sky images, it seems possible to forecast the evolution of atmospheric disturbances – and deduce from it k_c – in several simple steps:

1) clouds/clear-sky segmentation;

2) estimation of the cloud speed and direction;

3) forecasting of the position of clouds at the considered forecasting horizon.

However, there are finally few studies on DNI forecasting using a sky imager in the literature, a significant part of which originate in the University of California (see, for example, Chu et al. 2013; Marquez and Coimbra 2013; Quesada-Ruiz et al. 2014; Chu et al. 2015a, 2015b; Li et al. 2016), among other existing works (e.g. Bone et al. 2018; Nou et al. 2018). Indeed, while the above-mentioned first stage is quite well mastered (Chauvin et al. 2015; Karout et al. 2020), the last two are not. The main difficulty lies in the assessment and forecasting of cloud motion. For clouds whose altitude is approximately similar and that are globally moving through a translational motion, the process may yield very good results (see Figure 2.18). When there are several layers of clouds, at various altitudes, moving in various directions and at various speeds, the picture becomes significantly more complicated. Finally, in the case of non-rigid transformations of the cloud cover (fusion or split-off) or of sudden appearance or disappearance phenomena (depending on the humidity, temperature and pressure conditions), the forecasting becomes extremely difficult. The impact on DNI forecasting of errors on the estimation of cloud motion being strong (Li et al. 2016), the results deteriorate significantly.

NOTE. Since the focus is the proper forecasting of ramp phenomena and other sudden changes, most of the works consider a binary transmittance: zero for the clouds and 1 otherwise, which leads to errors in the amplitude of DNI forecasting, as shown in Figure 2.18 around 12h45, for example. Indeed, the clouds, depending on their thickness, do not always generate the same irradiance attenuation. These inaccuracies may, however, be addressed by estimating their transmittance, which is possible based on cameras (see Li et al. 2016; Nouri et al. 2019a).

Concerning the spatiotemporal forecasting of DNI, several exploratory works exist (Blanc et al. 2017; Kuhn et al. 2017; Nouri et al. 2019b). They rely on a multi-camera system allowing a three-dimensional reproduction of the atmospheric conditions directly above the instrumented site. However, the complexity of the processes to be implemented, both from a physical perspective (synchronization of cameras, identification of their optimal position, instrumentation for the validation of results, etc.) and from a logistics perspective (high sensitivity to the quality of images (reflections and saturations) and to the estimation of the altitude of clouds, overshadowing problems in the presence of several layers of clouds, estimation of cloud displacement depending on altitude class, projection of the shadow on the ground, etc.), seems burdensome to this date. The partial results obtained are nevertheless encouraging.

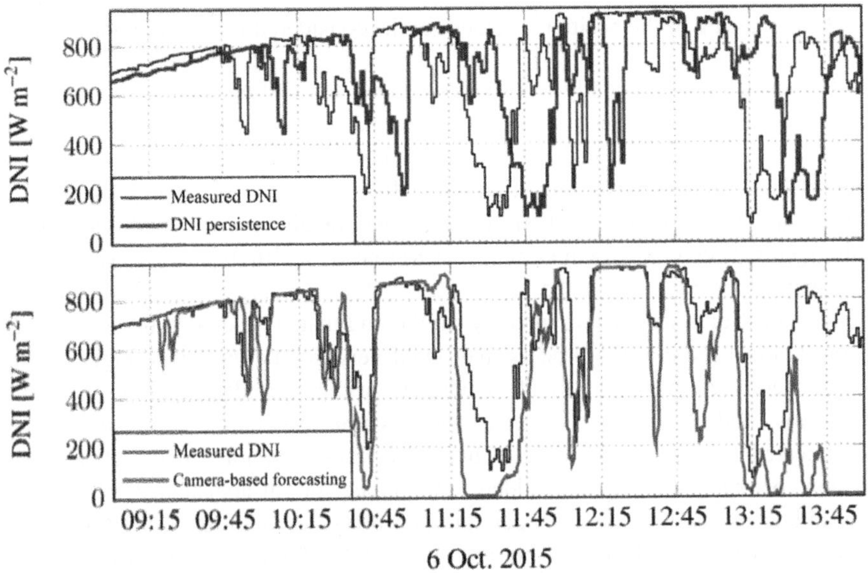

Figure 2.18. *Forecasting examples of DNI at* $\Delta t = 15\,min$*, using a DNI persistence (upper image) and a sky imager (lower image), in the case of a good forecasting of the cloud motion (sampling of measurements: 1 min; temporal resolution step of the forecasting: 20 s). For a color version of this figure, see www.iste.co.uk/flamant/energy.zip*

Generally speaking, these various works suggest that there is room for improvement and that reaching a robust intra-hour DNI forecasting is possible. This statement is notably explained by the complex nature of the behavior of clouds and reinforces the idea that it may be futile to attempt to precisely estimate the local motion in order to improve the forecasting. Referring to this point, models relying on tools involving artificial intelligence seem promising, but there are many remaining questions related to the relevant input parameters in order to improve DNI forecasting.

2.3. Conclusion

Solar resource assessment inevitably involves in-depth comprehension of solar radiation and its interactions with the atmosphere. The first part of this chapter focused on the various notions important for its assessment and measurement. DNI, in particular, is a function of a solid angle that must be specified: its measurement, as recommended by WMO, may differ from the actual radiation perceived by the solar field of a CSP plant. Consequently, an overestimation of the power plant generation should be avoided when the scattering phenomena in the solar corona are significant. It may be preferable to take into account the radial distribution of the solar flux in the circumsolar region.

Furthermore, the DNI can be split into two components: clear-sky direct normal irradiance and clear-sky index. The first one represents the DNI when no cloud overshadows the Sun. It is the higher limit of irradiance available for a CSP plant. The second term reflects the influence of clouds on this clear-sky irradiance. Evaluating these two components is essential for the power plant operator, indicating the available room for maneuver.

The second part of this chapter focused on the forecasting of the solar resource used by CSP plants, namely the direct normal irradiance. After a review of the essential notions to be known, the various existing approaches for making a DNI forecasting are presented, differentiating them according to the expected spatial–temporal horizon. Several examples of intra-hour DNI forecasting are then provided, once more taking advantage of the separation of DNI into clear-sky direct normal irradiance and clear-sky index. Though successful for short forecasting horizons or when DNI variability is low, persistence models proved to be unsuitable for ramp forecasting or for sudden variations of short duration, which are generally highly significant events for CSP. The forecasting based on a camera with a wide angle of view was then approached, highlighting the challenges that remain to be met. Indeed, while the results obtained up to now are promising, there is still room for progress, notably in terms of robustness of forecasting related to the atmospheric situation: cloud behavior forecasting is not always an easy task. One possible path is to elaborate hybrid models, depending on the complexity of the atmospheric situation. Simple methods are sufficient when the sky is fully clear, fully covered or when cloud displacement can be easily forecasted. However, during more complex atmospheric phenomena, this displacement may prove difficult to forecast. Advanced methods using both weather information and the generalization and inference capacity specific to techniques derived from artificial intelligence may help in doing without fine modeling.

2.4. Nomenclature

Symbol	Parameter	Unit
CSR	Circumsolar Ratio	–
DHI	Diffuse Horizontal Irradiance	$W{\cdot}m^{-2}$
DNI	Direct Normal Irradiance	$W{\cdot}m^{-2}$
GHI	Global Horizontal Irradiance	$W{\cdot}m^{-2}$
NWP	Numerical Weather Prediction	–
SAA	Solar/Azimuth Angle	°
SZA	Solar/Zenith Angle	°
WMO	World Meteorological Organization	–
I	Variable representing DNI	$W{\cdot}m^{-2}$
I_0	Extra-atmospheric illuminance for Earth/Sun Distance D_0	$W{\cdot}m^{-2}$
$\hat{I}$	Variable representing a DNI forecast	$W{\cdot}m^{-2}$
I_{cs}	Variable representing DNI in clear-sky conditions	$W{\cdot}m^{-2}$
I_{UA}	Solar constant	$W{\cdot}m^{-2}$
k_c	Indice ciel clair	–
$\hat{k}_c$	Estimation or forecast of the clear-sky index	–
m	Relative optical sir mass	m
Δt	Temporal forecast horizon	min
T_L	Turbidity coefficient	–
τ	Attenuation coefficient	–
τ_{cs}	Attenuation coefficient in clear-sky conditions	–
ξ_{CSP}	Angular sensitivity of solar concentrators	°
ξ_{Sun}	Apparent radius of the solar disk from Earth	°
ξ_{WMO}	Half-aperture angle recommended by the World Meteorological Organization for DNI measurement	°

2.5. References

Aguiar, L.M., Pereira, B., David, M., Díaz, F., Lauret, P. (2015). Use of satellite data to improve solar radiation forecasting with Bayesian artificial neural networks. *Solar Energy*, 122, 1309–1324.

Alonso-Montesinos, J. and Batlles, F. (2015). Solar radiation forecasting in the short- and medium-term under all sky conditions. *Energy*, 83, 387–393.

Berk, A., Bernstein, L., Anderson, G., Acharya, P., Robertson, D., Chetwynd, J., Adler-Golden, S. (1998). MODTRAN cloud and multiple scattering upgrades with application to AVIRIS. *Remote Sensing of Environment*, 65(3), 367–375.

Bernecker, D., Riess, C., Angelopoulou, E., Hornegger, J. (2014). Continuous short-term irradiance forecasts using sky images. *Solar Energy*, 110, 303–315.

Blanc, P. and Wald, L. (2012). The SG2 algorithm for a fast and accurate computation of the position of the sun for multi-decadal time period. *Solar Energy*, 86(10), 3072–3083.

Blanc, P., Espinar, B., Geuder, N., Gueymard, C., Meyer, R., Pitz-Paal, R., Reinhardt, B., Renné, D., Sengupta, M., Wald, L. et al. (2014). Direct normal irradiance related definitions and applications: The circumsolar issue. *Solar Energy*, 110, 561–577.

Blanc, P., Massip, P., Kazantzidis, A., Tzoumanikas, P., Kuhn, P., Wilbert, S., Schüler, D., Prahl, C. (2017). Short-term forecasting of high resolution local DNI maps with multiple fish-eye cameras in stereoscopic mode. *AIP Conference Proceedings*, 1850, 140004.

Blanco-Muriel, M., Alarcón-Padilla, D.C., López-Moratalla, T., Lara-Coira, M. (2001). Computing the solar vector. *Solar Energy*, 70(5), 431–441.

Boehm, J., Niell, A., Tregoning, P., Schuh, H. (2006). Global mapping function (GMF): A new empirical mapping function based on numerical weather model data. *Geophysical Research Letters*, 33(7).

Bone, V., Pidgeon, J., Kearney, M., Veeraragavan, A. (2018). Intra-hour direct normal irradiance forecasting through adaptive clear-sky modelling and cloud tracking. *Solar Energy*, 159, 852–867.

Bretagnon, P. and Francou, G. (1988). Planetary theories in rectangular and spherical variables. VSOP 87 solutions. *Astronomy and Astrophysics*, 202, 309–315.

Buie, D. and Monger, A. (2004). The effect of circumsolar radiation on a solar concentrating system. *Solar Energy*, 76(1–3), 181–185.

Buie, D., Monger, A., Dey, C. (2003). Sunshape distributions for terrestrial solar simulations. *Solar Energy*, 74(2), 113–122.

Champion, K., Cole, A., Kanto, A. (1985). Standard and reference atmospheres. In *Handbook of Geophysics and the Space Environment*, Jursa, A.S. (ed.). National Technical Information Service, Fort Belvoir, VA.

Chauvin, R., Nou, J., Thil, S., Grieu, S. (2015a). Modelling the clear-sky intensity distribution using a sky imager. *Solar Energy*, 119, 1–17.

Chauvin, R., Nou, J., Thil, S., Traoré, A., Grieu, S. (2015b). Cloud detection methodology based on a sky-imaging system. *Energy Procedia*, 69, 1970–1980.

Chauvin, R., Nou, J., Eynard, J., Thil, S., Grieu, S. (2018). A new approach to the real-time assessment and intraday forecasting of clear-sky direct normal irradiance. *Solar Energy*, 167, 35–51.

Chauvin, R., Nou, J., Thil, S., Grieu, S. (2019a). System for measuring components of solar radiation. Patent WO2019053232.

Chauvin, R., Nou, J., Thil, S., Grieu, S. (2019b). Système de mesure des composantes du rayonnement solaire. Brevet FR1758549.

Chow, C.W., Urquhart, B., Lave, M., Dominguez, A., Kleissl, J., Shields, J., Washom, B. (2011). Intra-hour forecasting with a total sky imager at the UC San Diego solar energy testbed. *Solar Energy*, 85(11), 2881–2893.

Chu, Y., Pedro, H.T., Coimbra, C.F. (2013). Hybrid intra-hour DNI forecasts with sky image processing enhanced by stochastic learning. *Solar Energy*, 98(Part C), 592–603.

Chu, Y., Pedro, H.T., Li, M., Coimbra, C.F. (2015a). Real-time forecasting of solar irradiance ramps with smart image processing. *Solar Energy*, 114, 91–104.

Chu, Y., Urquhart, B., Gohari, S.M., Pedro, H.T., Kleissl, J., Coimbra, C.F. (2015b). Short-term reforecasting of power output from a 48 MWe solar PV plant. *Solar Energy*, 112, 68–77.

Fu, C.-L. and Cheng, H.-Y. (2013). Predicting solar irradiance with all-sky image features via regression. *Solar Energy*, 97, 537–550.

Geuder, N. and Quaschning, V. (2006). Soiling of irradiation sensors and methods for soiling correction. *Solar Energy*, 80(11), 1402–1409.

Geuder, N., Affolter, R., Kraas, B., Wilbert, S. (2014). Long-term behavior, accuracy and drift of LI-200 pyranometers as radiation sensors in rotating shadowband irradiometers (RSI). *Energy Procedia*, 49, 2330–2339.

Grena, R. (2008). An algorithm for the computation of the solar position. *Solar Energy*, 82(5), 462–470.

Grena, R. (2012). Five new algorithms for the computation of sun position from 2010 to 2110. *Solar Energy*, 86(5), 1323–1337.

Grether, D., Evans, D., Hunt, A., Whalig, M. (1980). Measurements and analysis of circumsolar radiation. Technical report, Lawrence Berkeley Laboratory, University of California, CA.

Gueymard, C.A. (1993). Development and performance assessment of a clear sky spectral radiation model. *Proceedings of the 22nd ASES Conference*, Washington D.C., 433–438.

Gueymard, C.A. (2001). Parameterized transmittance model for direct beam and circumsolar spectral irradiance. *Solar Energy*, 71(5), 325–346.

Gueymard, C.A. (2010). Spectral circumsolar radiation contribution to CPV. *6th International Conference on Concentrating Photovoltaic Systems.*

Gueymard, C.A. (2012). Temporal variability in direct and global irradiance at various time scales as affected by aerosols. *Solar Energy*, 86(12), 3544–3553.

Gueymard, C.A. (2018). A reevaluation of the solar constant based on a 42-year total solar irradiance time series and a reconciliation of space borne observations. *Solar Energy*, 168, 2–9.

Hamill, T.M. and Nehrkorn, T. (1993). A short-term cloud forecast scheme using cross correlations. *Weather and Forecasting*, 8(4), 401–411.

Hammer, A., Heinemann, D., Hoyer, C., Kuhlemann, R., Lorenz, E., Müller, R., Beyer, H.G. (2003). Solar energy assessment using remote sensing technologies. *Remote Sensing of Environment*, 86(3), 423–432.

Ineichen, P. and Perez, R. (2002). A new airmass independent formulation for the Linke turbidity coefficient. *Solar Energy*, 73(3), 151–157.

Inman, R.H., Pedro, H.T., Coimbra, C.F. (2013). Solar forecasting methods for renewable energy integration. *Progress in Energy and Combustion Science*, 39(6), 535–576.

ISO (1999). Solar energy – Vocabulary, ISO 9488. International Organization for Standardization.

Karout, Y., Thil, S., Eynard, J., Grieu, S. (2020). Cloud/sky segmentation from ground camera based on supervised machine learning approach. *33rd International Conference on Efficiency, Cost, Optimization, Simulation and Environmental Impact of Energy Systems (ECOS 2020)*, Osaka.

Kasten, F. (1980). A simple parameterization of two pyrheliometric formulae for determining the Linke turbidity factor. *Meteorologische Rundschau*, 33, 124–127.

Kasten, F. and Young, A.T. (1989). Revised optical air mass tables and approximation formula. *Applied Optics*, 28(22), 4735–4738.

Kivalov, S.N. (2007). Improved ray tracing air mass numbers model. *Applied Optics*, 46(29), 7091–7098.

Kopp, G. (2014). An assessment of the solar irradiance record for climate studies. *Journal of Space Weather and Space Climate*, 4.

Kopp, G. and Lean, J.L. (2011). A new, lower value of total solar irradiance: Evidence and climate significance. *Geophysical Research Letters*, 38(1).

Kuhn, P., Wilbert, S., Schüler, D., Prahl, C., Haase, T., Ramirez, L., Zarzalejo, L., Meyer, A., Vuilleumier, L., Blanc, P. et al. (2017). Validation of spatially resolved all sky imager derived DNI nowcasts, 1850.

Kyle, H., Hoyt, D., Valette, B., Hickey, J., Maschhoff, R. (1993). Nimbus-7 earth radiation budget calibration history – Part I: The solar channels. Technical report, National Aeronautics and Space Administration, Washington D.C.

Lauret, P., Lorenz, E., David, M. (2016). Solar forecasting in a challenging insular context, *Atmosphere*, 7(2), 18.

Law, E.W., Prasad, A.A., Kay, M., Taylor, R.A. (2014). Direct normal irradiance forecasting and its application to concentrated solar thermal output forecasting – A review. *Solar Energy*, 108, 287–307.

Lean, J.L. and Rind, D.H. (2009). How will Earth's surface temperature change in future decades? *Geophysical Research Letters*, 36(15), L15708.

Li, M., Chu, Y., Pedro, H.T., Coimbra, C.F. (2016). Quantitative evaluation of the impact of cloud transmittance and cloud velocity on the accuracy of short-term DNI forecasts. *Renewable Energy*, 86, 1362–1371.

Linke, F. (1922). Transmisions-koeffizient und trübungsfaktor. *Beiträge zur Physik der Atmosphäre*, 10, 91–103.

Lopes, F.M., Silva, H.G., Salgado, R., Cavaco, A., Canhoto, P., Collares-Pereira, M. (2018). Short-term forecasts of GHI and DNI for solar energy systems operation: Assessment of the ECMWF integrated forecasting system in southern Portugal. *Solar Energy*, 170, 14–30.

Marquez, R. and Coimbra, C.F. (2013). Intra-hour DNI forecasting based on cloud tracking image analysis. *Solar Energy*, 91, 327–336.

Meeus, J. (1999). *Astronomical Algorithms.* Willmann-Bell, Richmond, VA.

Mendes, V., Prates, G., Pavlis, E., Pavlis, D., Langley, R. (2002). Improved mapping functions for atmospheric refraction correction in SLR. *Geophysical Research Letters*, 29(10), 1414.

Michalsky, J.J. (1988). The astronomical Almanac's algorithm for approximate solar position (1950–2050). *Solar Energy*, 40(3), 227–235.

Neumann, A. and von der Au, B. (1997). Sunshape measurements at the DLR solar furnace site in Cologne, Germany. *Proceedings of the International Solar Energy Conference*, Washington, D.C., 163–170.

Nou, J., Chauvin, R., Eynard, J., Thil, S., Grieu, S. (2018). Towards the intrahour forecasting of direct normal irradiance using sky-imaging data. *Heliyon*, 4(4), e00598.

Nouri, B., Wilbert, S., Segura, L., Kuhn, P., Hanrieder, N., Kazantzidis, A., Schmidt, T., Zarzalejo, L., Blanc, P., Pitz-Paal, R. (2019a). Determination of cloud transmittance for all sky imager based solar nowcasting. *Solar Energy*, 181, 251–263.

Nouri, B., Kuhn, P., Wilbert, S., Hanrieder, N., Prahl, C., Zarzalejo, L., Kazantzidis, A., Blanc, P., Pitz-Paal, R. (2019b). Cloud height and tracking accuracy of three all sky imager systems for individual clouds. *Solar Energy*, 177, 213–228.

Pape, B., Batlles, J., Geuder, N., Piñero, R.Z., Adan, F., Pulvermueller, B. (2009). Soiling impact and correction formulas in solar measurements for CSP projects. *15th International Conference on Concentrating Solar Power and Chemical Energy Systems* (SolarPACES 2009).

Quesada-Ruiz, S., Chu, Y., Tovar-Pescador, J., Pedro, H., Coimbra, C. (2014). Cloudtracking methodology for intra-hour DNI forecasting. *Solar Energy*, 102, 267–275.

Rabl, A. (1976). Comparison of solar concentrators. *Solar Energy*, 18, 93–111.

Rabl, A. and Bendt, P. (1982). Effect of circumsolar radiation on performance of focusing collectors. *Journal of Solar Energy Engineering*, 104(3), 237–250.

Ramírez, L. and Vindel, J.M. (2017). Forecasting and nowcasting of DNI for concentrating solar thermal systems. In *Advances in Concentrating Solar Thermal Research and Technology*, Blanco, M.J. and Santigosa, L.R. (eds). Woodhead Publishing, Amsterdam.

Rapp-Arrarás, I. and Domingo-Santos, J. M. (2011). Functional forms for approximating the relative optical air mass. *Journal of Geophysical Research*, 116, D24308.

Reda, I. and Andreas, A. (2004). Solar position algorithm for solar radiation applications. *Solar Energy*, 76(5), 577–589.

Rigollier, C., Bauer, O., Wald, L. (2000). On the clear sky model of the ESRA – European solar radiation atlas – with respect to the Heliosat method. *Solar Energy*, 68(1), 33–48.

Rigollier, C., Lefèvre, M., Wald, L. (2004). The method heliosat-2 for deriving shortwave solar radiation from satellite images. *Solar Energy*, 77(2), 159–169.

Schmidt, T., Kalisch, J., Lorenz, E., Heinemann, D. (2015). Evaluating the spatiotemporal performance of sky imager based solar irradiance analysis and forecasts. *Atmospheric Chemistry and Physics*, 16(5), 3399–3412.

Sengupta, M., Habte, A., Kurtz, S., Dobos, A., Wilbert, S., Lorenz, E., Stoffel, T., Renné, D., Gueymard, C., Myers, D. et al. (2015). *Best Practices Handbook for the Collection and Use of Solar Resource Data for Solar Energy Applications*. National Renewable Energy Laboratory, Golden, CO.

Stefferud, K., Kleissl, J., Schoene, J. (2012). Solar forecasting and variability analyses using sky camera cloud detection & motion vectors. *2012 IEEE Power and Energy Society General Meeting*, 1–6.

Stoffel, T., Renné, D., Myers, D., Wilcox, S., Sengupta, M., George, R., Turchi, C. (2010). Concentrating solar power: Best practices handbook for the collection and use of solar resource data. Technical Report, National Renewable Energy Laboratory, Golden, CO.

Tolba, H., Dkhili, N., Nou, J., Eynard, J., Thil, S., Grieu, S. (2020). Multi-horizon forecasting of global horizontal irradiance using online Gaussian process regression: A kernel study, *Energies*, 13, 4184.

Voyant, C., Notton, G., Kalogirou, S., Nivet, M.-L., Paoli, C., Motte, F., Fouilloy, A. (2017). Machine learning methods for solar radiation forecasting: A review. *Renewable Energy*, 105, 569–582.

West, S.R., Rowe, D., Sayeef, S., Berry, A. (2014). Short-term irradiance forecasting using skycams: Motivation and development. *Solar Energy*, 110, 188–207.

Wilbert, S. (2014). Determination of circumsolar radiation and its effect on concentrating solar power. PhD Thesis, Fakultaät für Maschinenwesen der Rheinisch-Westfälischen Technischen Hochschule Aachen.

Wilbert, S., Pitz-Paal, R., Jaus, J. (2012). Circumsolar radiation and beam irradiance measurements for focusing collectors. *COST ES 1002 WIRE Workshop on Remote Sensing Measurements for Renewable Energy*.

Wilbert, S., Reinhardt, B., de Vore, J., Röger, M., Pitz-Paal, R., Gueymard, C., Buras, R. (2013). Measurement of solar radiance profiles with the sun and aureole measurement system. *Journal of Solar Energy Engineering*, 135(4), 041002.

WMO (2010). Guide des instruments et des méthodes d'observation météorologiques. Guide, Organisation météorologique mondiale, Geneva.

Yang, H., Kurtz, B., Nguyen, D., Urquhart, B., Chow, C.W., Ghonima, M., Kleissl, J. (2014). Solar irradiance forecasting using a ground-based sky imager developed at UC San Diego. *Solar Energy*, 103, 502–524.

3

Optics of Concentrating Systems

François HÉNAULT[1], Benjamin GRANGE[2] and Quentin FALCOZ[3]
[1]*IPAG-CNRS, Grenoble, France*
[2]*PROMES-CNRS, Font-Romeu-Odeillo-Via, France*
[3]*PROMES-CNRS, UPVD, Perpignan, France*

3.1. Introduction

Solar concentrating systems are a very original and little-known part of optical sciences. Drawing inspiration from Archimede's famous "burning mirrors" (see section 3.2), their reflective surfaces are not optical surfaces in the common sense of the term, as their manufacturing cost would be exorbitant. The technologies employed rather rely on industrial glass manufacturing processes, and the implementation of mirrors is often conducted under mechanical constraint. Another particularity is that given the dimensions to be covered, the mirrors are generally composed of hundreds or thousands of individual facets ("segmented"). Finally, the precision of optical surfaces is significantly less restrictive compared to that of classical optical systems, which is measured in fractions of wavelength: here, the manufacturing precision is rather expressed in tenths of millimeters, the imagery capacity being therefore lost. As it will be seen, high-concentration performances can nevertheless be attained, and certain important concepts of classical optics are still applicable (section 3.3). Finally, sections 3.4 and 3.5 offer a more detailed presentation of the two types of the most commonly used concentrating field.

Concentrating Solar Thermal Energy,
coordinated by Gilles FLAMANT. © ISTE Ltd 2022.

3.2. History

3.2.1. *From Archimedes to 19th century*

The Greek Archimedes is often referred to as the inventor of concentrated solar power. The legend has it that during the second Punic war between the Romans and the Carthaginians, he instructed the Carthaginian soldiers to use their bronze shields to reflect the solar rays on the sails of the Roman fleet, setting them on fire. These events allegedly happened in the Syracuse port in 215 BC, and they are illustrated by famous paintings and engravings, such as that shown in Figure 3.1.

The question naturally arises as to whether it was actually possible that Archimedes accomplished this feat. The first historic note on these "burning mirrors" is from Anthemius of Tralles in the 6th century AD, hence 800 years after the event. The Roman historians narrating this war did not mention them. It therefore appears to be a legend, though Archimedes will always be remembered as the forefather of optical concentration.

Figure 3.1. Archimedes' Burning Mirrors, *fresque by Giulio Parigi (1571–1635), Uffizi Gallery, Florence. For a color version of this figure, see www.iste.co.uk/flamant/energy.zip*

In the history of France, "burning mirrors" are present at the court of King Louis XIV: in 1669, the Sun King paid 7,000 pounds for a one-meter diameter bronze mirror, manufactured by François Villette, an engineer, optician and fireworks expert. It was used to melt various objects and materials, and to conduct optical experiments for teaching purposes (Figure 3.2, upper left). It can still be seen nowadays in the storage area of Versailles palace. Those were the times when the subject also attracted the interest of scientists: Georges-Louis Leclerc, Count of Buffon (1707–1788), made a mirror composed of 168 facets for melting tin, lead and silver (Figure 3.2, upper right) and thus reached a temperature above 1,000°C. According to his *Natural History*: "I have also melted, or volatilized, by a mirror many matters which neither the fire, nor reverberatory furnace, nor the most powerful bellows furnace could cause to run." Later on, Antoine Laurent de Lavoisier (1743–1794) built a solar concentrator using two convergent lenses "with water" (liquid water confined between two convex glass slides; Figure 3.2, lower left). This is how he melted platinum at 1,750°C.

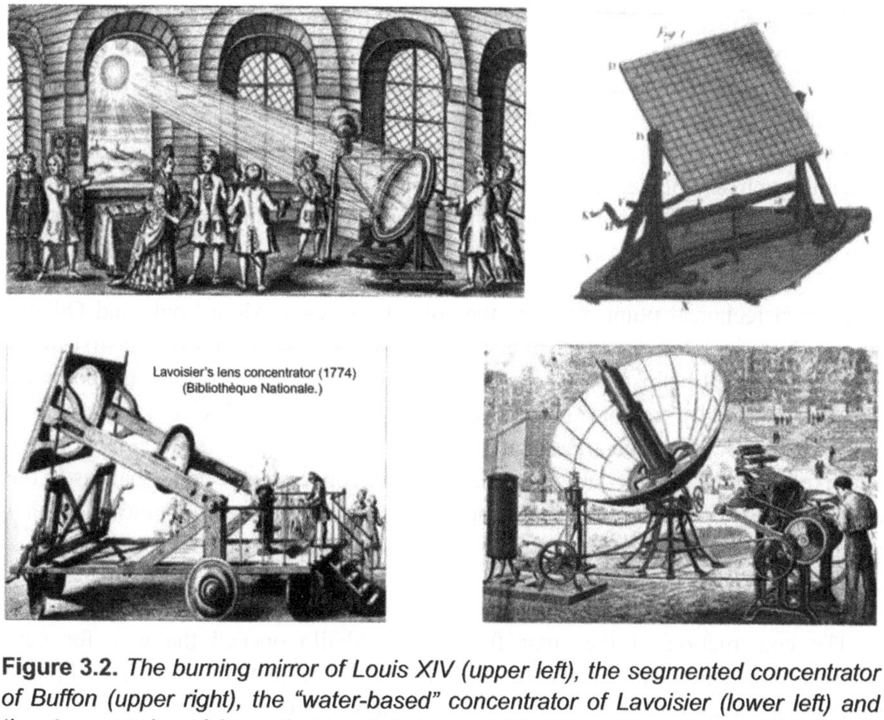

Figure 3.2. *The burning mirror of Louis XIV (upper left), the segmented concentrator of Buffon (upper right), the "water-based" concentrator of Lavoisier (lower left) and the steam engine of Augustin Mouchot (lower right). For a color version of this figure, see www.iste.co.uk/flamant/energy.zip*

Finally, Augustin Mouchot (1825–1912) was the pioneer of industrial solar applications, opening the way for the large installations of the 20th century (see the next section). He was the first to couple a solar concentrator with a steam engine, and presented it at the World Exhibition in Paris, in 1878. A forerunner of modern renewable energies, his main objective was to save coal-based natural resources.

3.2.2. *1950–1980: First industrial installations*

The pioneer in the field of large solar radiation concentrating installations of the 20th century is undoubtedly the French Félix Trombe (1906–1985). After having built the first prototype of double reflection solar furnace using a mirror belonging to the anti-aircraft defense at the Observatory of Meudon in 1946, he settled down in the Eastern Pyrenees in order to conduct more ambitious projects, starting with a 90 m^2 and 50 kW solar furnace in the Mont-Louis Citadel (Trombe 1955). But his focus was already on building the largest solar furnace in the world near the neighboring village of Odeillo, an installation concentrating 1,000 kW power, which was finally put into service in 1968 (Trombe 1973). Its 54 x 40 m parabolic concentrator is illuminated by 63 flat heliostats of 45 m^2 each (see Figure 3.3) and can reach temperatures above 3,000°C on an area 20 cm in diameter (Royère 1981). There is still nowadays a large research infrastructure of CNRS intended to explore domains such as the testing of space components, the measurement of optical properties at high temperatures, the thermodynamic systems and engines for the conversion of solar energy, and the solar thermochemistry.

From a technical point of view, the solar furnaces at Mont-Louis and Odeillo rely on the principle of double reflection (see Figure 3.4). They involve first of all one or several flat heliostats that serve for redirecting the solar rays along a common direction, generally a North–South horizontal axis. The reflected rays are then collected by a fixed concentrator that leads them to converge to a single focal point. The fixed geometry of the latter enables excellent concentration performances in the vicinity of the focal point, despite the energy loss resulting from the additional reflection. This explains the capability of the large solar furnace of Odeillo to reach concentrations of the order of 10,000.

The construction of the solar furnace at Odeillo opened the way for new industrial developments, such as the THEMIS project of the first French experimental solar power plant of 2.5 MW (Drouot 1984), built by CNRS, EDF and AFME (nowadays ADEME), finally put into service in 1983 on the neighboring site of Targasonne. It is composed of a field of 201 heliostats of 54 m^2 each, focusing the solar radiation to a $4 \times 4 \times 4 = 64$ m^3 cubic solar receiver installed at 86 m height (Figure 3.5). The solar power plant was closed in 1986 after a first promising phase of evaluation, and then was reconverted into astronomical observatory for the observation of cosmic rays. The tower concentrator (see section 3.5) was put into service in 2005 for conducting researches on new solar receiver design. Meanwhile, many other experimental or industrial solar thermal power plants were built throughout the world (Spain, the United States, Australia, etc.). The reader is referred to Chapter 1 of this book and to Li et al. (2016) for a recent synthesis.

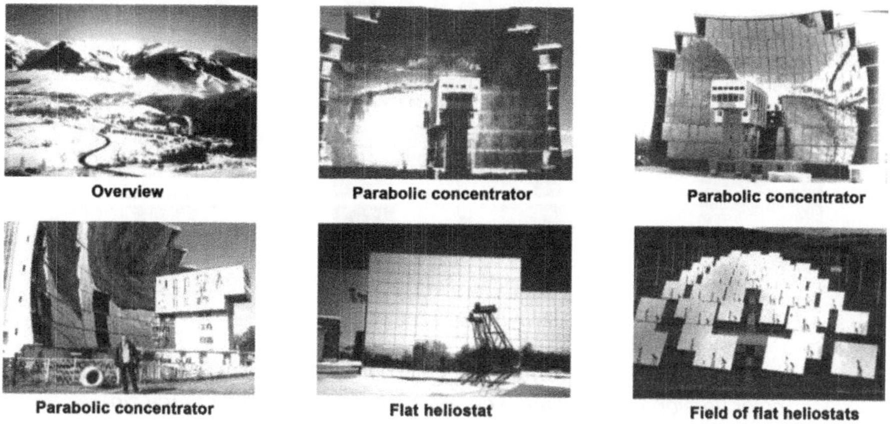

Figure 3.3. *Large 1,000 kW solar furnace at Odeillo: overview, images of the parabolic concentrator and images of the field of flat heliostats. For a color version of this figure, see www.iste.co.uk/flamant/energy.zip*

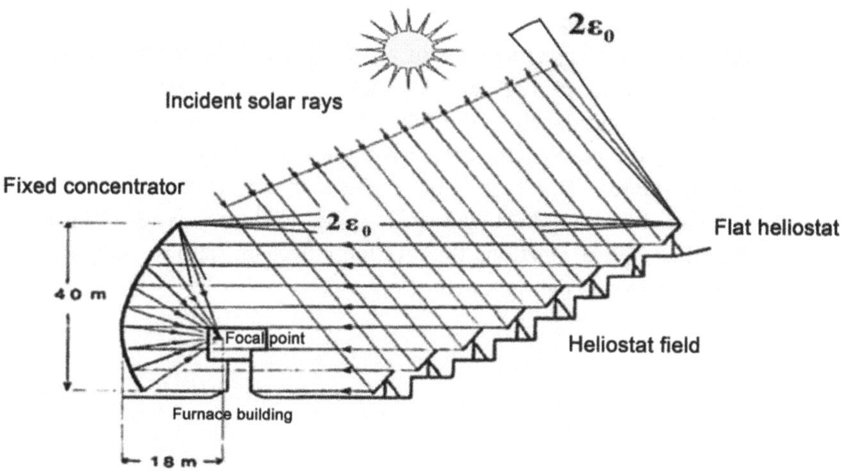

Figure 3.4. *Principle of the double reflection solar furnace*

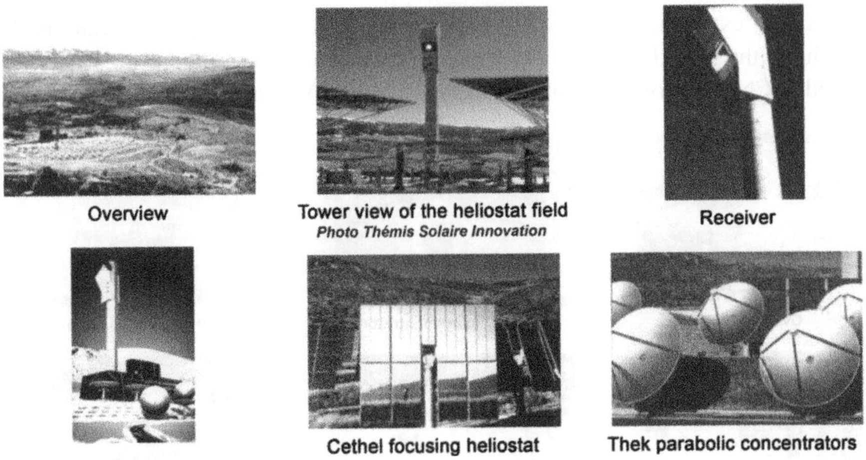

Figure 3.5. *The various components of Themis solar power tower. For a color version of this figure, see www.iste.co.uk/flamant/energy.zip*

3.3. Performances and limitations

3.3.1. *Specification of a solar concentrator*

The main scientific notations employed in this chapter are summarized in Table 3.1. The elementary specifications of a solar power concentrating system are schematically represented in Figure 3.6. They are in fact reduced to several fundamental characteristics, such as the total collected power Φ, the concentration factor C related to the receiver temperature and the spatial uniformity at its surface (left part of the figure). The first two notions can be replaced by equivalent specifications related to the collecting surface S and to the receiving surface S' by means of simplified optical relations that are the object of the next section. These relations are summarized in the right part of Figure 3.6.

Symbol	Parameter	Unit
Φ	Total collected power	W
DNI	Direct Normal Irradiance	W/m^2
T_A	Atmospheric transmission	–
E_0	Solar irradiance outside the atmosphere at the level of the terrestrial orbit, sometimes referred to as "solar constant"	W/m^2
S	Collecting surface	m^2
S'	Receiving surface (boiler)	m^2
D	Distance between the collecting and receiving surfaces, ideally equal to the focal length f of the concentrating system	m
i	Solar incidence angle on the collecting surface	rad
C_0	Theoretical concentration factor	–(*)
C	Actual concentration factor	–(*)
η	Ratio of the actual and theoretical concentration factors $\eta = C/C_0$	–
$L(\varepsilon)$	Formula for the variation of radiance at the surface of the solar disk beyond the atmosphere, depending on the angle ε with respect to the center of the Sun	$W/m^2/sr$
L_0	Radiance at the center of the Sun beyond the atmosphere	$W/m^2/sr$
ε_0	Angular radius of the Sun	rad
E'(M')	Irradiance distribution on the receiving surface, depending on the position of point M' on this surface	W/m^2
$\langle E' \rangle$	Average irradiance on the receiving surface	W/m^2
E'_{Max}	Maximal theoretical irradiance on the receiving surface	W/m^2
Ω	Solid angle of the Sun seen from the collecting surface	sr
Ω'	Solid angle of the collecting surface seen from the receiving surface	sr
T	Equivalent temperature of the black body or gray body	K
$\Delta(M)$	Local shape errors of the collecting surface	m
W(P)	Errors of the wave front seen from the receiving surface	m
δ_s	Standard deviation of the local surface errors	rad
δ_r	Standard deviation of the adjustment errors	rad
δ_p	Standard deviation of the pointing errors	rad
δ	Quadratic sum of errors on the reflected rays	rad
	(*) Sometimes expressed in equivalent number of Suns at the ground.	

Table 3.1. *List of parameters employed*

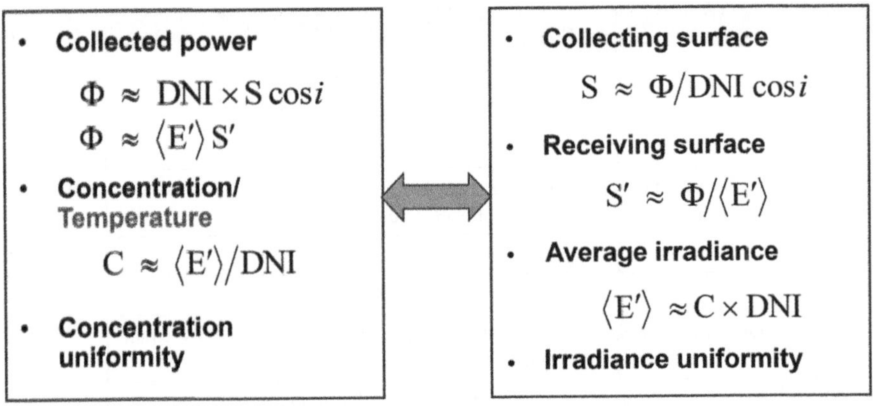

Figure 3.6. *Elementary specifications of solar concentrators*

3.3.2. *Collected power*

In first approximation, the collected power Φ is equal to the product of the total collecting surface S multiplied by the solar constant beyond the atmosphere E_0 and the losses due to the atmospheric transmission T_A. This result can be obtained by integration of the elementary geometrical extent over the collecting surface S and the receiving surface S′ (see Figure 3.7):

$$\Phi \approx T_A \iint_{S,S'} L(\varepsilon)\frac{dSdS'}{D^2} \qquad [3.1]$$

where $L(\varepsilon)$ is the law of radiance at the surface of the solar disk beyond the atmosphere, function of the angle ε with respect to the center of the Sun. For a Sun of uniform radiance, $L(\varepsilon)$ is constant:

$$L(\varepsilon) = L_0 = E_0/\pi \sin^2 \varepsilon_0 \approx E_0/\pi\varepsilon_0^2 \qquad [3.2]$$

where ε_0 is the angular radius of the Sun. But for an "ideal" optical system with no aberrations and manufacturing defects, the dimensions of the receiving surface are usually limited to the geometrical image of the Sun:

$$S' \approx \pi(D\varepsilon_0)^2 \qquad [3.3]$$

Combining the previous relations yields:

$$\Phi \approx T_A \frac{E_0}{\pi \varepsilon_0^2} \frac{SS'}{D^2} \cos i \approx T_A E_0 S \cos i$$
$$\approx \text{DNI} \times S \cos i \qquad [3.4]$$

where DNI = T_A E_0 is the direct solar irradiance at the Earth's surface, depending on many parameters such as the day of the year and the hour of the day, the geographical coordinates of the place (latitude, longitude and altitude) and the weather conditions (see Chapter 2). It is important to note the presence of a multiplicative factor cos*i* often referred to as "cosine effect" in solar applications. If these factors are considered negligible, it can be concluded that for an ideal optical system, the power depends only on the total collecting surface S and not on the receiving surface S′.

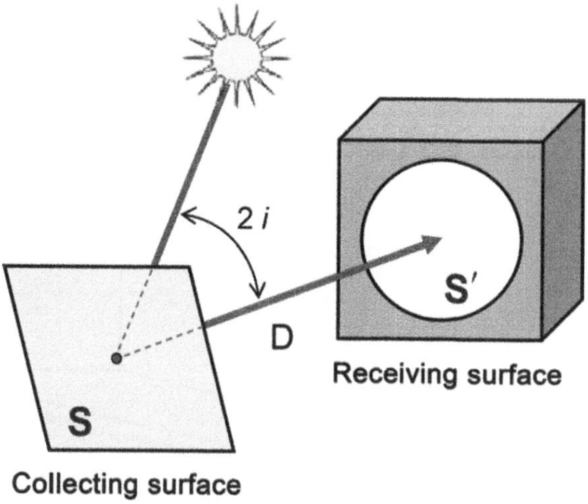

Figure 3.7. *Fundamental parameters of the collecting and receiving surfaces. For a color version of this figure, see www.iste.co.uk/flamant/energy.zip*

3.3.3. *Three definitions of concentration*

The scientific literature proposes several definitions of the concentration factor of an optical system, among which the three most practical ones are presented in the following sections.

3.3.3.1. *Ratio of surfaces*

According to one of the most famous references in the field of non-imaging optics (Winston et al. 2005), the average concentration factor C can be defined as the ratio of the collecting surface S and the receiving surface S′ (see Figure 3.8):

$$C=S/S' \qquad [3.5]$$

It is important to note that the cosine factor cos*i* is not present in this definition. If it is introduced, and the law of conservation of the geometrical extent is applied, then the total collected energy in relation 3.4 is written as:

$$\Phi \approx \mathrm{DNI} \times \mathrm{S} \cos i \approx \langle \mathrm{E}' \rangle \, \mathrm{S}' \qquad [3.6]$$

where $\langle E' \rangle$ is the average irradiance on the receiving surface. The expression of the concentration factor in Figure 3.6 is then obtained:

$$\mathrm{C} = \frac{\mathrm{S}\cos i}{\mathrm{S}'} = \frac{\langle \mathrm{E}' \rangle}{\mathrm{DNI}} \qquad [3.7]$$

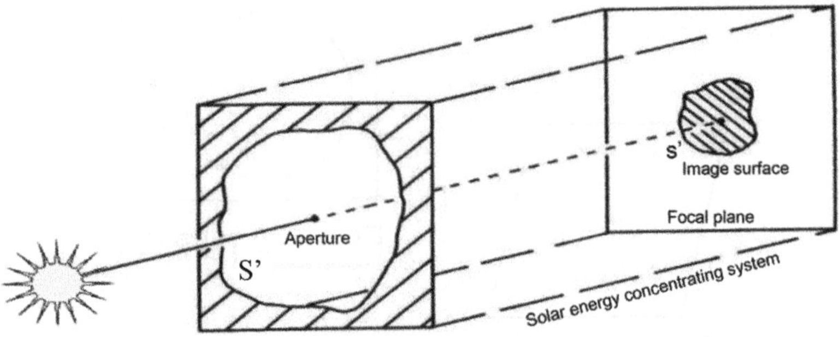

Figure 3.8. *Definition of the concentration factor C with respect to the collecting and receiving surfaces. For a color version of this figure, see www.iste.co.uk/flamant/energy.zip*

3.3.3.2. *Ratio of irradiances*

In practice, the concentration factor C depends on the distribution of irradiance E′(M′) formed on the receiving surface, variable depending on the spatial position of M′ points, which can be defined by their Cartesian coordinates (x′,y′) or polar coordinates (r′,θ′). An example of distribution E′(M′) is illustrated in false colors and in sectional view along X′ axis in Figure 3.9, where the black curve represents the actual irradiance distribution. This curve can be framed in two different ways:

– In the case of a concentrating system that is perfectly free of aberrations and manufacturing defects, the receiving surface is reduced to the geometrical image of the Sun, which is a disk of radius $D\varepsilon_0$ (see relation [3.3]). The irradiance is then maximal and constant over all the receiving surfaces (red curves in Figure 3.9), and the theoretical concentration factor C_0 is defined as the ratio of the maximal irradiance E'_{Max} to DNI:

$$C_0 = E'_{Max}/DNI \quad [3.8]$$

– From a practical point of view, the receiving surface S′ is similar to that of the receiver (blue curves in Figure 3.9). E'_{Max} is then replaced by the average irradiance $\langle E' \rangle$ on the receiving surface, and the expression of the actual concentration factor C is identical to that of equation [3.7].

The "canonical" definition of section 3.3.3.1 is equally applicable to both previous cases. Finally, it is important to note that the profile of the black curve in Figure 3.9 allows the consideration of a "local" concentration factor C(M′) proportional to the irradiance distribution (Hénault 1987), equal to:

$$C(M') = E(M')/DNI \quad [3.9]$$

and expressed in the number of equivalent suns at the ground.

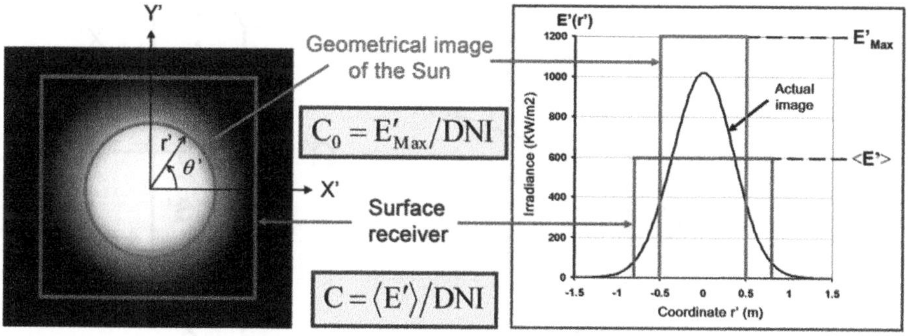

Figure 3.9. *Definition of the concentration factor C using the ratio of irradiances. For a color version of this figure, see www.iste.co.uk/flamant/energy.zip*

3.3.3.3. *Ratio of solid angles*

A more rigorous definition of the theoretical concentration factor C_0 relies on the ratio of solid angles Ω and Ω' at the inlet and outlet of the concentrating system (Hénault 2015). In Figure 3.10, Ω is the solid angle of the Sun seen from the Earth's surface, while Ω' is the solid angle of the useful surface of the concentrator *S. cosi* seen from the receiving surface S′. Knowing that $\Omega = \pi \sin^2 \varepsilon_0 \approx \pi \varepsilon_0^2$ and that Ω' is approximately equal to $S \cos i / D^2$, where D is the distance |SS′|, an expression of C_0 equivalent to that of equation 3.8 can be found:

$$C_0 = \frac{\Omega'}{\Omega} \approx \frac{S \cos i}{\pi \varepsilon_0^2 D^2} \approx \frac{E'_{Max}}{DNI} \qquad [3.10]$$

The main advantages of this definition are the following:

– it takes into account the actual geometry of the concentrating system and therefore yields more precise results;

– it establishes a clear relation with the laws of geometrical optics, since C_0 is the squared power of the angular magnification of the optical system;

– according to these laws, the concentration factors are mutually multiplied in the case of multistage concentrating systems.

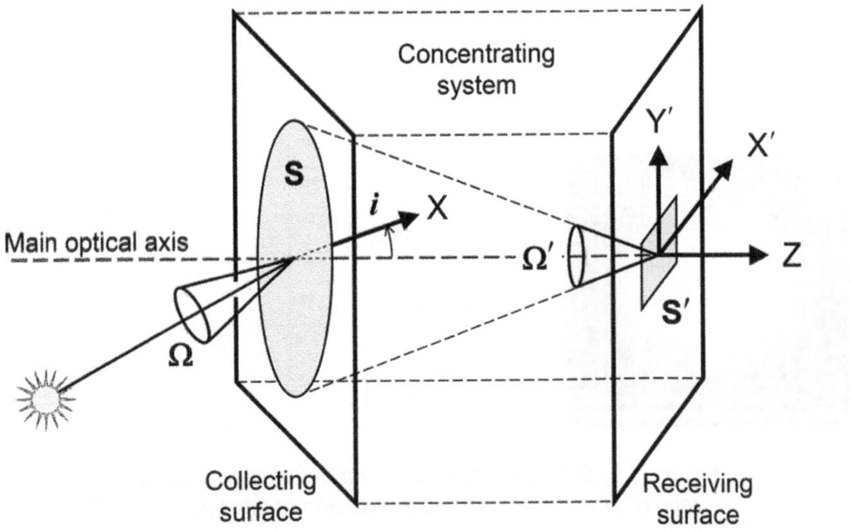

Figure 3.10. *Definition of the concentration factor C by the ratio of solid angles. For a color version of this figure, see www.iste.co.uk/flamant/energy.zip*

3.3.4. *Maximal concentration – Stefan's law*

According to the previous definitions of the theoretical concentration factor C_0 (relations [3.8] and [3.10]), the maximal concentration at the focal point of a solar concentrating system depends essentially on the maximal irradiance E'_{Max} (the atmospheric transmission factor T_A can be omitted, as it applies to both direct solar radiation and concentrated radiation). In classical photometry, E'_{Max} is calculated by the integration of elementary solid angles $d\alpha'$ over the angular aperture of the system α'_m (see Figure 3.11):

$$E'_{Max} = 2\pi\, L_0 \int_0^{\alpha'_m} \sin\alpha' \cos\alpha'\, d\alpha' = \pi\, L_0 \sin^2\alpha'_m \qquad [3.11]$$

hence the usual expression of C_0:

$$C_0 \approx \sin^2\alpha'_m / \varepsilon_0^2 \qquad [3.12]$$

For a flat receiving surface, the aperture α'_m of the concentrator cannot exceed 90 degrees. Knowing that the average angular radius of the Sun is 16 minutes of arc, a maximal theoretical concentration factor of 46,200 suns is obtained (Trombe 1971).

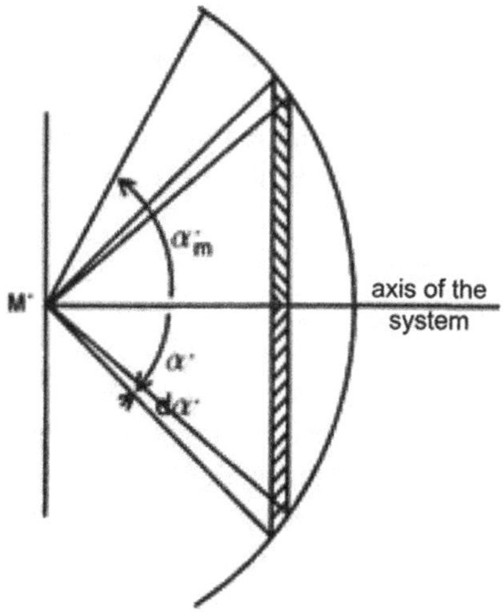

Figure 3.11. *Calculation of the maximal irradiance at the focal point of a concentrator. For a color version of this figure, see www.iste.co.uk/flamant/energy.zip*

For the optical engineer, the relation between the concentration factor (either theoretical or actual) and the temperature attainable on the receiving surface can be established using the Stefan–Boltzmann law. This law establishes the relation between the temperature T of a black or gray body and the irradiance received on its surface:

$$\sigma T^4 = \varepsilon E'_{Max} = \varepsilon C_0 \, DNI \cos\gamma \qquad [3.13]$$

where σ is the Stefan–Boltzmann constant equal to 5.67×10^{-8} W/m^2.K^4, ε is the emissivity of the gray body (equal to 1 for a black body) and γ is the average incidence angle of the rays on the receiving surface being considered negligible in what follows. An approximate expression of the maximal temperature attainable at the focal point of the solar concentrator is deduced:

$$T \approx \sqrt[4]{\frac{\varepsilon C_0 \, DNI}{\sigma}} \qquad [3.14]$$

This estimation should by no means replace the detailed thermodynamic models of the receiver (see Chapter 4). It can nevertheless be used to rapidly calculate the orders of magnitude of the concentrator performances. Hence, the high-concentration systems such as the solar furnaces require a concentration above 10,000 in order to exceed 3,000°C, while a concentration of approximately 200 or 300 is theoretically sufficient for reaching 1,000°C for the new generation of solar tower power plants. Finally, the linear or photovoltaic systems require only concentrations of the order of a dozen or several dozens.

NOTE 3.1.– Using a more general definition of the geometrical extent of an optical system, Welford and Winston showed that relation [3.12] is actually written as (Welford 1978):

$$C_0 \approx n^2 \sin^2\alpha'_m / \varepsilon_0^2 \qquad [3.15]$$

where n is the refractive index of the image medium, in which the receiver is immersed. It is then possible to exceed the limit of 46,200 suns, since the index is generally below 1. This was experimentally proven with the construction of a two-stage concentrating system (see Figure 3.12) equipped with a secondary hollow conical mirror filled with oil, thus establishing a “concentration world record” of the order of 80,000 suns (Gleckman 1988). However, it appears that this system was not the subject of subsequent applications.

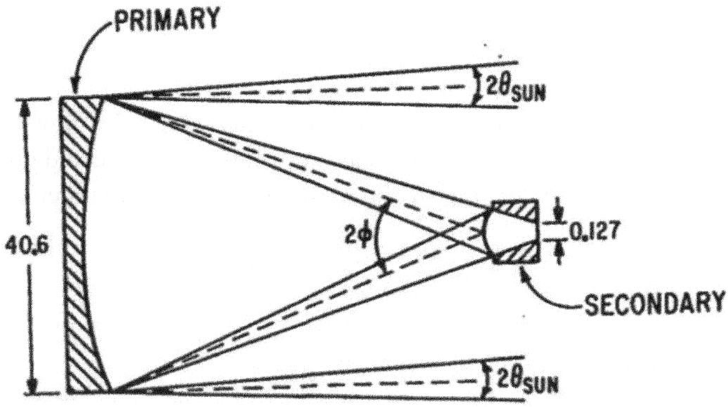

Figure 3.12. *Two-stage energy concentrating system (Gleckman 1988). For a color version of this figure, see www.iste.co.uk/flamant/energy.zip*

NOTE 3.2.– In the case of linear concentrating systems described in section 3.4, the concentration is only achieved along a single axis, and the expression of C_0 given in relation [3.12] must be modified as follows:

$$C_0 \approx \sin\alpha'_m / \varepsilon_0 \quad [3.16]$$

which yields a maximal theoretical concentration of 215 suns, and therefore lower maximal temperatures, according to the Stefan–Boltzmann law.

3.3.5. *Solar concentrator-specific errors*

One of the main problems faced by the optical engineer when designing an energy concentrating system is to establish the relation between theoretical and actual concentration factors C_0 and C defined in section 3.3.3.3, which will be the object of section 3.3.6. First, the error sources specific to solar installations must be defined. There are four such sources (Hénault 1987, 2015), as represented in Figure 3.13:

– local surface errors, represented by their angular standard deviation δ_S;

– adjustment errors of segments or individual modules of the concentrator, represented by their angular standard deviation δ_r;

– mirror pointing errors, represented by their angular standard deviation δ_p;

– microscopic surface errors, represented by their angular standard deviation δ_m.

Each of these types of error is presented and discussed in the following sections. It is important to note that the first three ones are similar in nature and can be added up in a quadratic sum, while the fourth one must be dealt with separately.

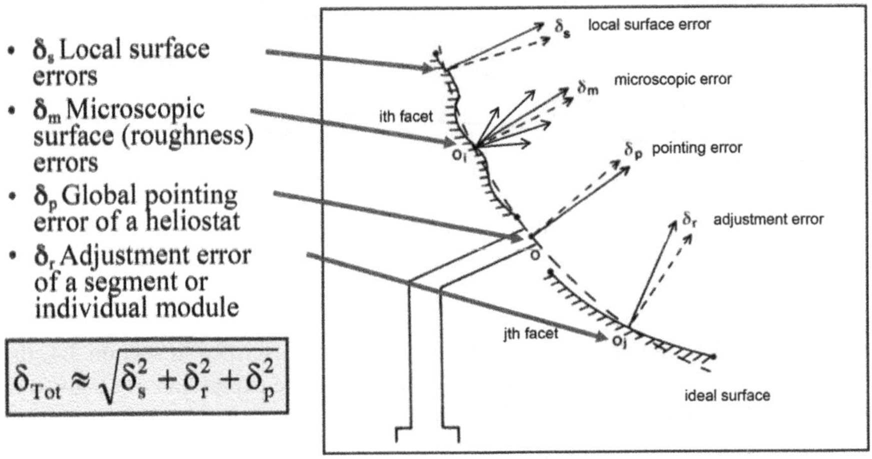

Figure 3.13. *Error sources specific to solar concentrators. For a color version of this figure, see www.iste.co.uk/flamant/energy.zip*

3.3.5.1. *Local surface errors*

Despite their much higher amplitudes (of the order of 0.1 mm compared to 0.001 mm for conventional polishing), the shape errors of solar concentrator mirrors follow the laws of classical geometrical optics. The latter are briefly summarized below and illustrated in Figure 3.14. They broadly rely on the wave front concept:

1) In classical optics, according to the Malus theorem, wave fronts are always perpendicular to light rays. Hence, for a sun reduced to a point at infinity (which assumes that $\varepsilon_0 = 0$), the rays incident on the concentrator are parallel and the incoming wave front is plane.

2) After reflection on the mirror centered in O, all the rays are concentrated in a single point O′ generally situated at the center of the receiver. For this purpose, the optical surface must have a special and unique shape, that of a section of parabola often referred to as "off-axis parabola". Then, all the rays effectively converge to O′

and the outgoing wave front is a sphere centered in O′ (see the upper part of Figure 3.14).

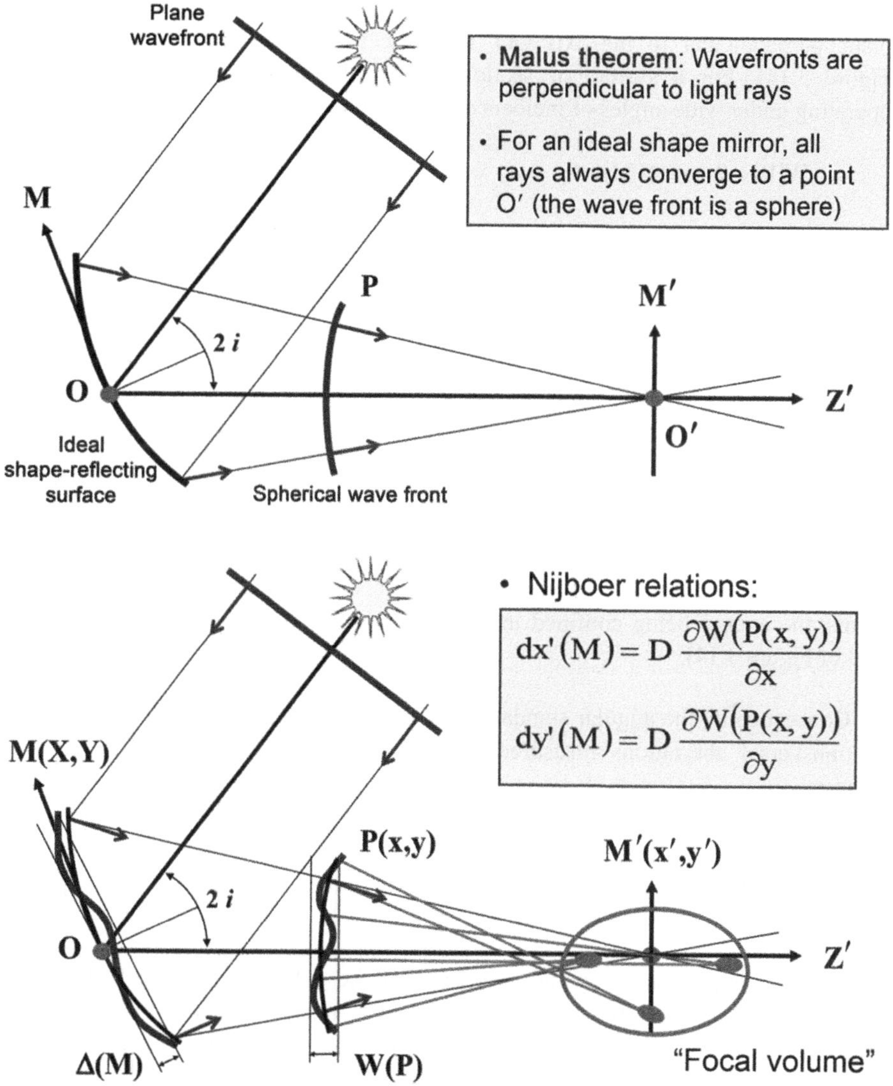

Figure 3.14. *Wave fronts reflected by an ideal shape mirror (upper image) and an actual concentrator (lower image). For a color version of this figure, see www.iste.co.uk/flamant/energy.zip*

3) For various reasons, this actual shape of the mirror differs from its ideal shape. Let us denote by Δ(M) these local surface errors, where M designates a point on the reflecting surface, and by W(P) the resulting wave front errors, seen in a plane perpendicular to the axis OO′ from the receiving surface (lower part of Figure 3.14). For a system or sections of a solar concentrating system often operating under wide angles of incidence i, the relation between Δ(M) and W(P) is:

$$W(P) \approx 2\cos i\, \Delta(M) \qquad [3.17]$$

According to the Malus theorem, the rays no longer converge to point O′. Their points of impact defined by their Cartesian coordinates (dx′(M),dy′(M)) in the receiving plane are obtained from the Nijboer relations of geometrical optics:

$$\begin{aligned} dx'(M) &= D\frac{\partial W(P(x,y))}{\partial x} \\ dy'(M) &= D\frac{\partial W(P(x,y))}{\partial y} \end{aligned} \qquad [3.18]$$

These relations remain valid for very high amplitude errors, such as those affecting solar concentrating mirrors. In this case, a real image of the Sun no longer forms, the energy being confined in a "focal volume" centered in point O′ (lower part of Figure 3.14).

Consequently, the angular standard deviation δ_s of local surface errors is related to transversal aberrations measured in the receiving plane by the approximate relation:

$$\delta_s \approx \frac{\sqrt{\left\langle dx'(M)^2 + dy'(M)^2 \right\rangle_M}}{2D\cos i} \qquad [3.19]$$

where $\langle \ldots \rangle_M$ represents a spatial average over the collecting surface S. Relation [3.19] is very general, and can also be applied to mirror adjustment and pointing errors (sections 3.3.5.2 and 3.3.5.3). It is employed by traditional methods for the measurement of standard deviations δ_s , δ_r and δ_p.

From a practical point of view, local surface errors depend essentially on mirror manufacturing and shaping methods. Figure 3.15 shows various types of surface errors Δ(M) commonly present in solar concentrators, from left to right:

– mirror curvature errors resulting from manufacturing defects and/or the fact that the focal distance of the mirror is not exactly adjusted to the distance D. They generate a defocus defect with respect to the receiving surface;

– deformations of slightly higher order, such as astigmatism, often due to mirror deformation under mechanical constraint;

– periodical defects of highest spatial frequencies due to float glass manufacturing process (rippling effect);

– finally, random errors of various origins: manufacturing, shaping, operating environment.

The presence and characteristics of these errors may significantly vary from one mirror to another, which in the end justifies their statistical approach.

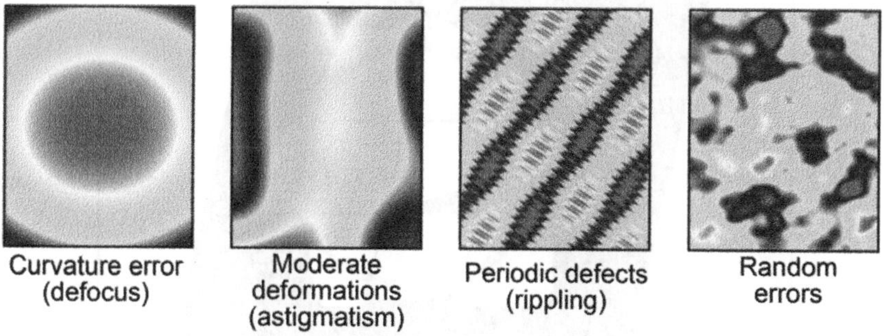

Figure 3.15. *False color representation of various types of local surface errors. For a color version of this figure, see www.iste.co.uk/flamant/energy.zip*

3.3.5.2. *Adjustment errors*

The adjustment errors or those of alignment of various facets or segments of the concentrating system (canting errors) are present in most solar concentrating systems. These are essentially mirror orientation errors, as their defects of lateral or longitudinal positioning along the axes X, Y and Z of the concentrator have a negligible influence on the concentration factor C. They are statistical in nature and are characterized by their angular standard deviation δ_r, which essentially depends on the mirror (heliostats and/or concentrator) adjustment procedures. Aging effects due to day/night and summer/winter alternation of these systems operating outdoors should also be taken into account. Wind force exerted on the supporting structure and the settlement under its own weight, itself variable with the orientation of the gravity vector, also contributes to mirror maladjustment. Consequently, the actual value of δ_r varies with time.

Figure 3.16 shows a didactic illustration of the adjustment errors of one of the flat heliostats of Odeillo solar furnace: from a distance, the Sun image observed through the correctly adjusted heliostat should reproduce a virtual image of the Sun delineated by the red circle. In the presence of adjustment defects, it shows dead zones inside the useful zone and spillages outside of it, which is useful information for determining the value of the standard deviation δ_r (Hénault 1989).

Figure 3.16. *Visualization of the adjustment errors of a flat heliostat (Hénault 1989). For a color version of this figure, see www.iste.co.uk/flamant/energy.zip*

3.3.5.3. *Pointing errors*

In the case of a system involving several dozens or hundreds of heliostats, such as a solar furnace (section 3.2.2) or a tower power plant (section 3.5), the pointing errors of various mirrors are also of statistical nature and are characterized by their angular standard deviation δ_p. The value of this parameter strongly depends on the type of guidance of mirror mounting, which can be open loop (section 3.3.5.3.1) or closed loop (section 3.3.5.3.2). These two control modes are illustrated in Figure 3.17. There is currently a trend towards using mixed systems, with a closed-loop guidance during strong insolation hours and open-loop guidance during weak insolation (cloud presence) or in order to bring back the concentrator to its initial position, referred to as "parking".

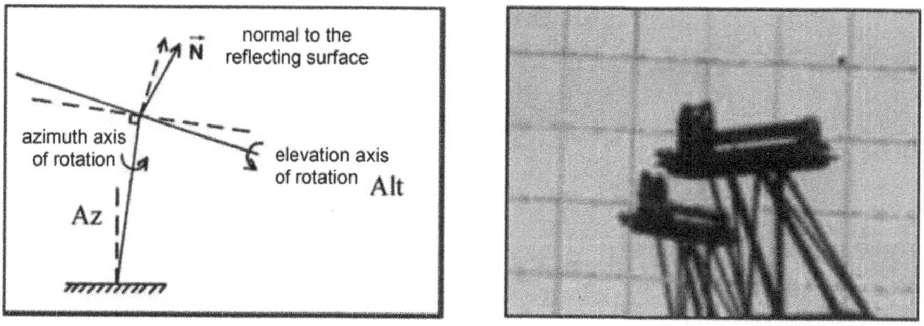

Figure 3.17. *Illustration of the open-loop control mode (on the left) or closed-loop control mode (on the right). For a color version of this figure, see www.iste.co.uk/flamant/energy.zip*

3.3.5.3.1. Open-loop control

This type of control involves a software control system instructing the heliostat to align the vector normal to its surface $\vec{N}_0$ so that it reflects the Sun vector $\vec{S}_0$ along a target vector $\vec{R}_0$ pointing to the receiver, according to the vector relation of Descartes reflection (see Figure 3.18):

$$\vec{N}_0 = \left(\vec{S}_0 + \vec{R}_0\right) \Big/ \sqrt{2\left(1 + \vec{S}_0 \vec{R}_0\right)} \qquad [3.20]$$

In the widespread case of altazimuth mount, the main error causes are illustrated in the left image in Figure 3.17, as follows:

– a defect of verticality of the azimuth axis of rotation;

– a perpendicularity error of the height and azimuth axes of rotation.

An additional cause is the uncertainty on the exact geometry of the system (precise knowledge of vectors $\vec{S}_0$ and $\vec{R}_0$), precise knowledge on the initial parking position, the absolute resolution and precision of the coders, and their refresh frequency (Escobar-Toledo et al. 2014). It is important to note that some of these errors can be calibrated and partially compensated by the software control of the heliostat.

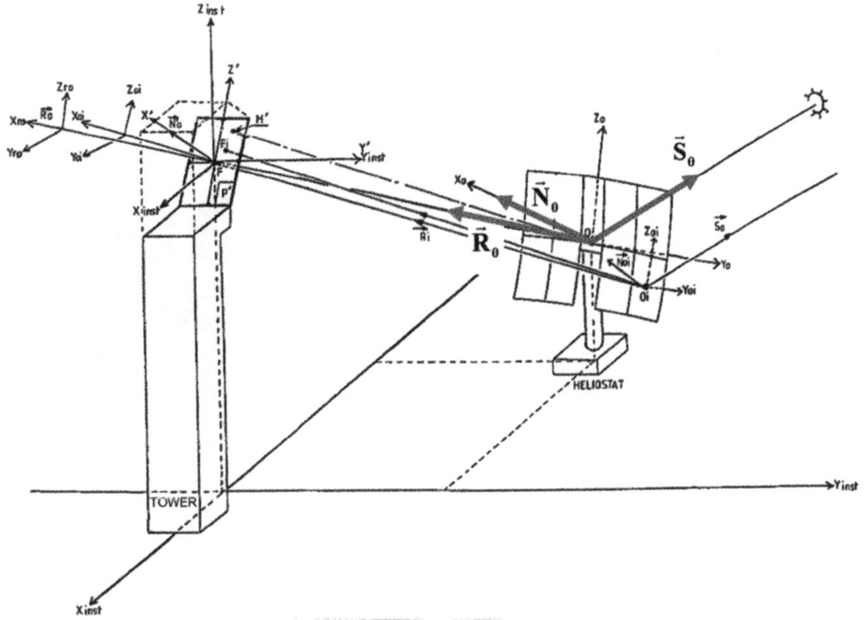

Figure 3.18. *Open-loop control of a heliostat of a tower power plant. For a color version of this figure, see www.iste.co.uk/flamant/energy.zip*

3.3.5.3.2. Closed-loop control

In the closed-loop systems, the vector normal to the concentrator $\vec{N}_0$ is controlled by the error signals provided by a fixed guiding telescope (see the image on the right of Figure 3.17), whose optical axis is parallel to the target $\vec{R}_0$. Here, the errors are essentially reduced to defective alignment between the target vector $\vec{R}_0$ and the real optical axis of the telescope, as well as to its optoelectronic design, on which the guidance precision depends. The simplest version of a guiding telescope uses a four-quadrant detector placed at the focal point of a focalization optical system. Nowadays, there are many improved designs, a vast subject that is beyond the scope of this book.

3.3.5.4. *Microscopic surface errors*

Microscopic surface errors, sometimes referred to as roughness or "waviness", are the result of glass manufacturing or polishing processes at the scale of several micrometers. Figure 3.19 shows an illustration of these errors and one of their standard measurement means (Silvestri et al. 2010). They essentially generate diffusion throughout the image medium, so that a small proportion of the diffused rays finally reach the receiving surface. Their influence on the concentration factor is therefore negligible, being essentially related to the total collected power Φ. As such, they can be considered as an additional attenuation factor similar to the atmospheric transmission TA in relation 3.1. According to the classical formula of "Total Integrated Scattering" (TIS), the approximate expression of this factor is (Harvey et al. 2012):

$$T_m \approx 1-(4\pi \cos i\, \sigma_m/\lambda)^2 \quad [3.21]$$

where σ_m is the standard deviation of microscopic surface errors and λ is the wavelength of the radiation. It is important to note here an essential difference with respect to the three previous types of error: here, the standard deviation σ_m relates to the mirror surface errors, while the standard deviations δ_s, δ_r, δ_p relate to their angular errors, and therefore to the slopes of these same surfaces.

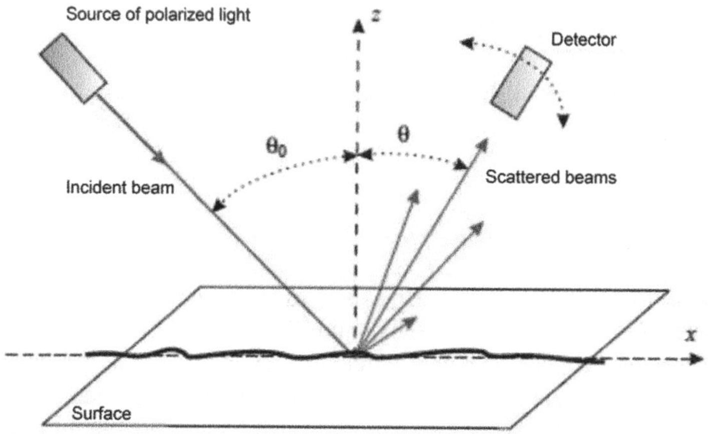

Figure 3.19. *Microscopic surface errors and their measurement means (Silvestri et al. 2010). For a color version of this figure, see www.iste.co.uk/flamant/energy.zip*

3.3.6. *Concentration losses and the "golden rule" of solar concentration*

In a reference article of 1970s (Lipps 1976), F. Lipps established that k_a, distribution of irradiance E′(M′) formed in the receiving plane, can be expressed as a convolution product of the "ideal" geometric image of the Sun with a geometric pulse response function that would be observed in the same plane if the Sun were reduced to a single point in the sky (which means fixing a zero angle ε_0). This convolution relation can be generalized by adding to it statistical error cones corresponding to various defects defined in sections 3.3.5.1–3.3.5.3, and projected in the receiver plane (Hénault 1989). Assuming that Gaussian laws of error are applicable and using several elementary properties of the convolution product, a simplified expression of the ratio η of the actual concentration factor C and its theoretical value C_0 in the form of a Lorentz function is obtained (Hénault 2015):

$$\eta = \frac{C}{C_0} = \frac{1}{1+\delta^2/\varepsilon_0^2} \qquad [3.22]$$

where δ is the quadratic sum of angular errors on the rays reflected by the concentrator, which is expressed as a function of standard deviations δ_s, δ_r, δ_p:

$$\delta \approx 2\cos i\sqrt{\delta_s^2+\delta_r^2+\delta_p^2} \qquad [3.23]$$

Then, the expression of the black body equivalent temperature in relation 3.14 becomes:

$$T \approx \sqrt[4]{\frac{\varepsilon\, \eta\, C_0\, DNI}{\sigma}} \qquad [3.24]$$

Relations [3.22] and [3.24] can be used to draw the curves of the effective concentration factor C and of the temperature T depending on angular errors δ for various values of the theoretical concentration C_0 (Figure 3.20). It can be noted, for example, that the concentration C decreases by about 50% if δ is equal to ε_0 and by 80% if it is equal to $2\varepsilon_0$. These losses are reflected by temperature drops of 17.5% and 35%, respectively. These relations can be considered as "golden rules" of the optical concentration defining the tolerances on the slopes of optical surfaces and/or on the deviations of the reflected rays, in order to reach the required concentration and temperature performances. Hence, a mirror slope error of 1 mrad RMS is sufficient to ensure a drop in temperature below 4.5%, while a more demanding specification of 0.5 mrad RMS guarantees a loss in temperature of about 1% (see Table 3.2). Obviously, these simple rules for the optical engineer cannot replace a detailed thermodynamic study of the receiving system.

Surface slope RMS error (mrad)	0.5	1
Reflected rays RMS error (mrad)	1	2
Concentration loss (%)	4.4	15.6
Temperature loss (%)	1.1	4.1

Table 3.2. *Golden rule of solar concentration*

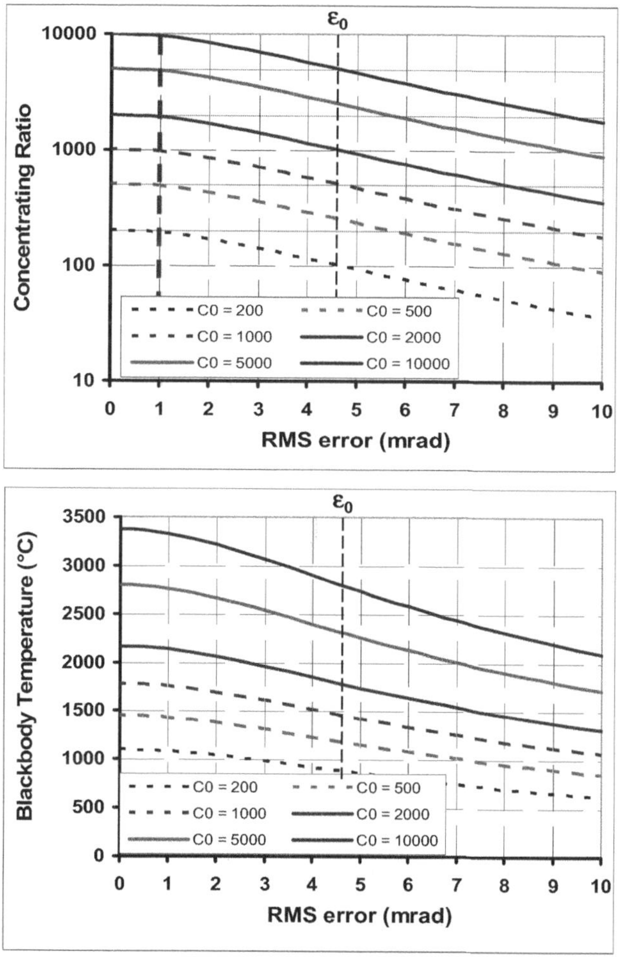

Figure 3.20. *Concentration and temperature error depending on angular errors δ (Hénault 2015). For a color version of this figure, see www.iste.co.uk/flamant/energy.zip*

As indicated in Chapter 1, the parabolic trough concentrators (linear concentration) and tower concentrators (point concentration) are the most developed at the industrial scale. The following sections are therefore dedicated to a presentation of these devices.

3.4. Optical qualification of parabolic trough concentrators

3.4.1. *Definitions*

The optical qualification of a parabolic trough concentrator involves the experimental determination of its global optical efficiency. This global optical efficiency reflects the ratio of the solar power received by the parabolic trough concentrator (P_{sol}) and the power incident on the solar receiver (P_{inc}).

By definition:

$$\eta_{opt,glob} = \frac{P_{inc}}{P_{sol}} \qquad [3.25]$$

The global optical efficiency reflects the power attenuation of the solar radiation between the inlet and the outlet of the parabolic trough concentrator. Following the path of a light ray, various physical phenomena can be identified as causing this attenuation, as schematically represented in Figure 3.21.

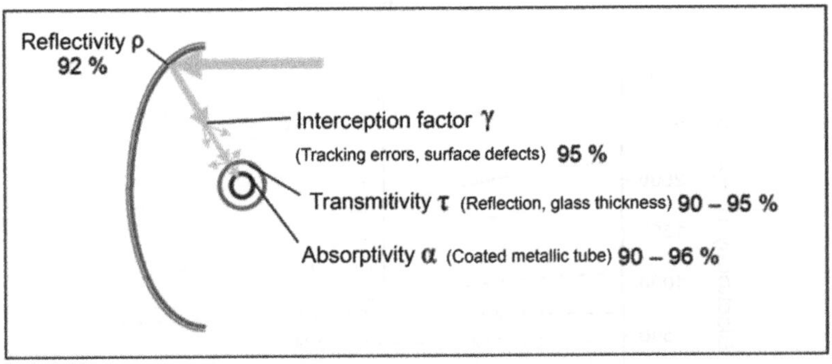

Figure 3.21. *Schematic representation of the physical phenomena affecting global optical efficiency. For a color version of this figure, see www.iste.co.uk/flamant/energy.zip*

The following are noted:

– the reflection of the incident ray on the reflecting surface, characterized by reflectivity ρ;

– the probability that the reflected ray is intercepted by the solar receiver at the focal point, characterized by the interception factor γ. It particularly reflects tracking errors, surface defects, etc.;

– the part of the reflected light traveling through the glass tube that protects the solar receiver, characterized by transmittivity τ;

– the proportion of the reflected ray absorbed by the absorber, characterized by the absorptivity of the material it is composed of, α.

The global optical efficiency can finally be expressed as the combination of these various phenomena according to the following equation:

$$\eta_{opt,glob} = \rho.\gamma.\tau.\alpha. \quad [3.26]$$

3.4.2. *Methodology*

The experimental determination of the global optical efficiency of a parabolic trough collector involves the measurement of powers in the energy balance of the system (Fasquelle et al. 2017). It is important to note here that the thermal losses undergone by the receiver (P_{losses}) must be known in advance. Therefore, the following quantities are employed:

The solar power P_{sol} at the entry of the collector, which is calculated based on the DNI and the collecting surface:

$$P_{sol} = DNI \cdot S_{coll} \quad [3.27]$$

The useful power received by the fluid P_u which is calculated from the experimental measurements of flow rates and temperatures at the system inlet and outlet:

$$P_u = \dot{m}_f \cdot C_f \cdot (T_{outlet} - T_{inlet}) \quad [3.28]$$

The energy balance of the solar receiver finally yields:

$$\eta_{opt,glob} . P_{sol} = P_u + P_{losses} \quad [3.29]$$

$$\eta_{opt,glob} = \frac{P_u + P_{losses}}{P_{sol}} = \frac{\dot{m}_f \cdot C_f \cdot (T_s - T_e) + P_{losses}}{DNI \cdot S_{coll}} \quad [3.30]$$

The global optical efficiency can therefore be determined experimentally from the measurement of thermal losses of the receiver, the solar power at the collector entry and the useful power provided by the transfer fluid.

It is important to note, among other cases, the case of zero incidence angle of the rays (and consequently maximal optical efficiency) (Fasquelle et al. 2017). For this purpose, an attenuation coefficient k_θ is defined, and this coefficient is calculated for each incidence angle θ_i and allows the comparison of the optical efficiency at each instant with the maximal optical efficiency $\eta_{opt,max}$

$$k_\theta = \frac{\eta_{opt}}{\eta_{opt,max}} \quad [3.31]$$

The attenuation coefficient k_θ is therefore a factor ranging between 0 and 1 that depends on the incidence angle. It is sometimes referred to as the incidence angle modifier (IAM). It makes it possible to account for the optical losses due to the fact that the collecting surface is not perpendicular to the incident solar rays. It takes into consideration the cosine effect. A description of the set of potential optical errors that might be affected by the incidence angle is provided by Güven et al. (1986).

The principle of optical qualification and determination of the attenuation coefficient k_θ is represented in Figure 3.22.

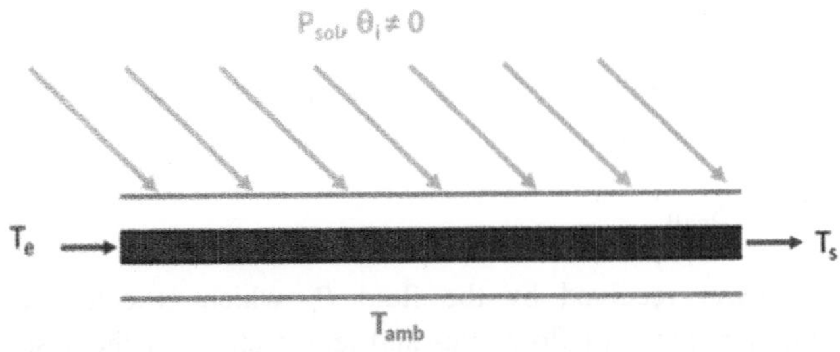

Figure 3.22. *Optical qualification test, evaluation of k_θ (Fasquelle et al. 2017). For a color version of this figure, see www.iste.co.uk/flamant/energy.zip*

Therefore, by conducting optical qualification tests at several times of the day for several incidence angles of the solar rays reaching the collector, it is possible to determine the variation of the attenuation coefficient k_θ depending on the incidence angle and to determine the global optical efficiency.

3.4.3. *Example*

In order to illustrate the previously described method, this section provides an example of optical qualification conducted on a 72 m long parabolic trough concentrator qualified (Lüpfert et al. 2003; Valenzuela et al. 2014) at CIEMAT (*Centro de Investigaciones Energéticas, Medioambientales y Tecnológicas*), on the PSA (*Plataforma Solar de Almeria*) site.

Parameters of the parabolic trough collector	
Length	72 m
Aperture	5.76 m
Maximal optical efficiency	0.768

Table 3.3. *Characteristics of the collector used for the optical qualification (Valenzuela et al. 2014)*

The maximal optical efficiency, measured when the incidence angle is zero, is 0.768. Having measured this maximal efficiency, the authors studied the influence of the incidence angle, through the evolution of the attenuation coefficient. These results are represented in Figure 3.23. This figure shows that k_θ decreases very rapidly for angles above 40°, which explains the low optical efficiency of these systems during winter, in Europe, for the North–South-oriented collectors.

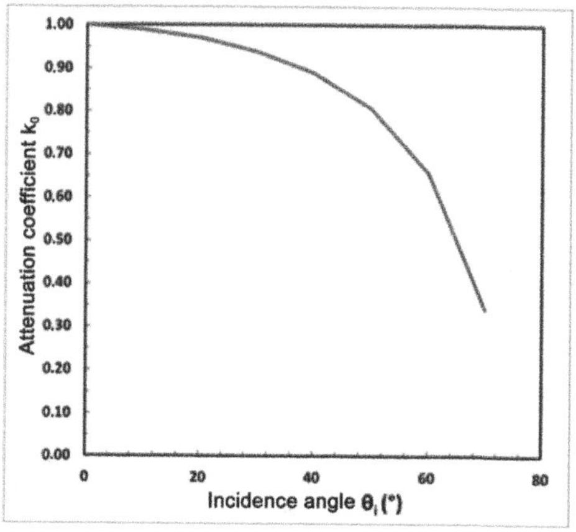

Figure 3.23. *Example of optical qualification (Valenzuela et al. 2014). For a color version of this figure, see www.iste.co.uk/flamant/energy.zip*

3.5. The heliostat fields of tower power plants

3.5.1. *Description*

"Heliostat" comprises the Greek words "helio" for sun and "stat" for stationary, reflecting the stationary character of the reflected spot throughout the day. Heliostats send the concentrated solar flux to the receiver, generally placed on top of a tower. The power required on the receiver allows the dimensioning of the field of heliostats, which is a difficult task, as it must take into account the position of the Sun, which varies throughout the day and depending on the season. The field display must be optimized in order to limit the heliostat–receiver distance. Indeed, the farther the heliostats, the higher the optical losses. The position of the heliostats in the field also depends on the height of the tower.

The main components of a heliostat are a slightly concave reflecting surface, a support structure and a sun-tracking mechanical system, comprising motor systems and control systems. Figure 3.24 presents a standard heliostat design.

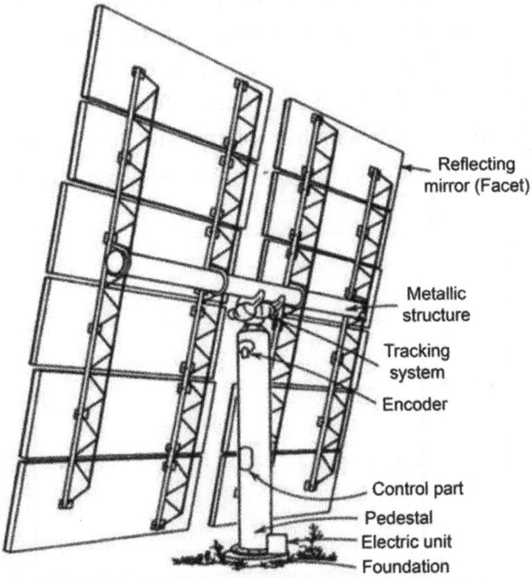

Figure 3.24. *Standard design of a heliostat*

Over 50 years of development led to enhanced performances of heliostats, particularly in terms of mirror reflectivity, wind resistance (mechanical structure) and optical precision. The size of a heliostat varies greatly depending on the manufacturer. Some prefer large heliostats (surface area greater than 100 m^2) to limit their number and ease the control, while others focus on small size heliostats (surface area below 10 m^2) as a means to improve optical quality, facilitate adjustment, reduce the wind load and facilitate their installation.

On modern heliostats, the mirror surface is composed of fine back-silvered glass plates. This glass has a low iron content, and therefore high transparence, to provide a better global reflectivity (transmission of the incident radiation, reflection on the silver layer and transmission of the reflected radiation). A 93–96% reflectivity is commercially obtained. Corrosion and oxidation are prevented by protecting the back-silver coating with, for example, a resin (Chapter 7). This makes it possible to increase the service life of the mirrors. The mirrors are placed on a supporting structure. The holding unit and the shape of the support maintain the curvature of the mirrors. In order to have the best concentration on the receiver at the top of the tower, a thorough study of the mutual positioning of the mirrors is required.

3.5.2. *Optical losses*

The performances of a heliostat field depend on its optical efficiency, which is calculated with the following equation:

$$\eta_{opt} = \rho_{mirror}.\left(1 - f_{cos(\alpha)}\right).\left(1 - f_{sb}\right).\left(1 - f_{at}\right).\left(1 - f_{sp}\right) \qquad [3.32]$$

It is affected by various optical losses. The first optical loss is caused by mirror reflectivity ρ_{mirror} ranging between 90% and 96% (depending on the technologies and on the soiling level) which amounts to an optical loss of 4–10%. The second, the cosine loss $f_{cos(\alpha)}$, is the dominant optical loss. As shown in Figure 3.25, the wide incidence angle of the solar ray on heliostat B generates a lower incident power on the receiver than that for heliostat A.

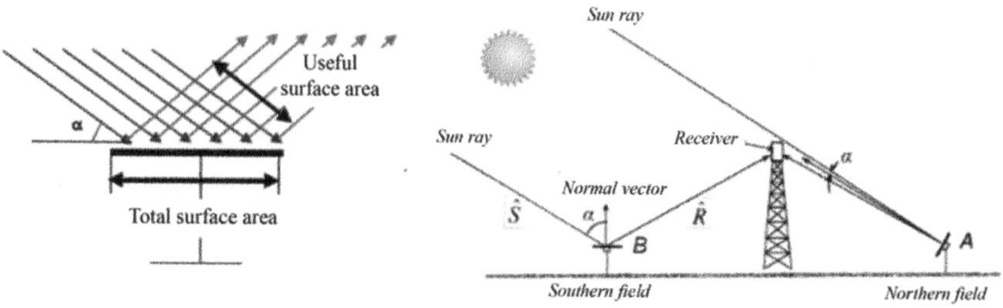

Figure 3.25. *Schematic representation of optical losses due to cosine effect. For a color version of this figure, see www.iste.co.uk/flamant/energy.zip*

There are also shading losses (one heliostat shades another, as illustrated in Figure 3.26) and blocking losses (the rays reflected by a heliostat are blocked by another heliostat placed on the path of the reflected rays towards the receiver, illustrated in Figure 3.27), represented by f_{sb} in equation [3.32].

Figure 3.26. *Optical losses due to shading. For a color version of this figure, see www.iste.co.uk/flamant/energy.zip*

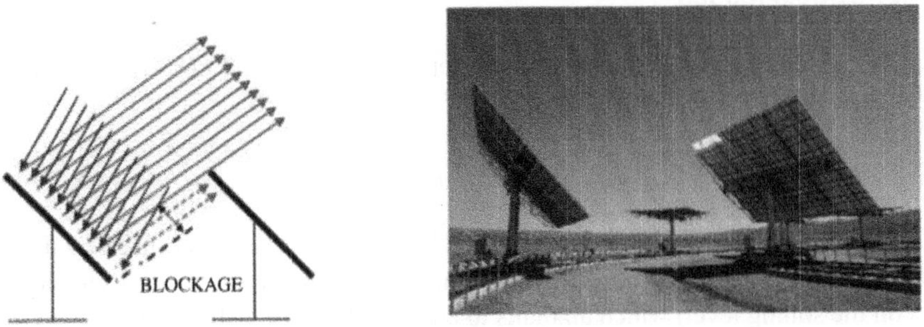

Figure 3.27. *Optical losses by blocking. For a color version of this figure, see www.iste.co.uk/flamant/energy.zip*

Figure 3.28 illustrates the optical losses due to atmospheric attenuation f_{at}. These are directly related to the distance between the heliostat and the receiver, and depend on air quality.

Finally, the spillage losses at the solar receiver f_{sp} depend on the optical errors of the heliostats, the heliostat/receiver distance and the dimensions of the receiver. They are represented in Figure 3.29.

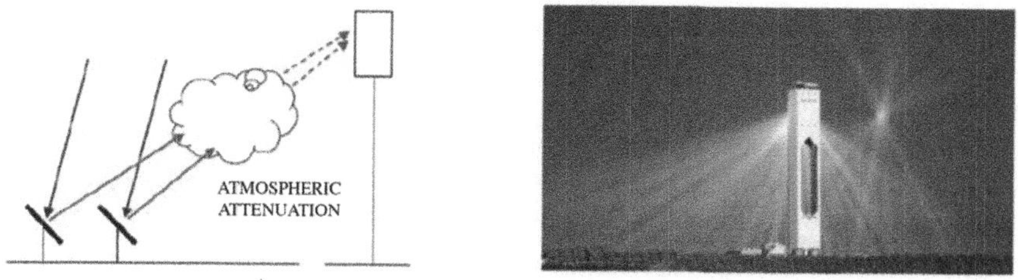

Figure 3.28. *Optical losses due to atmospheric attenuation. For a color version of this figure, see www.iste.co.uk/flamant/energy.zip*

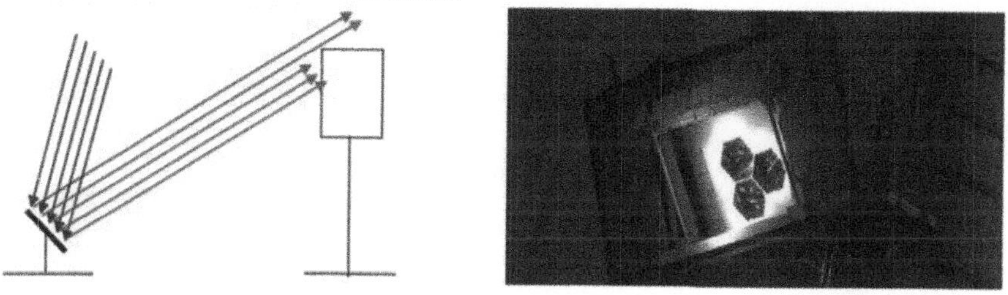

Figure 3.29. *Optical losses due to spillage. For a color version of this figure, see www.iste.co.uk/flamant/energy.zip*

The heliostat field contributes 30–50% of the total investment cost and 40% of the total energy losses. It is therefore essential to optimize its design in order to reduce its cost and improve the overall performances of the solar power plant. Many algorithms were developed in order to optimize the configuration of heliostat fields (Barberena 2016). It is important to take into account the nature of the solar receiver, the latitude of the site, the size of heliostats, the expected concentration and incident power.

Depending on the latitude of the site, there are two configurations of heliostat fields: Northern or Southern field (depending on the hemisphere) or surrounding field, as shown in Figure 3.30.

At an equal surface of mirrors, a surrounding field yields better efficiency, as heliostats are closer to the tower, but, on the other hand, many heliostats have significant cosine losses. At low latitudes, the Sun is very high in the sky throughout the year, which favors the use of a surrounding field. On the contrary, for latitudes above 30–35°, a Northern/Southern field has a better annual efficiency. The optimum display must therefore be identified for each site depending on technological limitations. There is a connection between the field display and the optical system of the receiver entry. For example, the Gemasolar heliostat field illuminates a cylindrical receiver at 360°. For a cavity receiver placed on top of the tower, a surrounding field is useless, as the receiver is oriented in a single direction.

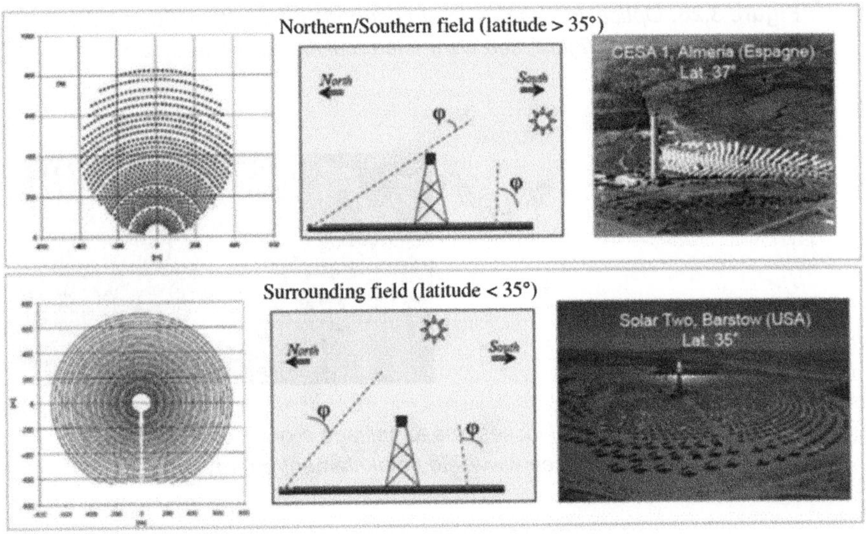

Figure 3.30. *Northern/Southern fields and surrounding fields. For a color version of this figure, see www.iste.co.uk/flamant/energy.zip*

A further possible configuration is a combination of a heliostat field with a second reflection at the top of the tower redirecting solar rays towards the base of the tower, referred to as beam-down. Figure 3.31 shows a schematic representation of this technology.

Though the beam-down adds to the optical losses, this configuration significantly reduces the length of pipelines. The generation components are on the ground. Furthermore, this power plant concept allows the coupling of a cavity receiver with a surrounding field. However, given that the surface of the reflector on top of the tower must be 5–100 times smaller than the total surface of the field, it must tolerate a flux density that is 5–100 times higher than the density of the incident solar radiation, which may lead to beam-down mirror overheating. Finally, this design cannot be used for high power systems.

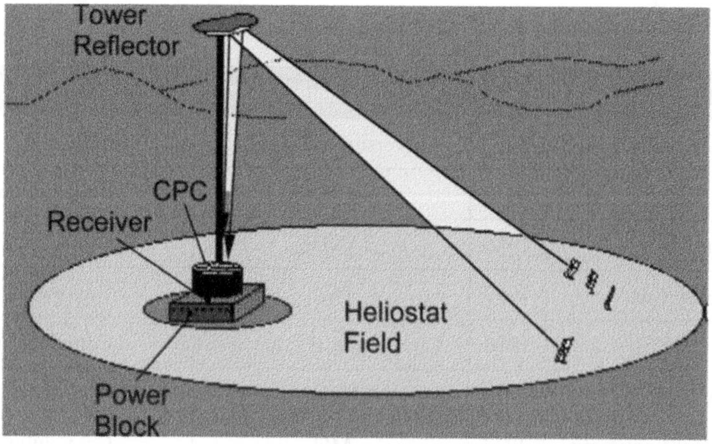

Figure 3.31. *Schematic representation of the principle of a beam-down tower power plant (Romero 2002)*

3.5.4. *Simulations of heliostat fields*

Many optical models were developed in order to predict the distribution of flux reaching the solar receiver (Li 2016). Two approaches are generally considered: the ray-tracing method and the convolution method. The principle of the ray-tracing method, also known as the statistical or Monte Carlo method, is to randomly choose a beam of rays from surface 1 and to determine which ones arrive on surface 2. The irradiation of an elementary surface is proportional to the number of rays impacting it. The precision and the calculation time increase with the number of rays and the complexity of the geometry. The ray-tracing software programs are the most robust instruments for modeling the optical systems and the radiative transfers in the concentrated solar power technologies (Asselineau 2015). Indeed, these statistical methods can be used for the integration of spatial, angular, spectral and temporal variations of the energy intensity, as well as of directional and spectral properties for opaque and transparent surfaces (Howell 1998). They provide good precision, but require a lot of calculation time.

In the convolution methods (or cone optics), the reflected solar radiation is modeled by a convolution product of normal Gaussian distributions corresponding to each one of the errors, the one of the sun shape and the one of the heliostat. These methods are less precise, as they do not take into account the optical aberrations, but they are faster and are therefore generally used for optimization problems.

Table 3.4 presents a non-exhaustive list of existing tools allowing the simulation of the optics of concentrating solar fields.

Optical simulation tool	Year	Developer	Method
CAMPO	2011	University of Zaragoza	-
DELSOL	1979	SANDIA	Convolution
winDELSOL	1986	SANDIA	Convolution
Fiat_Lux	1997	PSA	Ray-tracing
HFLCAL	1986	DLR	Convolution
MIRVAL	1978	SANDIA	Ray-tracing
MIT code/Biomimetic	2012	MIT	Convolution
SolarPILOT	2018	NREL	Convolution and Ray-tracing
SOLSTICE	2016	CNRS PROMES and Meso-Star SAS	Ray-tracing
SolTRACE	2003	NREL	Ray-tracing
STRAL	2009	DLR	Ray-tracing
Tonatiuh	2009	CENER	Ray-tracing
VEGAS	2010	ETH Zurich	Ray-tracing

Table 3.4. *Non-exhaustive list of tools available for the optical modeling of concentrating solar systems (adapted from Li 2016)*

As an example, Figure 3.32 illustrates the optimal layout of a heliostat field calculated by SolarPILOT software for a 100 MW_{th} average thermal power incident on the receiver. It can be noted, on the one hand, that the optical efficiency of heliostats located near the tower is the highest (above 85%) and, on the other hand, that the distance between the rows increases with the distance to the tower (limitation of shading and blocking losses). For the simulated power, the last row of heliostats is at about 1 km from the tower.

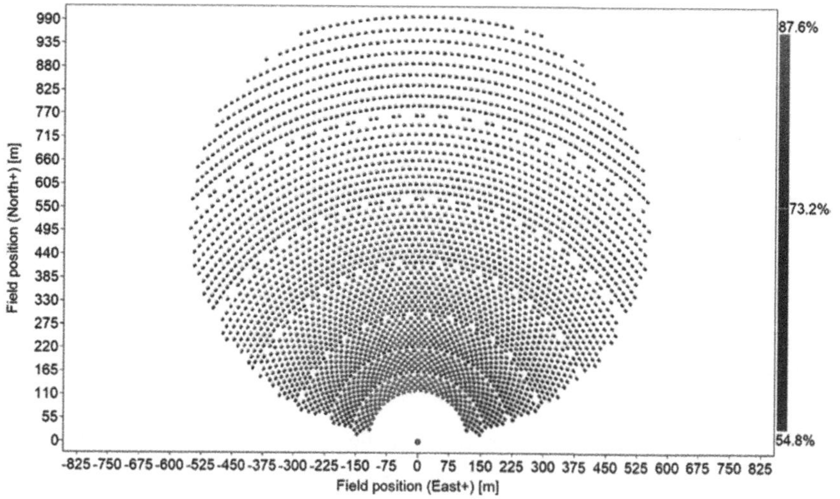

Figure 3.32. *Heliostat field simulated by SolarPILOT software for a 100 MWth average power incident on the solar receiver. The tower height is 120 m, and the surface area of heliostats is 49 m*^{2}*. The right scale reflects the optical efficiency of heliostats. For a color version of this figure, see www.iste.co.uk/flamant/energy.zip*

3.6. Conclusion

As a key component of concentrated solar power, the optical system integrates multiscale scientific and technological issues ranging from the submicronic for the mirror surface defects to the kilometric for the collecting fields, not to mention the metric scale of the concentrator facets. The search for high efficiencies, and therefore high operating temperatures, places higher demands on the optical quality of concentrators. Conversely, the push for lower costs should not lead to a decrease in the quality of the service life of the components subjected to sometimes extreme climate variations. In order to reach the DOE goals for 2030 (50 USD/MWh for the base load generation power plants and 100 USD/MWh for the peak generation power plants; see Chapter 1), the cost of heliostats must be around 75 USD/m^2, which means a 35–40% reduction with respect to the state of the art. Research and innovation challenges are therefore significant, but the objective is attainable, considering the cost reduction observed in recent years for a modest deployment of concentrated solar power.

3.7. References

Asselineau, C.-A., Zapata, J., Pye, J. (2015). Integration of Monte-Carlo ray tracing with a stochastic optimization method: Application to the design of solar receiver geometry. *Optics Express*, 23(11), A437–A443.

Barberena, J.G., Mutuberria Larrayoz, A., Sánchez, M., Bernardos, A. (2016). State-of-the-art of heliostat field layout algorithms and their comparison. *Energy Procedia*, 93, 31–38.

Drouot, L.P. and Hillairet, M.J. (1984). The Themis program and the 2500-KW Themis solar power station at Targasonne. *Journal of Solar Energy Engineering*, 106, 83–89.

Escobar-Toledo, M., Arancibia-Bulnes, C.A., Iriarte-Cornejo, C., Waissman, J., Riveros-Rosas, D., Cabanillas, R.E., Estrada, C.A. (2014). Heliostat image drift behavior for different error sources. *Journal of Renewable and Sustainable Energy*, 6, 023117.

Fasquelle, T. (2017). Modélisation et caractérisation expérimentale d'une boucle solaire cylindro-parabolique intégrant un stockage de type thermocline. PhD Thesis, University of Perpignan Via Domitia, Perpignan.

Fasquelle, T., Falcoz, Q., Neveu, P., Lecat, F., Flamant, G. (2017). A thermal model to predict the dynamic performances of parabolic trough lines. *Energy*, 141, 1187–1203.

Gleckman, P. (1988). Achievement of ultrahigh solar concentration with potential for efficient laser pumping. *Applied Optics*, 27, 4385–4391.

Güven, H.M. and Bannerot, R.B. (1986). Determination of error tolerances for the optical design of parabolic troughs for developing countries. *Solar Energy*, 36(6), 535–550.

Harvey, J.E., Schröder, S., Choi, N., Duparré, A. (2012). Total integrated scatter from surfaces with arbitrary roughness, correlation widths, and incident angles. *Optical Engineering*, 51, 013402.

Hénault, F. (1987). Concentration du rayonnement solaire par simple et double réflexion : contribution aux méthodes de réglage et de contrôle des surfaces réflectrices à facettes. PhD Thesis, Paris-Saclay University, Paris.

Hénault, F. (2015). Fast computation of solar concentrating ratio in presence of opto-mechanical errors. *Solar Energy*, 112, 183–193.

Hénault, F. and Bonduelle, B. (1989). Modèle de calcul des flux au foyer du four solaire de 1000 kW d'Odeillo. Un outil pour la recherche et le développement en thermique des matériaux. *Entropie*, 146147, 81–92.

Hénault, F. and Royère, C. (1989). Concentration du rayonnement solaire : analyse et évaluation des réponses impulsionnelles et des défauts de réglage de facettes réfléchissantes. *Journal of Optics*, 20, 225–240.

Howell, J.R. (1998). The Monte Carlo method in radiative heat transfer. *Journal of Heat Transfer*, 120(3), 547–560.

Jafrancesco, D., Cardoso, J.P., Mutuberria, A., Leonardi, E., Les, I., Sansoni, P., Francini, F., Fontani, D. (2018). Optical simulation of a central receiver system: Comparison of different software tools. *Renewable and Sustainable Energy Reviews*, 94, 792–803.

Li, L., Coventry, J., Bader, R., Pye, J., Lipiński, W. (2016). Optics of solar central receiver systems: A review. *Optics Express*, 24(14), A985–A1007.

Lipps, F. (1976). Four different views of the heliostat flux density integral. *Solar Energy*, 18, 555–560.

Lüpfert, E., Zarza-moya, E., Geyer, M., Nava, P., Langenkamp, J., Schiel, W., Esteban, A., Osuna, R., Mandelberg, E. (2003). Euro Trough collector qualification complete – Performance test results from PSA. *ISES Solar. World Congress*, Gothenburg.

Romero, R., Buck, R., Pacheco, M.E. (2002). An update on solar central receiver systems, projects and technologies. *Solar Energy Engineering*, 124, 98–108.

Royère, C. (1981). Les applications du four solaire de 1000 kW du CNRS à Odeillo. *Entropie*, 97, 147–160.

Silvestri, Z., Du Colombier, D., Macé, S., Pinot, P. (2010). Moyens de caractérisation des surfaces et leurs applications en métrologie. *Revue française de métrologie*, 2010-3, 11–25.

Trombe, F. (1955). Le grand Four solaire de Mont-Louis. *L'Astronomie*, 69, 45–60.

Trombe, F. and Le Phat Vinh, A. (1971). *Cahiers de l'AFEDES*, n°3.

Trombe, F. and Le Phat Vinh, A. (1973). Thousand kW solar furnace, built by the national center of scientific research, in Odeillo (France). *Solar Energy*, 15, 57–61.

Valenzuela, L., López-Martín, R., Zarza, E. (2014). Optical and thermal performance of large-size parabolic-trough solar collectors from outdoor experiments: A test method and a case study. *Energy*, 70, 456–464.

Welford, W. and Winston, R. (1978). *The Optics of Non-Imaging Concentrators*. Academic Press, Cambridge, MA.

Winston, R., Miñano, J.C., Benítez, P., Shatz, N., Bortz, J.C. (2005). *Nonimaging Optics*, 1st edition. Elsevier, Burlington, MA.

4

Solar Receivers

Benjamin Grange
PROMES-CNRS,
Font-Romeu-Odeillo-Via, France

4.1. Introduction

The objective of a solar receiver is to absorb concentrated solar radiation and transfer the energy to the heat transfer fluid (see Chapter 5). Depending on the considered collector field (linear or point concentration), there are two types of receivers. Section 4.2 presents the use of absorber tubes when the concentration is linear, particularly the parabolic trough concentrators, while section 4.3 describes the solar receivers used in tower power plants (point concentration).

4.2. Absorber tubes for linear concentrators

4.2.1. *Description*

Absorber tubes are composed of two concentric tubes: an internal metal tube through which the transfer fluid flows and an external glass tube, as shown in Figure 4.1. The external tube is made of low-iron borosilicate glass to increase its solar radiation transmittance. This tube can be used to create reduced pressure (≈ 5–10 mbar) in the annular space and thus limit thermal losses. A selective coating is applied on the metal tube in order to increase its absorption in the visible range and reduce its emission in the infrared range (see Chapter 7). The glass and metal tubes are connected by means of thermal expansion bellows. Besides tolerating the expansion of the two tubes, the latter offer a tight annular space between the two

Concentrating Solar Thermal Energy,
coordinated by Gilles Flamant. © ISTE Ltd 2022.

tubes to create the vacuum. The chemical getters in the ring between the two tubes allow the absorption of gas molecules in order to preserve the low pressure.

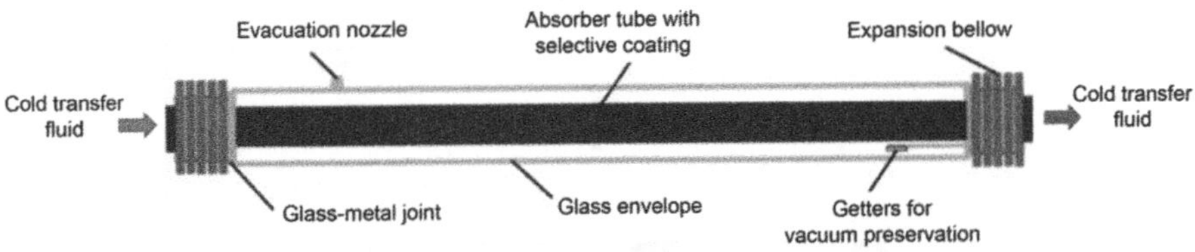

Figure 4.1. *Schematic representation of an absorber tube and its components (Espinosa-Rueda et al. 2016). For a color version of this figure, see www.iste.co.uk/flamant/energy.zip*

The commonly used transfer fluids for absorber tubes are oil and water–steam, which are discussed in Chapter 5.

Table 4.1 presents the technical characteristics of commercial absorber tubes manufactured by Rioglass (2020).

	PTR 70-4G	PTR 80-4G	PTR 90-4G
Solar absorptivity	≥ 94%	≥ 95.5%	≥ 95.5%
Transmittance	≥ 97%	≥ 96.5%	≥ 97%
Emissivity	≤ 9.5% at 400°C	≤ 9.5% at 400°C	≤ 9.5% at 400°C
Inner/outer diameters of the metal tube	66/70 mm	76/80 mm	84/89 mm
Thermal losses	-	125 W/m at 300°C 185 W/m at 350°C 260 W/m at 400°C	-
Maximal temperature of the fluid	400°C	400°C	400°C

Table 4.1. *Technical characteristics of the absorber tubes commercialized by Rioglass (2020)*

Due to manufacturing constraints, the maximal length of a tube is below 5 m; consequently, several tubes are connected in series in order to reach the desired total length.

4.2.2. *Thermal losses*

As shown in Figure 4.2, there are many thermal phenomena inside an absorber tube. One part of the concentrated solar radiation is reflected or absorbed by the glass envelope. The major part flowing through is then absorbed by the metal tube, particularly by selective coating. The temperature of the tube drives the emission of infrared radiation, which translates into thermal losses. Convective losses are also present, but they are minimized, thanks to the presence of vacuum between the two tubes.

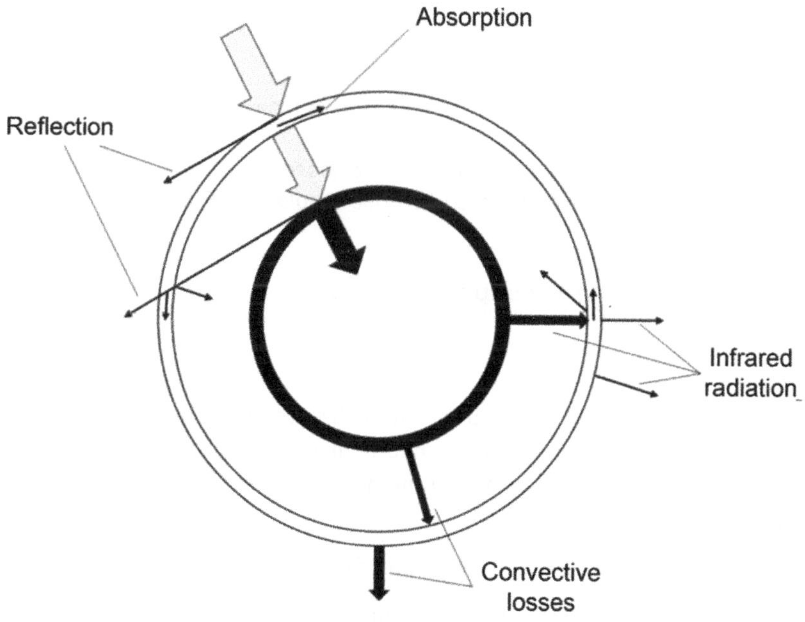

Figure 4.2. *Thermal phenomena in an absorber tube. For a color version of this figure, see www.iste.co.uk/flamant/energy.zip*

To find the thermal performances of the tube, the manufacturers provide a correlation giving the linear losses of the absorber tubes. Figure 4.3 presents the curve of linear thermal losses of Rioglass PTR 80-4G tubes depending on the temperature of the tube.

There are also models for the evaluation of the thermal performances of absorber tubes (Vasquez Padilla et al. 2011).

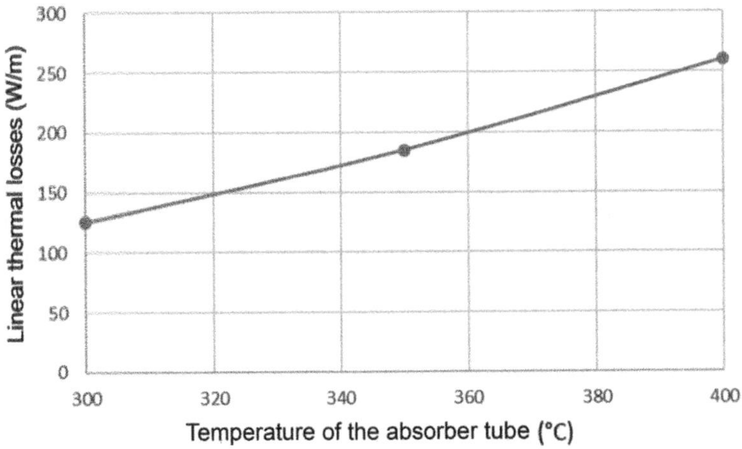

Figure 4.3. *Evolution of linear thermal losses depending on the temperature of the absorber tube Rioglass PTR 80-4G*

4.3. Solar receivers for tower power plants

4.3.1. *Description*

Solar receivers play a role similar to that of absorber tubes, but they are generally placed on top of a tower and must tolerate higher flux densities, ranging between 400 kW/m² and 1,000 kW/m², coming from the heliostat fields. The configuration of solar receivers depends on:

– the transfer fluid employed (see Chapter 5);

– the nature of the absorber (metal, ceramics);

– its geometry (flat or with cavity).

Flat receivers, presented in Figure 4.4, are adapted to surrounding fields mentioned in Chapter 3; they are easy to manufacture, but their radiative and convective losses are not negligible. They are well adapted to fluids, allowing the use of high solar flux densities ($\geq$ 800 kW/m^2) in order to limit the losses.

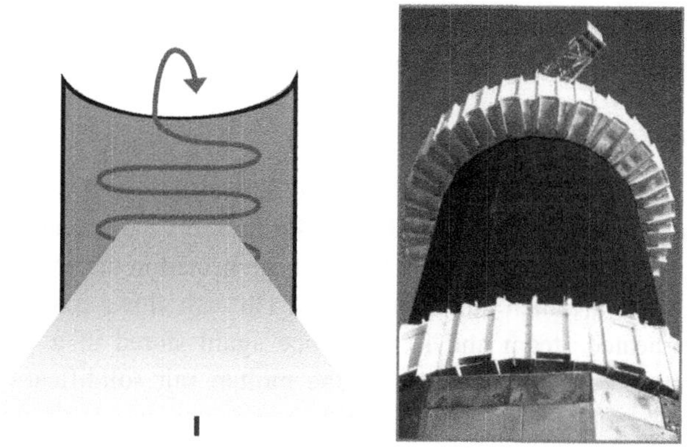

Figure 4.4. *Diagram and image of a circular flat receiver. For a color version of this figure, see www.iste.co.uk/flamant/energy.zip*

Cavity receivers, presented in Figure 4.5, can only be used with Northern/Southern fields. However, the presence of a cavity makes it possible to significantly reduce convective and radiative losses, which are preferred in applications at very high temperatures (~1,000°C).

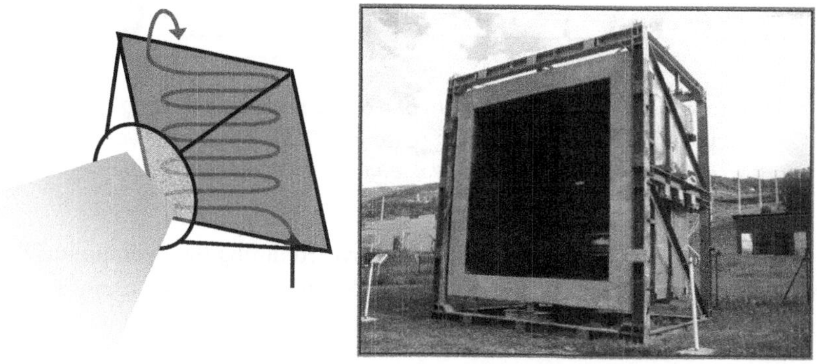

Figure 4.5. *Diagram and image of a cavity receiver. For a color version of this figure, see www.iste.co.uk/flamant/energy.zip*

4.3.2. *Receivers in the commercial tower power plants*

The technology of a tower power plant is mainly defined by the heat transfer fluid and the thermodynamic conversion cycle employed. The availability of high-capacity thermal energy storage is an essential element to be considered for evaluating various technologies. The two heat transfer fluids present in commercial power plants are molten salts and water–steam.

4.3.2.1. *Molten salt receivers*

The advantage of solar receivers with molten salt is that the heat transfer fluid can be used also as storage medium. Figure 4.6 presents the operating principle of a power plant using molten salts. After having been heated in the solar receiver on top of the tower, salts are then stored in a tank. Hot salt flows into an exchanger to generate superheated steam and is then once again stored in a cold tank whose temperature must be maintained above the molten salt solidification temperature (between 120°C and 260°C depending on their composition, see Chapter 5).

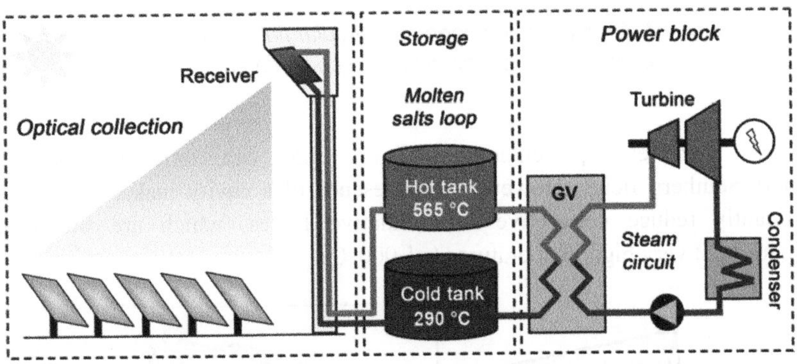

Figure 4.6. *Schematic diagram of a tower power plant using molten salt as the heat transfer fluid (Grange 2012). For a color version of this figure, see www.iste.co.uk/flamant/energy.zip*

The salt-induced corrosion problems discovered during the operation of the first generations of this type of power plant (see Chapter 1) led to the use of new alloys (Inconel 625, Inox 316Ti; Alvarez-Lara 2009).

In the 1980s, THEMIS power plant (Bonduelle et al. 1989) employed this technology to generate 2.5 MW_{el} on Targasonne site (France). Since then, Solar Two power plant, generating 12.4 MW_{el} (Cohen et al. 1999), started operating (1996–1999) at Barstow in the United States. These installations, whose purpose was to validate the technology, confirmed the proper operation of various components (heliostat fields, receiver, storage, steam generator). However, particular attention should be paid to salt temperature evolution inside the system, as it may solidify at certain points (valve, pipeline, for example) and thus lead to power plant shutdown. Since then, startup and shutdown strategies were established thanks to the operation of solar power plants such as Gemasolar.

The annual electricity generation of the Gemasolar power plant developed by Torresol Energy (Lata et al. 2010) is equivalent to that of a 50 MW_{el} solar thermal power plant without storage. However, the considered turbine is of about 20 MW_{el} since the objective is to generate electricity independently of the insolation conditions, thus reducing the generation cost. Using this technology, the power plant can generate electricity for 15 hours without insolation, which amounts to 740 MWh_{th} storage capacity (Burgaleta et al. 2011). Table 4.2 presents the main characteristics of this power plant, while Figure 4.7 shows an image of the tower with the illuminated solar receiver.

Number of heliostats	2,650
Total surface area of the mirrors	306,658 m^2
Tower height	140 m
Receiver power	120 MW_{th}
Turbine power	19.9 MW_{el}
Storage capacity (number of operating hours without the sun)	15 h
Storage tank size (height × diameter)	10.5 × 23 m
Quantity of salts	8,500 t
Generated electrical power	110,000 MWh/year

Table 4.2. *Main characteristics of Gemasolar (Burgaleta et al. 2011)*

Figure 4.7. *Gemasolar, tower power plant with molten salt near Seville, Spain (Ho 2017). For a color version of this figure, see www.iste.co.uk/flamant/energy.zip*

4.3.2.2. *Water–steam receivers*

As shown in Figure 4.8, unlike molten salts, water–steam heat transfer fluid can be used directly in the thermodynamic cycle of the turbine. This avoids an additional stage of thermal exchange.

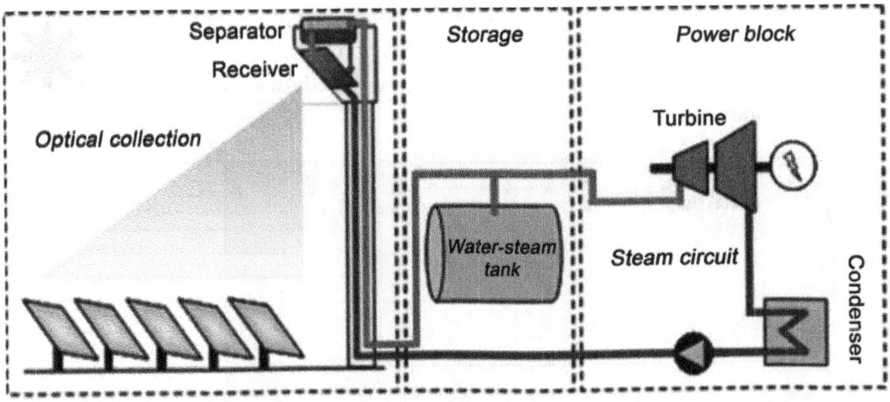

Figure 4.8. *Schematic diagram of a tower power plant with water–steam as the heat transfer fluid (Grange 2012). For a color version of this figure, see www.iste.co.uk/flamant/energy.zip*

The excellent heat transfer offered by saturated water steam allows operation at high solar concentration. However, the local changes in thermal transfer between the liquid phase and the steam phase may generate strong thermal gradients, thereby resulting in the deformation of the walls of the tubes.

The first two tower power plants using water–steam as the heat transfer fluid are Planta Solar 10 (PS10) and Planta Solar 20 (PS20), which were built in the south of Spain and put into service in 2007 and 2009, respectively (Figure 4.9). They generate an electric power of 11 MW_e and 20 MW_e, respectively. The receivers have a cavity containing several tube panels for water–steam heating. They have a one-hour storage capacity at half-power thanks to systems for the thermal storage of pressurized water.

In 2014, the solar power plant of Ivanpah became operational in Southern California, which involves three units, as shown in Figure 4.10. Each of these units generates superheated steam at 540°C and 160 bar. The total electric power is 377 MW_e.

Figure 4.9. *PS10 (left) and PS20 (right) near Seville, Spain. For a color version of this figure, see www.iste.co.uk/flamant/energy.zip*

Figure 4.10. *Ivanpah tower power plant (Ho 2017). For a color version of this figure, see www.iste.co.uk/flamant/energy.zip*

4.3.3. *Emerging designs*

The heat transfer fluids present in commercial solar power plants do not allow temperatures above 600°C. New solar receiver technologies are being developed to reach higher temperatures, which would increase the efficiency of the thermal–electrical conversion. However, the increase in the solar receiver temperature involves more convective and radiative losses. In order to avoid losing solar–thermal conversion efficiency what was gained in thermal–electrical conversion efficiency, it is essential to optimize the design of the solar receiver to limit thermal losses. This section presents several examples of emerging designs.

4.3.3.1. *Volumetric air receivers*

Volumetric receivers contain porous structures that absorb solar radiation in depth. The porous structure may be metallic or ceramic (e.g. silicon carbide – SiC), and the air can be heated from 850°C in the first case up to 1,000°C in the second case. Volumetric receivers have excellent performance, as they are theoretically able to capture radiation throughout the depth and thus strongly increase exchange surfaces, reduce the peaks of local fluxes and decrease radiative losses. Furthermore, the air temperature is very close to the wall temperature at the absorber outlet, as shown in Figure 4.11.

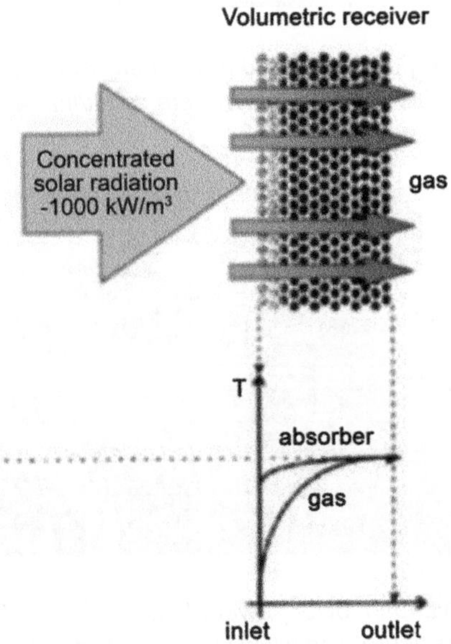

Figure 4.11. *Schematic representation of a receiver (Grange 2012). For a color version of this figure, see www.iste.co.uk/flamant/energy.zip*

The main drawback of these receivers is the necessity, in the case of a hybrid gas turbine technology, to integrate a shield window to contain the pressurized air. Indeed, in addition to high pressures (up to 15 bar), these shield windows are subjected to heating by absorbing a minimum fraction of concentrated solar radiation and especially by absorbing the thermal radiation emitted by the hot parts of the absorber. Although they resist at a small scale (several hundred kilowatts), their service life is much shorter in industrial applications.

The two main applications of volumetric air receivers are:

– open loop atmospheric air receiver associated with a Rankine cycle (steam). Design requiring an air–water/steam exchanger;

– closed-loop pressurized air receiver (presence of a shield window) for a Brayton cycle (gas).

Various designs of volumetric receivers are presented by Avila-Marin (Avila-Marin 2011), and several ceramic foams were tested by Mey et al. (2016).

4.3.3.2. *Surface pressurized air receivers*

In surface receivers, the problem of the shield window does not exist, as the transfer fluid flows between two walls at the cost of slightly lower performances. Indeed, the exchange surface is smaller than in the previous case, as the heat is absorbed by the illuminated surface and is then transmitted to the heat transfer fluid. The flux exchanged per insolated unit surface is irreparably lower and the acceptable solar fluxes are therefore limited (except for molten metals HTF). Indeed, as shown in Figure 4.12, the temperature gap between the wall and the fluid is far more significant than in the case of a volumetric receiver. In order to minimize this gap, it is therefore imperative to use a highly conductive material and to favor the thermal transfers between the material and the fluid (see Chapters 5 & 6). Thermal transfers can be improved, for example, by inserting fins in the tubes. Indeed, turbulence enables better thermal transfers as well as more significant pressure loss, leading to a drop in the efficiency of the turbine and therefore of the installation. Consequently, searching for a good compromise between the thermal transfers and the pressure loss is essential.

The operating principle of this technology, known as the solar hybrid gas turbine, is presented in Figure 4.13. The compressor, on the same shaft as the turbine, sucks and compresses the air before it gets into the solar receiver. In order to reach the temperature required at the turbine inlet (between 900°C and 1,300°C), the air is injected into the combustion chamber. A major drawback of (atmospheric or pressurized) air receivers is that it is difficult to store heat.

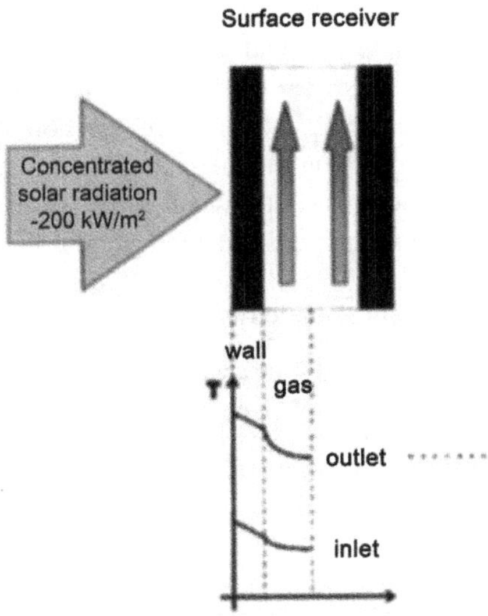

Figure 4.12. *Schematic representation of a surface receiver (Grange 2012). For a color version of this figure, see www.iste.co.uk/flamant/energy.zip*

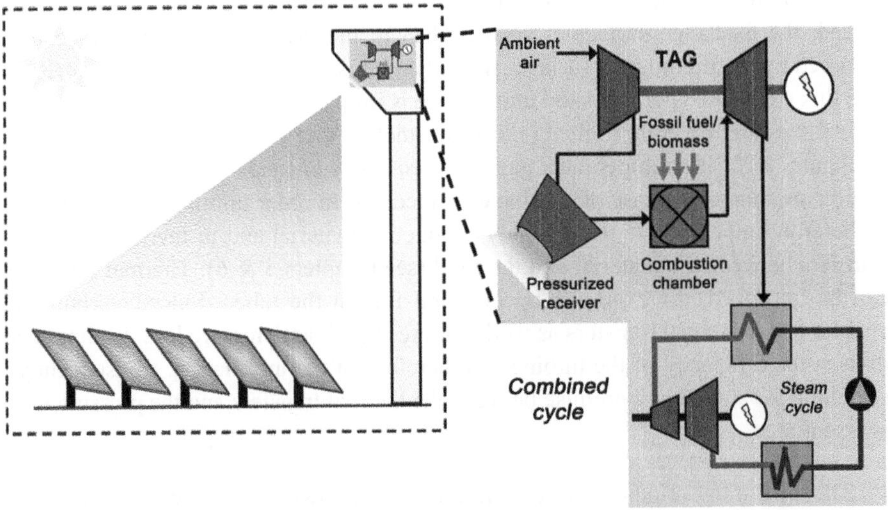

Figure 4.13. *Schematic diagram of a tower power plant with pressurized air as the heat transfer fluid coupled with a gas turbine (Grange 2012). For a color version of this figure, see www.iste.co.uk/flamant/energy.zip*

4.3.3.3. *Particle receivers*

The main advantage of particles compared to molten salts is the possibility of operating at temperatures above 1,000°C without any low temperature limit (as for molten salts). Several technologies were studied and tested and are presented in the following sections.

4.3.3.3.1. Receiver with an upward dense fluidized bed

The principle of the receiver with an upward dense fluidized bed is schematically presented in Figure 4.14. The particles are fluidized in vertical tubes subjected to concentrated solar radiation and thus capture the energy by conduction and infrared radiation in the tubes. Particles can be either stored or used in a heat exchanger in order to transmit their energy to a working fluid (pressurized air, supercritical CO_2, etc.; Benoit et al. 2015). The performance of this design applied to a medium power solar power plant (~ 20 MW_e) was simulated (Behar et al. 2020).

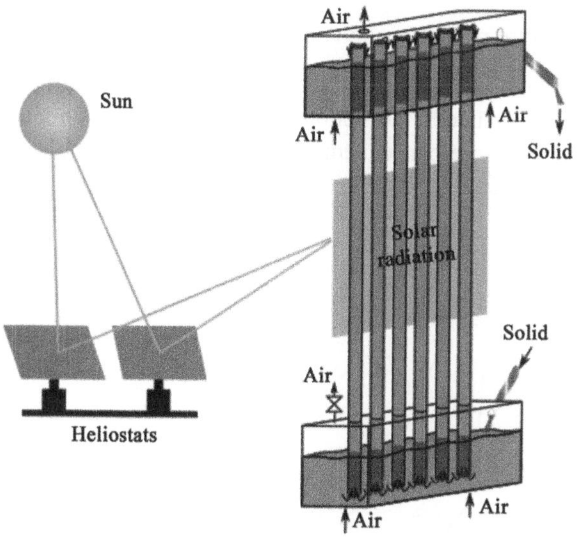

Figure 4.14. *Principle of a solar receiver with the vertical transport of dense fluidized gas–particle suspension subjected to concentrated solar radiation (Boissière 2015). For a color version of this figure, see www.iste.co.uk/flamant/energy.zip*

4.3.3.3.2. Particle curtain receiver

The particle curtain receiver has a cavity in which the particles are fed so that they form a curtain. Concentrated solar radiation illuminates the particles directly. In order to guarantee the significant increase in temperature, inserts must be installed

(Figure 4.15) to increase the residence time of the particles subjected to radiation (Ho et al. 2017).

Figure 4.15. *Images of inserts allowing the increase in the residence time of the particles subjected to concentrated solar radiation (Ho et al. 2017). For a color version of this figure, see www.iste.co.uk/flamant/energy.zip*

4.3.3.3.3. Centrifugal particle receiver

As shown in Figure 4.16, particles are introduced in the centrifugal receiver and travel along the walls under the action of the centrifugal force induced by the receiver rotation. They are consequently directly heated by the concentrated solar radiation (Wu et al. 2015).

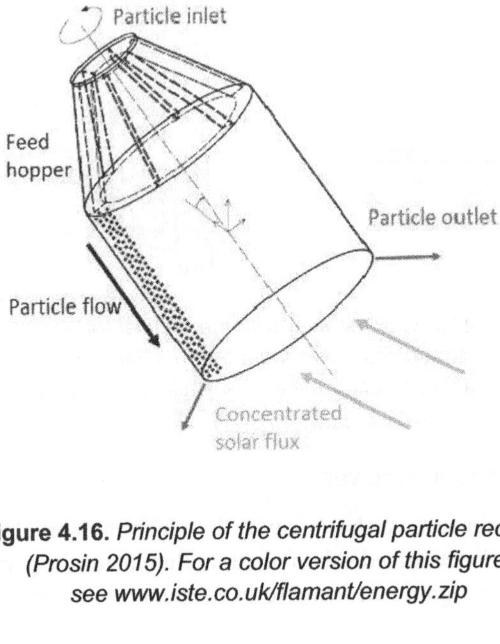

Figure 4.16. *Principle of the centrifugal particle receiver (Prosin 2015). For a color version of this figure, see www.iste.co.uk/flamant/energy.zip*

4.3.3.4. *Receivers with liquid metals*

The essential advantage of these receivers is that they can take in high solar fluxes (> 1 MW/m^2) due to the high thermal conductivity of liquid metals (Coventry et al. 2015). However, given its reactivity with water and its flammability with air, liquid sodium, which is the most studied, is a potentially dangerous choice in case of leakage.

4.3.3.5. *Balance of emerging technologies*

Many other solar receiver technologies are studied to heat a transfer fluid to higher temperatures and thus increase the overall efficiency of the solar power plant by increasing the efficiency of the thermodynamic cycle. These technologies are described in many scientific articles (Behar et al. 2013; Ho 2014; Ho 2017).

4.3.4. *Thermal losses*

Solar receivers are subjected to thermal losses, which are the focus of this section. These thermal losses are represented in Figure 4.17 for a cavity receiver.

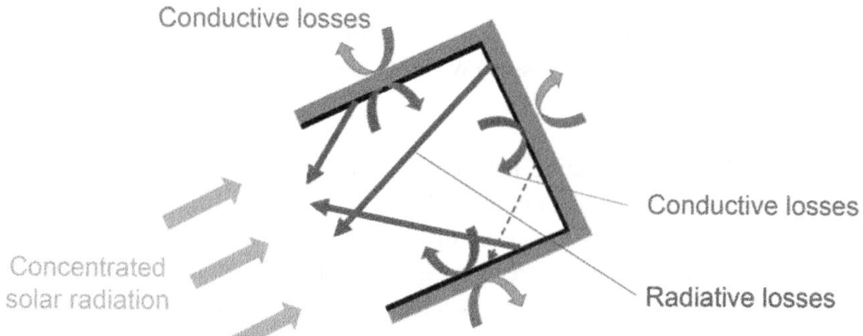

Figure 4.17. *Schematic representation of thermal losses in a solar cavity receiver. For a color version of this figure, see www.iste.co.uk/flamant/energy.zip*

In order to illustrate the main origins of thermal losses, the following section presents a simplified model of the solar receiver.

An efficient thermal insulation on the back face of the receiver generally minimizes conductive losses that become negligible compared to convective and radiative losses.

The thermal efficiency of the solar receiver can be written as:

$$\eta_{th} = \frac{\alpha Q_{inc} - Q_{losses}}{Q_{inc}} \quad [4.1]$$

where Q_{inc} is the incident power (W), Q_{losses} corresponds to thermal losses (W) and α is the absorptivity of the receiver (-).

The incident power is:

$$Q_{inc} = \eta_{field} IC \quad [4.2]$$

where η_{field} is the optical efficiency of the heliostat field (-), I is the direct solar radiation (W/m²) and C is the concentration factor.

Considering the radiative and convective losses, Q_{losses} becomes:

$$Q_{losses} = \varepsilon\sigma F_{view} T_{rec}^4 + f_{conv} h_{conv} (T_{rec} - T_{amb}) \quad [4.3]$$

where ε is the emissivity of the receiver (-), σ is the Stefan–Boltzmann constant ($W.m^{-2}.K^{-4}$), F_{view} is the view factor of the receiver towards the aperture (-), T_{rec} is the temperature of the receiver (K), f_{conv} is a multiplier of convective losses (-), h_{conv} is the coefficient of convective losses (W/m².K) and T_{amb} is the ambient temperature (K).

The multiplier of convective losses f_{conv} can be used to decrease or increase convective losses, thus reflecting the influence of the coefficient of convective losses on the thermal performances of the solar receiver. The coefficient of convective losses depends on the receiver geometry (flat or cavity) as well as on the wind velocity. The view factor of the receiver towards the aperture F_{shape} represents the part of the infrared radiation coming from the high temperature receiver that escapes through the aperture.

The thermal efficiency of the receiver can thus be written as:

$$\eta_{th} = \alpha - \frac{\varepsilon\sigma F_{view} T_{rec}^4 + f_{conv} h_{conv} (T_{rec} - T_{amb})}{\eta_{field} IC} \quad [4.4]$$

Considering an absorptivity α of 0.95, an emissivity ε of 0.85, an ambient temperature T_{amb} of 20°C, an optical efficiency of the heliostat field η_{field} of 0.6, a direct solar radiation of 800 W/m² and a coefficient of convective losses h_{conv} of 10 W/m².K, Figures 4.18–4.20 present the evolution of the thermal efficiency of a solar receiver depending on its temperature and the concentration factor, the view factor and the coefficient of convective losses, respectively.

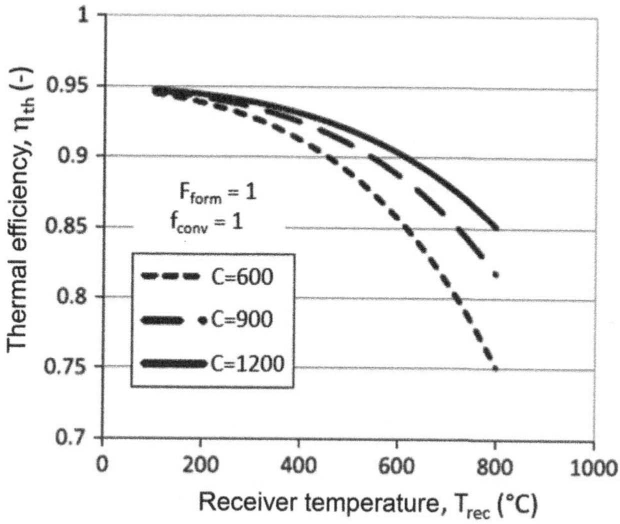

Figure 4.18. *Evolution of the thermal efficiency of a solar receiver depending on its temperature and concentration factor (Ho 2014)*

These curves show that the parameters that are the most important for preserving a high thermal efficiency are, on the one hand, a high concentration factor, and on the other hand, a low view factor. A high concentration factor requires a high transfer coefficient in the solar receiver to prevent the wall from overheating and thus damaging the receiver. As far as the view factor is concerned, it is reduced by the implementation of a cavity and decreases with the size of the aperture. However, it is essential to take into account the optical–thermal coupling, as for a very small aperture, spillage losses increase, and the optical efficiency is consequently affected.

As for the coefficient of convective losses, it influences to a lesser degree the thermal efficiency, but it should not be neglected, as its reduction leads to a gain of several points of thermal efficiency when operating at a very high temperature.

This study can be used to quantify the influence of various parameters on the thermal efficiency of a solar receiver. A more precise estimation of performances requires the development of a more complete model, which is the object of the following section.

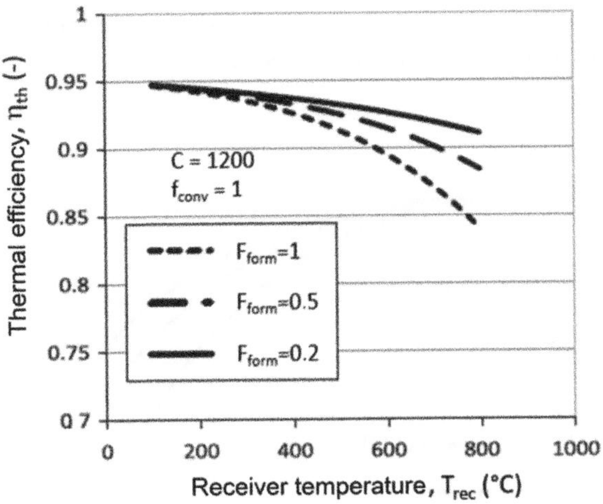

Figure 4.19. *Evolution of the thermal efficiency of a solar receiver depending on its temperature and the view factor (noted as F_{form} in the figure) of the receiver towards the aperture (Ho 2014)*

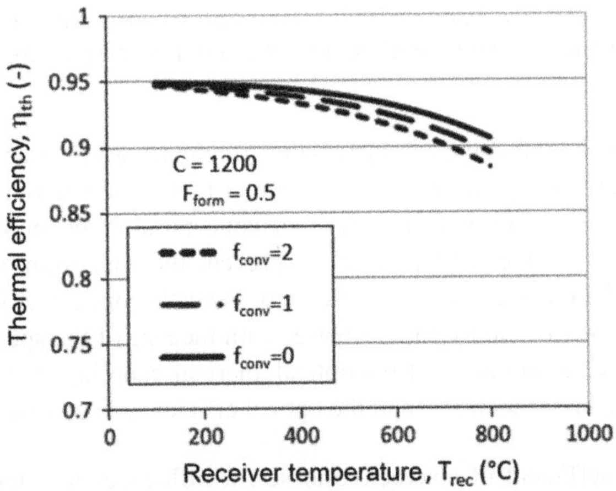

Figure 4.20. *Evolution of the thermal efficiency of a solar receiver depending on its temperature and the coefficient of convective losses (Ho 2014)*

4.3.5. *Thermal model of solar receivers*

There are various ways for modeling solar receivers, either through simple and rapid models, or through finite element models or thermal–hydraulic simulations that require longer computational time but are generally more precise.

This section presents a model based on solving a linear system of equations that can be used to evaluate the thermal performances of a solar receiver in steady state. For this purpose, the receiver must be discretized into elementary surfaces and the solar power incident on each surface must be evaluated using a code or optical software presented in Chapter 3. The thermal–optical properties of the components must also be known. Furthermore, all the view factors, dimensionless numbers representing the fraction of the total hemispherical radiative flux of the surface S_i reaching the surface S_j must be known. They are essential for the study of radiative balances in the receiver. For simple geometries, shape factors can be calculated using the equations proposed by Howell (2020). For more complex geometries, the Monte-Carlo method can be used to evaluate the shape factors, although it is quite burdensome (Vujicic et al. 2006).

S_i	Surface area of the element i (m^2)
F_{ij}	view factor of the surface element i to the surface element j (-)
J_i^S	Radiosity in the solar range of the surface element i (W/m^2)
J_i^{IR}	Radiosity in the infrared range of the surface element i (W/m^2)
ρ_i^S	Reflectivity in the solar range of the surface element i (-)
ρ_i^{IR}	Reflectivity in the infrared range of the surface element i (-)
φ_i^{inc}	Direct incident flux on the surface element i (W/m^2)
ε_i	Emissivity of the surface element i (-)
σ_{boltz}	Stefan–Boltzmann constant ($W/m^{-2}.K^{-4}$)
$T_i^{réc}$	Wall temperature of the surface element i (K)

Table 4.3. *Parameters involved in the net radiation method*

Using the previously calculated shape factors, a radiative balance is established by the net radiation method (Siegel 1992). It considers two ranges of wavelength: the solar spectral range (0.3 µm ≤ λ ≤ 3 µm) characterized by the exponent S, and the infrared spectral range (λ > 3 µm) characterized by the exponent IR. Hence, the net radiation method yields:

$$S_i J_i^S = S_i \rho_i^S \varphi_i^{inc} + \rho_i^S \sum_j S_j F_{ji} J_j^S \qquad [4.5]$$

$$S_i J_i^{IR} = S_i \varepsilon_i \sigma (T_i^{rec})^4 + \rho_i^{IR} \sum_j S_j F_{ji} J_j^{IR} \qquad [4.6]$$

Table 4.3 presents the parameters involved in relations [4.5] and [4.6].

The relation of reciprocity of view factors:

$$S_i F_{ij} = S_j F_{ji} \qquad [4.7]$$

leads to the following equations that are solved using a matrix system:

$$J_i^S = \rho_i^S \left(\varphi_i^{inc} + \sum_j F_{ij} J_j^S \right) \qquad [4.8]$$

$$J_i^{IR} = \varepsilon_i \sigma (T_i^{rec})^4 + \rho_i^{IR} \left(\sum_j F_{ij} J_j^{IR} \right) \qquad [4.9]$$

This data can be used to calculate the net power absorbed by the element i:

$$P_i^{net} = S_i \varphi_i^{inc} + \sum_j S_j F_{ji} J_j - S_i J_i \qquad [4.10]$$

where $J_i = J_i^S + J_i^{IR}$ is the total radiosity of the element i (W/m²).

Applying once more the relation of reciprocity [4.7], we have:

$$P_i^{net} = S_i \left(\varphi_i^{inc} + \sum_j F_{ij} J_j - J_i \right) \qquad [4.11]$$

Besides the radiative losses deduced from the radiative balance recalculated at each iteration, the model must take into account the convective losses (front face of the receiver) and conductive losses (back face of the receiver):

$$P_i^{convective} = h_{conv} S_i \left(T_i^{rec} - T_{amb} \right) \qquad [4.12]$$

$$P_i^{conductive} = U S_i \left(T_i^{rec} - T_{amb} \right) \qquad [4.13]$$

The coefficient of convective losses h_{conv} can be calculated using correlation (Samanes et al. 2015) or finite element models (Yuan et al. 2015), while the conductive–convective coefficient through the insulation is defined by:

$$U = \frac{1}{\left(\frac{1}{h_{conv}} + \frac{e_{insulation}}{\lambda_{insulation}}\right)} \quad [4.14]$$

where $e_{insulation}$ is the thickness of the insulation (m) and $\lambda_{insulation}$ is the thermal conductivity of the insulation (W/m.K).

A heat transfer coefficient h_i^{tr} must be known (measured experimentally or estimated by a correlation, see Chapter 5) to evaluate the power transmitted to the transfer fluid. Hence, the power transferred P_i^{tr} can be calculated as:

$$P_i^{tr} = h_i^{tr} S_i \left(T_i^{rec} - T_i^{avg}\right) \quad [4.15]$$

where T_i^{avg} is the average temperature of the transfer fluid in the element i (K).

This transferred power can be used to find the temperature of the transfer fluid leaving the element i:

$$T_i^s = T_i^e + \frac{P_i^{tr}}{\dot{m}_i c_p} \quad [4.16]$$

where T_i^e is the temperature of the transfer fluid at the entry (K), $\dot{m}_i$ is the flow rate of the transfer fluid (kg/s) and c_p is the specific heat capacity of the transfer fluid (J/kg.K).

Hence, a residual power P_i^{res} can be calculated as:

$$P_i^{res} = P_i^{net} - \left(P_i^{tr} - P_i^{losses}\right) \quad [4.17]$$

with $P_i^{losses} = P_i^{convective} + P_i^{conductive}$ (W).

This residual power is used to make an enthalpy balance on the elements of the receiver. A time interval Δt is introduced to calculate the increase in the temperature of the wall using the following relation:

$$T_{i,t+1}^{rec} = T_{i,t}^{rec} + \frac{P_i^{res} \Delta t}{m_i^{rec} c_{p,rec}} \quad [4.18]$$

where m_i^{rec} is the mass of the element i (kg) and $c_{p,rec}$ is the specific heat capacity of the receiver material (J/kg.K).

Finally, an iterative method can be used to find the temperatures of the fluid going in and out of each element, wall temperatures and thermal efficiency of the solar receiver.

Figure 4.21 presents the geometry of a pressurized air solar receiver, which is composed of a cavity through which the air flows. The preheated air is then introduced in absorber modules placed at the bottom of the cavity in two stages.

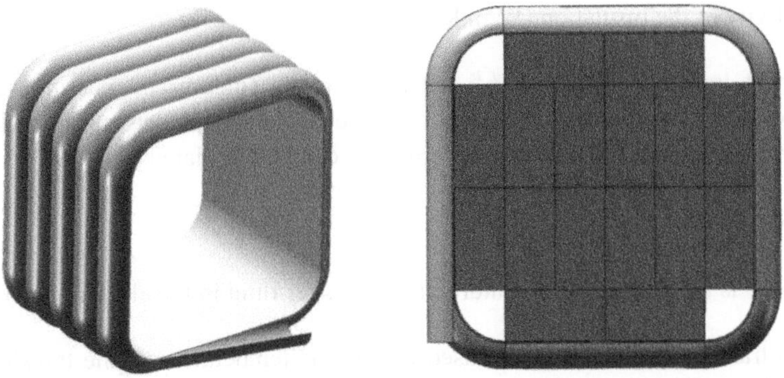

Figure 4.21. *Receiving cavity in front of the receiver, (a) isometric view, (b) front view with absorber modules (Verdier 2016)*

The air flow guided by a set of hydraulic connections at the back surface is shown in Figure 4.22.

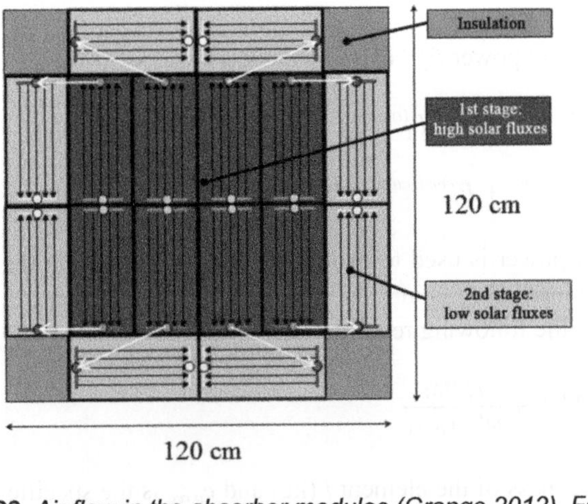

Figure 4.22. *Air flow in the absorber modules (Grange 2012). For a color version of this figure, see www.iste.co.uk/flamant/energy.zip*

Using an optical model to find out the incident solar fluxes (Chapter 3) and the thermal model presented in this section, it is possible to evaluate the air and wall temperatures of each element of the discretization, as shown in Figure 4.23.

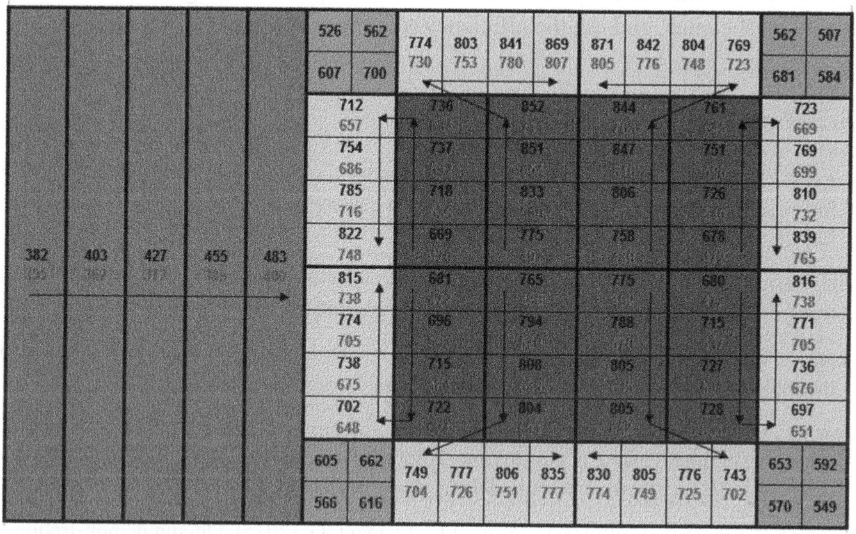

Figure 4.23. *Mapping of wall temperatures (upper part in black) and outgoing air temperatures (lower part in red) for each element of the discretization. Air flow is symbolized by arrows; temperatures in °C (Grange 2012). For a color version of this figure, see www.iste.co.uk/flamant/energy.zip*

This type of model can be used to evaluate the thermal performances of a solar receiver, but it has to be compared to finite element models or thermo-hydraulic simulations (Colomer et al. 2014; Hazmoune et al. 2020), which are more computational time intensive but more precise, as they allow the modeling of the transfer fluid flow in the receiver.

4.4. Conclusion

Solar receiver technology for linear concentrators has matured and reached quasi-optimal performances thanks to the development of materials with controlled optical properties and to an industrial feedback over several decades. This standardization level was not yet reached by the receivers for tower power plants due to wider ranges of technological and operating temperatures. Furthermore, the acceptable solar flux densities differ by an order of magnitude between the two technologies, which involves very different operational constraints. It is therefore clear that the innovation possibilities are stronger for receivers adapted to point concentrators. For this type of receiver, the optical model (Chapter 3) and the thermal model must be coupled, as the relevant property is the global solar energy collection efficiency, and not the efficiency of each component.

4.5. References

Alvarez-Lara, M. and Perosanz, F. (2009). Alloys selection for molten salts central receivers for solar power plants. *Proceedings of 15th SolarPACES International Symposium*, Berlin.

Avila-Marin, A.L. (2011). Volumetric receivers in solar thermal power plants with central receiver system technology: A review. *Solar Energy*, 85(5), 891–910.

Behar, O., Khellaf, A., Mohammedi, K. (2013). A review of studies on central receiver solar thermal power plants. *Renewable and Sustainable Energy Reviews*, 23, 12–39.

Behar, O., Grange, B., Flamant, G. (2020). Design and performance evaluation of a modular solar power plant using a gas turbine combined cycle with fluidized particle-in-tube receiver and direct thermal storage. *Energy Conversion & Management*, 220, 113108.

Benoit, H., Lopez, I.P., Gauthier, D., Sans, J.L., Flamant, G. (2015). On-sun demonstration of a 750 degrees C heat transfer fluid for concentrating solar systems: Dense particle suspension in tube. *Solar Energy*, 118, 622–633.

Boissière, B. (2015). Etude hydrodynamique et thermique d'un nouveau concept de récepteur solaire à suspensions denses gaz-particules. PhD Thesis, Université de Toulouse, Toulouse.

Bonduelle, B., Rivoire, B., Ferriere, A. (1989). La centrale expérimentale THEMIS : bilan et perspectives. *Revue de Physique Appliquee*, 24(4), 453–461. doi:10.1051/rphysap:01989002404045300.

Burgaleta, J., Arias, S., Ramirez, D. (2011). Gemasolar the first tower thermosolar commercial plant with molten salt storage system. *Proceedings of 17th SolarPACES International Symposium*, Grenada.

Cohen, G., Kearney, D., Kolb, G. (1999). Final report on the operation and maintenance improvement program for concentrating solar power plants. Sandia report SAND99-1290.

Colomer, G., Chiva, J., Lehmkuhl, O., Oliva, A. (2014). Advanced CFD&HT numerical modeling of solar tower receivers. *Energy Procedia*, 49, 50–59.

Coventry, J., Andraka, C., Pye, J., Blanco, M., Fisher, J. (2015). A review of sodium receiver technologies for central receiver solar power plants. *Solar Energy*, 122, 749–762.

Espinosa-Rueda, G., Navarro-Hermoso, J.L., Martínez-Sanz, N., Gallas-Torreira, M. (2016). Vacuum evaluation of parabolic trough receiver tubes in a 50 MW concentrated solar power plant. *Solar Energy*, 139, 36–46.

Grange, B. (2012). Modélisation et dimensionnement d'un récepteur solaire à air pressurisé pour le projet PEGASE. PhD Thesis, Université de Perpignan Via Domitia, Perpignan.

Hazmoune, M., Aour, B., Chesneau, X., Debbache, M., Ciupageanu, D.-A., Lazariou, G., Hadjiat, M.M., Hamidat, A. (2020). Numerical analysis of a solar tower receiver novel design. *Sustainability*, 12, 6957.

Ho, C.K. (2017). Advances in central receivers for concentrating solar applications. *Solar Energy*, 152, 38–56.

Ho, C.K. and Iverson, B.D. (2014). Review of high-temperature central receiver design for concentrating solar power. *Renewable and Sustainable Energy Reviews*, 29, 835–846.

Ho, C.K., Christian, J.M., Yellowhair, J., Jeter, S., Golob, M., Nguyen, C., Repole, K., Abdel-Khalik, S.I., Siegel, N., Al-Ansary, H. et al. (2017). Highlights of the high-temperature falling particle receiver project: 2012–2016. *Proceedings of 22nd SolarPACES International Symposium*, Abu Dhabi.

Howell, J.R., Siegel, R., Menguc, M.P. (2010). *Thermal Radiation Heat Transfer*, 5th edition. Taylor and Francis/CRC, New York [Online]. Available at: https://web.engr.uky.edu/rtl/Catalog/ [Accessed 29 December 2020].

Lata, J., Alcalde, S., Fernández, D., Lekube, X. (2010). First surrounding field of heliostats in the world for commercial solar power plants – Gemasolar. *Proceedings of 16th SolarPACES International Symposium*, Perpignan.

Mey-Cloutier, S., Caliot, C., Kribus, A., Gray, Y., Flamant, G. (2016). Experimental study of ceramic foams used as high temperature volumetric solar absorber. *Solar Energy*, 136, 226–235.

Pablo, D.T.-S., Krishna, D.P.N., Carlos, I.R.-S. (2020). Parabolic trough solar collectors: A general overview of technology, industrial applications, energy market, modeling, and standards. *Green Processing and Synthesis*, 9, 595–649. doi.org/10.1515/gps-2020-0059.

Prosin, T., Pryor, T., Creagh, C., Amsbeck, L., Buck, R. (2015). Hybrid solar and coal-fired power plant with air preheating using solid particle receiver. *Energy Procedia*, 69, 1371–1381.

Rioglass (2020). Rioglass [Online]. Available at: https://www.rioglass.com/ [Accessed 29 December 2020][1].

Samanes, J., García-Barberena, J., Zaversky, F. (2015). Modeling solar cavity receivers: A review and comparison of natural convection heat loss correlations. *Energy Procedia*, 69, 543–552.

1 See Pablo et al. (2020).

Siegel, R. and Howell, J.R. (1992). *Thermal Radiation Heat Transfer*, 3rd edition. Taylor & Francis, New York.

Vasquez Padilla, R., Demirkaya, G., Goswami, D.Y., Stefanakos, E., Rahman, M.M. (2011). Heat transfer analysis of parabolic trough solar receiver. *Applied Energy*, 88, 5097–5110.

Verdier, D. (2016). Stockage thermique de protection à chaleur latente intégré à un récepteur solaire à air pressurisé. PhD Thesis, Université de Perpignan Via Domitia, Perpignan.

Vujicic, M., Lavery, N.P., Brown, S.G.R (2006). View factor calculation using the Monte-Carlo and numerical sensitivity. *Communications in Numerical Methods in Engineering*, 22(3), 197–203.

Wu, W., Trebing, D., Amsbeck, L., Buck, R., Pitz-Paal, R. (2015). Prototype testing of a centrifugal particle receiver for high-temperature concentrating solar applications. *Solar Solar Energy Engineering*, 137(4), 041011.

Yuan, J.K., Ho, C.K., Christian, J.M. (2015). Numerical simulation of natural convection in solar cavity receivers. *Solar Energy Engineering*, 137(3), 031004.

5

Heat Transfer Fluids for Solar Power Plants

Gilles FLAMANT
PROMES-CNRS, Font-Romeu, France

5.1. Introduction

In a solar power plant, the heat transfer fluid (HTF) flows through the solar receiver and transfers heat to the heat storage system or for the conversion into the electricity system (power block). The heat transfer fluid differs from the working fluid. The latter is employed in a thermodynamic system that generates work, which is most often a steam turbine. In general, the thermal energy contained in the heat transfer fluid is transferred to the working fluid by a set of exchangers. Solar power plants with direct steam generation are a specific case in which the heat transfer fluid and the working fluid are one and the same. Steam flows through the solar receiver and is directly sent to the turbine. A steam storage system (more precisely pressurized water) can be integrated in the loop (see Chapter 8). The heat transfer fluid is most often a single phase (liquid or gas), but it can also be a two phase (water–steam or solid–gas).

This chapter begins by reviewing the macroscopic physics of fluid–wall heat transfers in section 5.2 and then presents the properties and domains of thermal stability of the most common heat transfer fluids in section 5.3. A detailed presentation of fluid–wall heat transfer coefficients is provided in section 5.4. Finally, current research on the new heat transfer fluids at high temperature is summarized in section 5.5.

Concentrating Solar Thermal Energy,
coordinated by Gilles FLAMANT.

5.2. Review of thermal transfer physics

The convection heat transfer coefficient, h (W/m^2.K), is defined by the ratio of the density of power or flux, q (W/m^2), and the wall–fluid temperature difference, $T_{wi}-T_{fb}$ (K) (equation [5.1]). T_{wi} is the internal temperature of the wall on the fluid side, and T_{fb} is the average temperature at the core of the fluid flow:

$$h = q/(T_{wi} - T_{fb}) \quad [5.1]$$

The absorbed net radiative flux, q, can be expressed as a conduction flux in the fluid near the wall, index w. Equation [5.2] expresses this flux with respect to the normal to the wall:

$$q = -\lambda_f \left(\partial T_f / \partial n\right)_w \quad [5.2]$$

where λ_f is the thermal conductivity of the fluid at the fluid temperature near the wall T_{fw}.

Combining equations [5.1] and [5.2] leads to

$$h = -\lambda_f \frac{\left(\partial T_f / \partial n\right)_w}{T_{wi} - T_{fb}} \quad [5.3]$$

Given the continuity of the heat flux, q is also equal to the conduction flux through the wall (equation [5.4]):

$$q = -\lambda_w (\partial T_w / \partial n)_w \quad [5.4]$$

Figure 5.1 shows a schematic representation of the previous mechanisms.

The previous relations can be used to describe the constraints and the compromise to be made when choosing the heat transfer fluid adapted to various concentrating installations. The first constraint concerns the wall temperature (equation [5.1]).

For a given material, there is a maximal operating temperature assuring mechanical properties adapted to the working conditions and a service life in conformity with that of the installation (generally 25–30 years). Hence, for a fixed T_w, the maximal temperature of the heat transfer fluid is determined by the heat transfer coefficient for a given absorbed net radiative flux. Conversely, there is a maximum acceptable radiative flux for a fixed h. On the contrary, equation [5.3] clearly shows the dependence of the transfer coefficient, on the one hand, on the heat transfer fluid conductivity and, on the other hand, on the local temperature gradient in the fluid. This latter characteristic is a function of flow conditions, in particular on the velocity of the fluid. In operational terms, the existence of a high velocity flow (potentially inducing a turbulent flow) is therefore favorable to thermal transfers. However, this operating mode involves a consumption of pumping energy (parasitic mechanical energy consumption) that may have a non-negligible influence on the net efficiency of the solar power plant. A compromise on the flow conditions should therefore be reached.

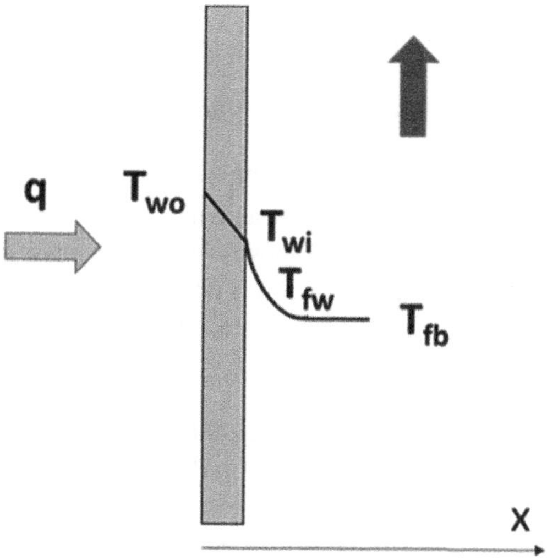

Figure 5.1. *Schematic representation of the continuity of the heat flux at the fluid–wall interface in 1D geometry. For a color version of this figure, see www.iste.co.uk/flamant/energy.zip*

Given the previous considerations, the estimation of the fluid–wall heat transfer coefficient is essential for dimensioning and modeling solar receivers. This estimation is conducted in the classical manner from correlations linking the Nusselt number *Nu* (transfer coefficient) to Reynolds number *Re* (speed, flow regime) and often to the Prandtl number *Pr*, with:

$$Nu = \frac{h\,d}{\lambda_f} \qquad [5.5]$$

$$Re = \frac{\rho_f v_f d}{\mu_f} \qquad [5.6]$$

$$Pr = \frac{c_{pf}\mu_f}{\lambda_f} \quad [5.7]$$

where d is the hydraulic diameter of the pipe, ρ_f, v_f, μ_f, c_{pf} are the density, velocity, dynamic viscosity and the heat capacity of the fluid, respectively. These properties depend on the nature of the fluid and on the temperature. It is therefore important to express the variation of these properties as a function of temperature before introducing the correlations enabling the calculation of transfer coefficients.

5.3. Fluids, stability and properties

5.3.1. *Thermal stability of heat transfer fluids*

The most commonly used heat transfer fluids in commercial solar power plants are liquids. Synthetic oil is employed in parabolic trough solar power plants, and *molten salt* is chosen to be used in the most recent solar tower power plants. It is important to note that unlike molten salt, oil is not used as the heat storage material. The only material currently used for heat storage in commercial power plants is *molten salt.*

The operating range of synthetic oil is 12–400°C. Its composition corresponds to the eutectic mixture 73.5% diphenyl oxide ($C_{12}H_{10}O$) and 26.5% diphenyl ($C_{12}H_{10}$) (Figure 5.2). This oil is commercialized by Solutia and Dow Chemical companies under the commercial names Therminol VP-1 and Dowtherm A, respectively.

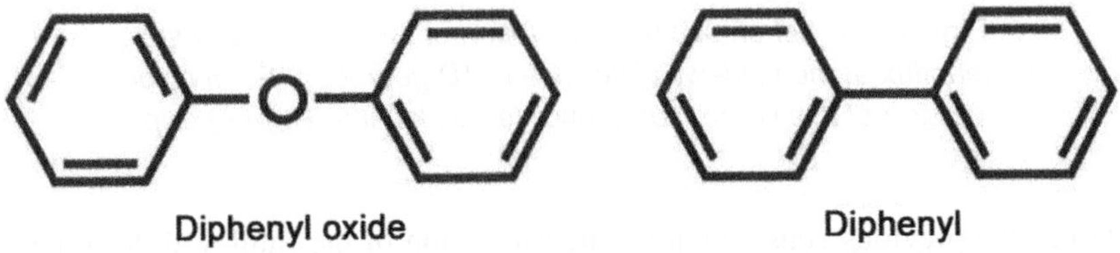

Figure 5.2. *Chemical compounds of synthetic oils*

Given the stability limit of synthetic oil, its working temperature is 390°C in solar power plants. Above 400°C, the oil gets degraded and releases hydrogen. Hydrogen diffuses through the metal constituting the absorber tube and slowly deteriorates the insulating properties of the annular space between the absorber tube and the surrounding glass tube. For this reason, hydrogen collectors are placed in this space in order to maintain the vacuum. As the boiling temperature of the oil is 257°C at atmospheric pressure, the circuit must be placed under pressure (approximately 15 atmospheres) in order to maintain oil under liquid form. To avoid any oxidation reaction, nitrogen is used for this purpose. Arkema Jarytherm synthetic oil is stable up to 350°C (very slow degradation below 330°C) and makes it possible to work at atmospheric pressure under neutral gas. Below 12°C, the viscosity of Therminol VP-1 and Dowtherm A oil is too high to allow pumping, with the low temperature recommended being 15°C. Consequently, putting into service a solar field under cold climate requires the existence of a heating system.

Solar salt (or molten salt) is a eutectic mixture of 60% sodium nitrate ($NaNO_3$) and 40% potassium nitrate (KNO_3) mass percentage or about 50/50 in molar percentage (Figure 5.3). Its melting temperature is 223°C and degrades above 600°C. Consequently, its operating range in solar power plants is between 290°C and 565°C.

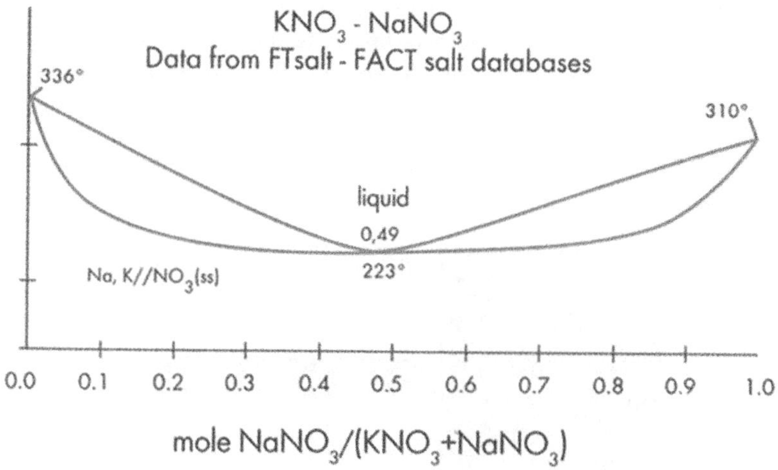

Figure 5.3. *Phase diagram of the KNO_3–$NaNO_3$ mixture (SQM 2020)*

The use of molten salts in solar tower power plants yields a double gain compared to synthetic oil in parabolic trough power plants. The increase in operating temperature generates an increase in the efficiency of the thermodynamic cycle of approximately +14% (relative) – see Chapter 1 – and the specific storage capacity is tripled. This factor corresponds to the increase in the temperature difference between the hot and cold tanks. Moreover, in tower power plants, the salt acts as the heat transfer and storage fluid, which avoids the use of a heat exchanger (oil–salt exchanger in parabolic trough power plants).

Other fluids are used, which were tested or are undergoing evaluation:

– molten salts: HITEC (7 wt% $NaNO_3$, 40 wt% $NaNO_2$, 53 wt% KNO_3), HITEC XL (7 wt% $NaNO_3$, 45 wt% KNO_3, 48 wt% $Ca(NO_3)_2$;

– liquid metals: sodium (Na), lead–bismuth eutectic (LBE), 44.5 wt% Pb, 55.5 wt% Bi;

– gas: air (at atmospheric pressure or under pressure), helium, hydrogen, carbon dioxide;

– water–steam.

In particular, the water–steam mixture is used in solar power plants with Fresnel linear concentrators and in tower power plants; however, it leads to expensive or complex storage solutions. Air is proposed as the transfer fluid in volumetric receivers implementing porous absorbers operating under atmospheric pressure (Mey et al. 2016) or compact absorbers operating under pressure (Grange et al. 2011). Helium and hydrogen are used in receivers for dish–Stirling solar systems (Reinalter et al. 2008).

The previously listed ranges of the working temperature of heat transfer fluids are summarized in Table 5.1.

Heat transfer fluids	Min–max temperature (°C)
Synthetic oil Therminol VP-1	12–400
Molten salts	
Solar salt	260–600
HITEC	140–530
HITEC XL	130–550
Liquid metals	
Na	98–883
LBE	125–1,670

Table 5.1. *Working temperature of liquid heat transfer fluids*

5.3.2. *Physical properties of heat transfer fluids*

The main physical properties of heat transfer fluids are specific heat capacity (c_p, $J.kg^{-1}.K^{-1}$), thermal conductivity (λ, $W.m^{-1}.K^{-1}$), density (ρ, $kg.m^{-3}$) and dynamic viscosity (μ, Pa.s) – the index f is omitted in this section. These properties vary with temperature. Table 5.2 presents the functions that can be used to determine the physical properties depending on this parameter (Benoit et al. 2016).

Equations	Validity domain
Liquids	
Therminol VP-1 (Solutia)	
$c_p = 2.82\,T + 716$	285 K < T < 673 K
$\lambda = 1.73 \times 10^{-7}\,T^2 + 7.62 \times 10^{-6}\,T + 0.14$	285 K < T < 673 K
$\rho = -7.61 \times 10^{-4}\,T^2 - 2.24 \times 10^{-1}\,T + 1{,}191$	285 K < T < 673 K
$\mu = (-2.3 \times 10^{-5}\,T^3 + 5.61 \times 10^{-3}\,T^2 - 19.89\,T + 1{,}822)^{-1}$	285 K < T < 673 K
Molten salts	
Solar salt (Zavoico 2001)	
$c_p = 1{,}443 + 0.172\,(T - 273.15)$	533 K < T < 873 K
$\lambda = 0.443 + 1.9 \times 10^{-4}\,(T - 273.15)$	533 K < T < 873 K
$\rho = 2{,}090 - 0.636\,(T - 273.15)$	533 K < T < 873 K
$\mu = 2.2714 \times 10^{-2} - 1.2 \times 10^{-4}\,(T - 273.15) + 2.281 \times 10^{-7}\,(T - 273.15)^2 - 1.474 \times 10^{-10}\,(T - 273.15)^3$	533 K < T < 873 K
HITEC (Siegel et al. 2011)	
$c_p = 1.56 \times 10^3$	415 K < T < 808 K
$\lambda = 0.411 + 4.36 \times 10^{-4}(T - 273.15) - 1.54 \times 10^{-6}(T - 273.15)^2$	415 K < T < 808 K
$\rho = -0.74 \times (T - 273.15) + 2{,}084$	415 K < T < 808 K
$\mu = 10^{2.7374} \times (T - 273.15)^{-2.104}$	415 K < T < 808 K
HITEC XL (Siegel et al. 2011)	
$c_p = -0.33\,T + 1{,}634$	403 K < T < 823 K
$\lambda = 0.519$	403 K < T < 823 K
$\rho = 2{,}240 - 0.827 \times (T - 273.15)$	403 K < T < 823 K
$\mu = 10^{6.1374} \times (T - 273.15)^{-3.36406}$	403 K < T < 823 K
Liquid metals	
Sodium (Fink and Leibowitz 1995)	
$c_p = 1658.2 - 0.84790\,T + 4.4541 \times 10^{-4}\,T^2 - 2.9926 \times 10^6\,T^{-2}$	371 K < T < 1,255 K
$\lambda = 124.67 - 0.11381\,T + 5.5226 \times 10^{-5}\,T^2 - 1.1842 \times 10^{-8}\,T^3$	371 K < T < 1,255 K
$\rho = 219 + 275.32\left(1 - \frac{T}{2503.7}\right) + 511.58\left(1 - \frac{T}{2503.7}\right)^{0.5}$	371 K < T < 1,255 K

$ln(\mu) = -6.4406 - 0.3958\, ln(T) + \frac{556.835}{T}$	371 K < T < 1,255 K
Lead–bismuth eutectic (Sobolev 2007)	
$c_p = 159 - 2.72 \times 10^{-2}\, T + 7.12 \times 10^{-6}\, T^2$	400 K < T < 1,500 K
$\lambda = 3.61 + 1.517 \times 10^{-2}\, T - 1.741 \times 10^{-6}\, T^2$	400 K < T < 1,100 K
$\rho = 1{,}1096 - 1.3236\, T$	400 K < T < 1,300 K
$\mu = 0.000494\, exp\left(\frac{6270}{RT}\right)$	400 K < T < 1,500 K
Gas	
Air (Lemmon and Jacobsen 2004; Green and Perry 2008)	
$c_p = 0.1805 \times T + 950.1$	500 K < T < 1,500 K
$\lambda = -9.72 \times 10^{-9}\, T^2 + 7.14 \times 10^{-5}\, T + 6.22 \times 10^{-3}$	500 K < T < 1,500 K
$\rho = \frac{P}{r_{air}\, T}$	500 K < T < 1,500 K
$\mu = \mu_r \left(\frac{T}{T_r}\right)^{1.54} \frac{T_r + B}{T + B}$ *where T_r = 273.15 K standard temperature, μ_r = 1.716 Pa.s standard viscosity and B = 110.4 K Sutherland's constant for air*	500 K < T < 1,500 K
Carbon dioxide (NIST)	
$c_p = 651 + 0.918\, T - 3.32 \times 10^{-4}\, T^2$	650 K < T < 1,100 K
$\lambda = -1.10 \times 10^{-2} + 9.74 \times 10^{-5}\, T - 1.57 \times 10^{-8}\, T^2$	650 K < T < 1,100 K
$\rho = \frac{p}{r_{CO_2}\, T}$	650 K < T < 1,100 K
$\mu = 5.94 \times 10^{-7} + 5.3 \times 10^{-8}\, T - 1.23 \times 10^{-11}\, T^2$	650 K < T < 1,100 K
Helium (NIST)	
$c_p = 5{,}183 + 8.97 \times 10^{-3}\, T - 2.58 \times 10^{-6}\, T^2$	500 K < T < 1,500 K
$\lambda = 7.08 \times 10^{-2} + 3.33 \times 10^{-4}\, T - 3.91 \times 10^{-8}\, T^2$	500 K < T < 1,500 K
$\rho = \frac{p}{r_{He}\, T}$	500 K < T < 1,500 K
$\mu = 8.64 \times 10^{-6} + 4.23 \times 10^{-8}\, T - 4.7 \times 10^{-12}\, T^2$	500 K < T < 1,500 K
Hydrogen (NIST)	
$c_p = 14{,}994 - 1.72\, T + 1.72 \times 10^{-3}\, T^2$	500 K < T < 1,000 K
$\lambda = 5.94 \times 10^{-2} + 4.32 \times 10^{-4}\, T + 3.6 \times 10^{-8}\, T^2$	500 K < T < 1,000 K
$\rho = \frac{p}{r_{H2}\, T}$	500 K < T < 1,000 K
$\mu = 3.69 \times 10^{-6} + 1.97 \times 10^{-8}\, T - 3.08 \times 10^{-12}\, T^2$	500 K < T < 1,000 K

Table 5.2. *Variation with the temperature of physical properties of heat transfer fluids*

5.4. Fluid–wall heat transfer coefficients

5.4.1. *Flow conditions*

The fluid–wall heat transfer coefficient is generally estimated from correlations such as $Nu = f(Re, Pr)$.

As indicated in section 5.2, the sought-for conditions for obtaining a high heat transfer coefficient are those of a developed turbulent flow obtained for Re > 4,000. This limit imposes a minimal fluid velocity in the receiver in order to avoid local overheating due to a decrease of the heat transfer coefficient that may lead to its degradation. As an example, Figure 5.4 represents the evolution of the Reynolds number as a function of temperature at 2 m/s for the previously mentioned liquid heat transfer fluids. It indicates that the flow is turbulent for all these fluids under standard working conditions. This condition is also verified for gases flowing at 15 m/s.

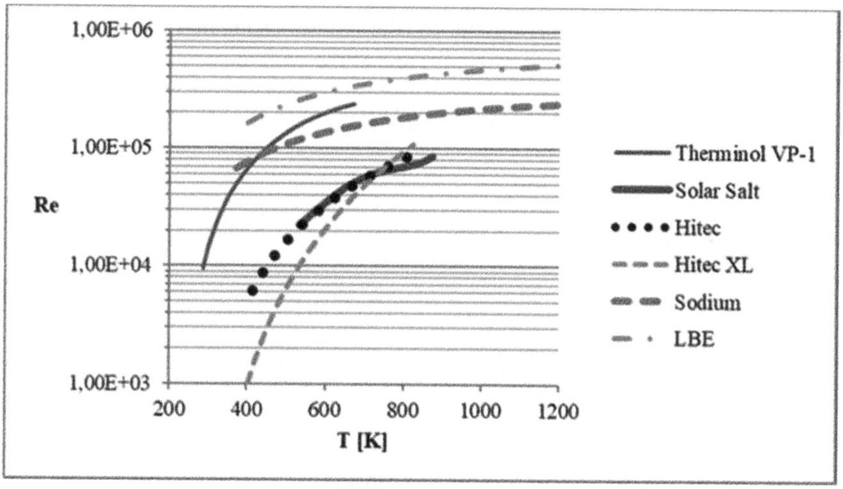

Figure 5.4. *Reynolds number depending on the temperature for liquid heat transfer fluids flowing at 2 m/s in a tube receiver (Benoit et al. 2016). For a color version of this figure, see www.iste.co.uk/flamant/energy.zip*

5.4.2. *Correlations*

A comparative study of various correlations is proposed in Benoit et al. (2016). In what follows, only the results of chosen correlations are presented.

For fluids other than molten salts and liquid metals, the Gnielinski correlation (Gnielinski 2013) is recommended:

$$Nu = \frac{(\xi/8)(Re_b - 1,000)Pr_b}{1 + 12.7\sqrt{\xi/8}\left(Pr_b^{2/3} - 1\right)} \times \left[1 + (d/L)^{2/3}\right]K \qquad [5.8]$$

with

$$\xi = [1.84\, log_{10}(Re_b) - 1.64]^{-2}$$

For liquids: $K = \left(\frac{Pr_b}{Pr_w}\right)^{0.11}$

For gases: $K = \left(\frac{T_w}{T_b}\right)^n$

and $n = -\left(log_{10}(T_w/T_b)\right)^{1/4} + 0.3$

The indices b and w correspond to the value calculated for the average temperature at the center of the flow (bulk) and at the wall, respectively.

For molten salts, the correlation proposed in Wu et al. (2012) was chosen:

$$Nu = 0.02948\, Re^{0.787} Pr^{1/3} \qquad [5.9]$$

$$10^4 \leq Re \leq 4.6 \times 10^4$$

$$1.6 \leq Pr \leq 23.9$$

For liquid metals, the correlation used is the following (Cheng and Tak 2006):

$$Nu_{10} = A + 0.018\, Pe^{0.8} \qquad [5.10]$$

$$A = \begin{cases} 4.5 & Pe \leq 1,000 \\ 5.4 - 9 \times 10^{-4} Pe & 1,000 \leq Pe \leq 2,000 \\ 3.6 & Pe \geq 2,000 \end{cases}$$

5.4.3. *Heat transfer coefficients*

The fluid–wall heat transfer coefficients that depend on the temperature are determined from the previous correlations and physical properties for a given fluid velocity. A velocity of 2 m/s for the liquids and 15 m/s for the gases was chosen to meet the turbulent flow criteria.

Figure 5.5 presents the results for the synthetic oil used in parabolic trough solar power plants.

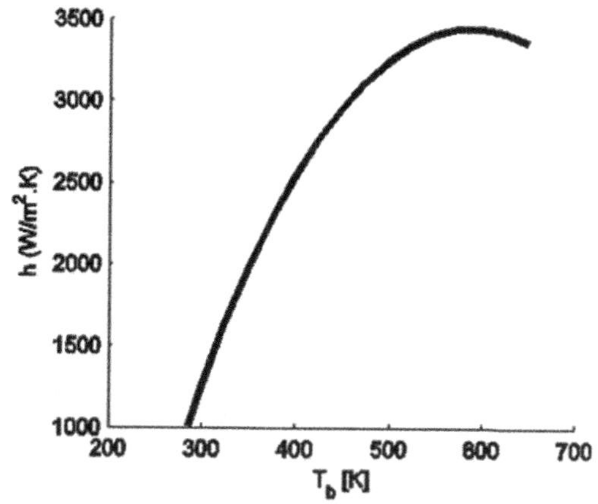

Figure 5.5. *Transfer coefficient for Therminol VP-1 (2 m/s) (Benoit et al. 2016)*

The heat transfer coefficients for molten salts, liquid metals and gases are represented in Figures 5.6, 5.7 and 5.8, respectively. These results are provided by Benoit et al. (2016).

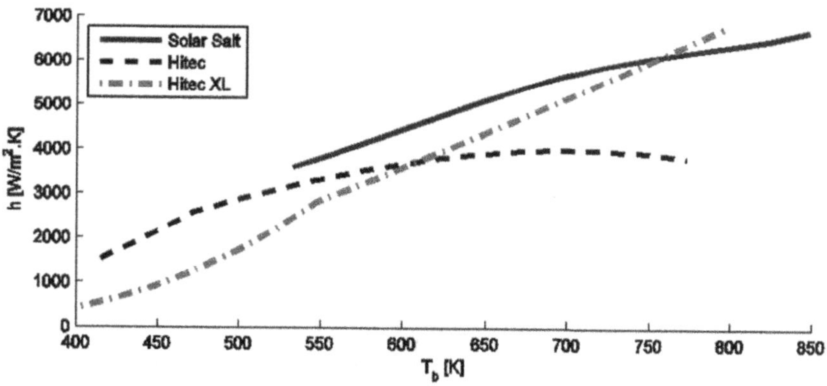

Figure 5.6. *Fluid–wall heat transfer coefficients for molten salts (2 m/s). For a color version of this figure, see www.iste.co.uk/flamant/energy.zip*

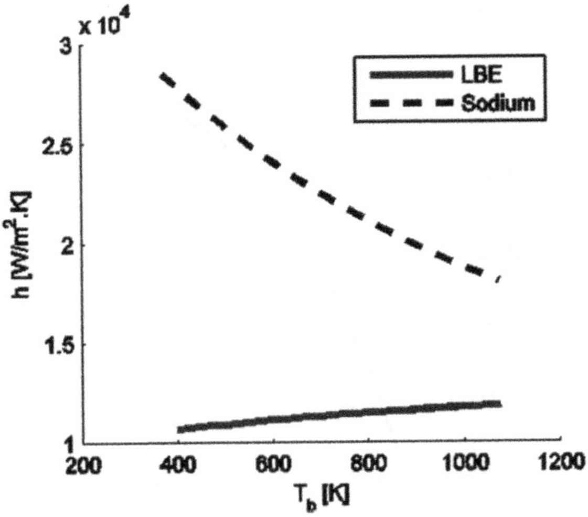

Figure 5.7. *Fluid–wall heat transfer coefficients for liquid metals (2 m/s). For a color version of this figure, see www.iste.co.uk/flamant/energy.zip*

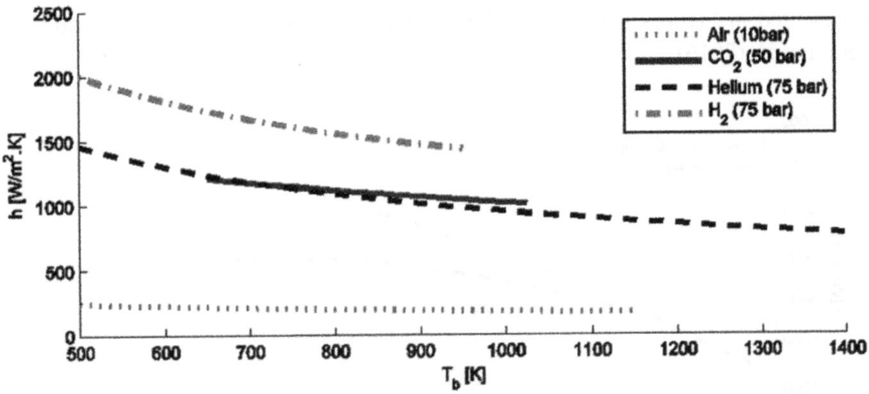

Figure 5.8. *Fluid–wall heat transfer coefficients for gases (15 m/s). For a color version of this figure, see www.iste.co.uk/flamant/energy.zip*

As expected considering the relative value of thermal conductivity of these fluids, the descending order of the heat transfer coefficients is the following:

Liquid metals > molten salts > synthetic oil > gas

For sodium, h varies between 28,500 and 18,000 $W.m^{-2}.K^{-1}$ within the considered temperature range. Due to this very high value, very compact solar receivers that are able to accept incident solar flux densities of the order of 2–3 MW/m^2 can be considered. The development of such a solution is hampered by safety problems related to the implementation of this metal that is highly reactive to air in an open environment. It nevertheless has been commercialized by the Australian company Vast Solar, starting with small modules (Vast 2020).

The heat transfer coefficient for solar salt is established within the range of 3,600–6,700 $W.m^{-2}.K^{-1}$ for the working temperature domain, which is approximately double the value for synthetic oil.

For hydrogen under 75 bars, h decreases from 2,000 $W.m^{-2}.K^{-1}$ to 1,440 $W.m^{-2}.K^{-1}$ between 500 and 1,000 K. Nevertheless, these values are close to those obtained for synthetic oil.

As far as water is concerned, the change of state leads to a complex evolution of the flow structure, which induces significant variations of the wall heat transfer phenomena, summarized in Figure 5.9 (Kandlikar 1997).

The detailed study of these phenomena is beyond the scope of this chapter. Nevertheless, the orders of magnitude are the following: for the liquid phase $h \sim 10{,}000$ $W.m^{-2}.K^{-1}$, in the boiling state $h \sim 50{,}000$–$100{,}000$ $W.m^{-2}.K^{-1}$ and for the superheated steam phase $h \sim 1{,}000$ $W.m^{-2}.K^{-1}$. It can be readily deduced that it is extremely difficult (and risky) to control the solar flux distribution applied to a solar receiver intended to generate superheated steam from liquid water in a single passage.

Taking into account the limit working temperatures of common liquid heat transfer fluids (synthetic oil and molten salt), the safety problems posed by the use of sodium, the strong corrosion of steels by liquid metals (in particular the LBE, Pacio and Wetzel 2013) and finally low heat transfer coefficients attainable by low pressure gases (except hydrogen), alternative solutions are being developed. These solutions are aimed, on the one hand, at the reduction of the melting temperature of salts to prevent the formation of solid plugs at cold points and, on the other hand, at the increase of the fluid working temperature so that they become compatible with high-efficiency thermodynamic cycles, which operate at 750°C and above, for the sCO_2 and combined cycles.

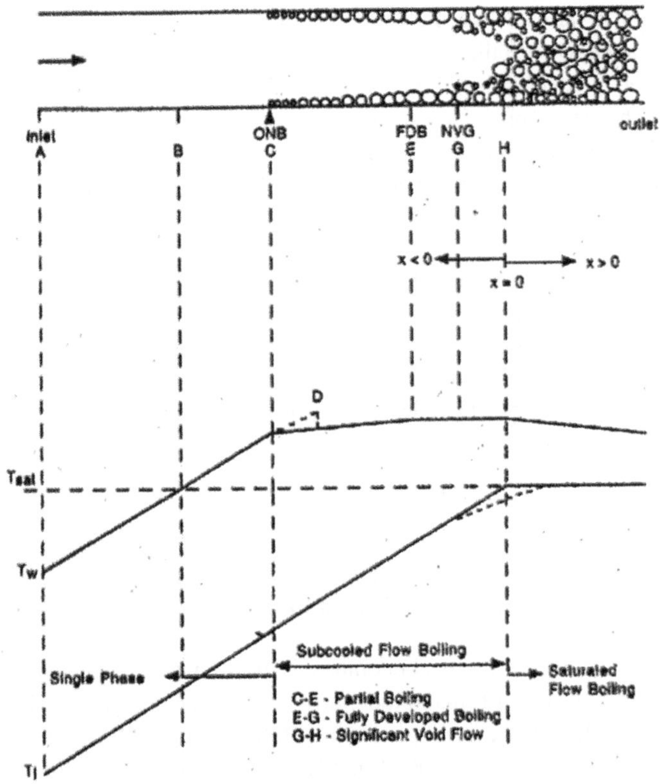

Figure 5.9. *Various types of flow during the liquid water–steam phase change (Kandlikar 1997)*

5.5. Solutions being developed

Current research focuses on extending the temperature range of salts and proposing solutions for exceeding the higher limit of solar salt use, which is about 600°C.

5.5.1. *Reduction of the melting temperature of salts*

The solidification of the binary compound $NaNO_3$–KNO_3 at approximately 220°C leads to the installation of pipework heating to prevent plug formation. This constraint motivated research on new compounds that can decrease the melting temperature of nitrate salts. Table 5.3, presented in Li et al. (2014), shows that several nitrate mixtures have a melting temperature below or equal to 100°C. The cost of these salts, significantly higher than that of solar salts, has prevented them from being marketed until now.

	Composition (% in weight)	T_f(°C)	T_m(°C)
Solar salt	60% $NaNO_3$/40% KNO_3	220	600
HITEC	53% KNO_3/7% $NaNO_3$/40% $NaNO_2$	142	454–538
HITEC XL	48% $Ca(NO_3)_2$/7% $NaNO_3$/45% KNO_3	133	500
M1	44% $Ca(NO_3)_3$/12% $NaNO_3$/44% KNO_3	127.6	622
M2	25.9% $LiNO_3$/20% $NaNO_3$/54.1% KNO_3	118	435
M3	30% $LiNO_3$/18% $NaNO_3$/52% KNO_3	120	550
M4	17.5% $LiNO_3$/14.2% $NaNO_3$/50.5% KNO_3/17.8% $NaNO_2$	99	500
M5	6% $NaNO_3$/23% KNO_3/8% $LiNO_3$/19% $Ca(NO_3)_2$/44% $CsNO_3$	65	561

Table 5.3. *Chemical composition of various nitrate-based salts according to Li et al. (2014). Tf = melting temperature, Tm = maximum working temperature*

5.5.2. *Increase of the maximum working temperature*

Besides the liquid metals already presented above, other options enabling the operation at temperatures above 600°C (for high temperature cycles, the threshold is $T \geq 700$°C) are the molten salts based on chlorides and carbonates, gases under high pressure and particles (Mehos et al. 2017).

For example, the chloride 37.5% $MgCl_2$, 62.5% KCl (% mass) is stable up to 1,400°C but melts at 426°C; the carbonate 33.4% Na_2CO_3, 34.5% K_2CO_3, 32.1% Li_2CO_3 has a melting temperature of 398°C and is stable up to about 750°C. These two examples show that with the salts, the increase in the maximum working temperature leads to a melting temperature far above that of solar salt.

The gases under consideration are air, carbon dioxide and helium. Employing supercritical CO_2 in the receiver is not realistic, given the working pressures (200–300 bars). The works are related to the use of these gases under 60–70 bars in compact receivers–exchangers (Li et al. 2011).

Particles have the advantage of meeting the high temperature stability constraint and offering the possibility of being used as the heat storage medium (similar to solar salt; see Kang et al. 2019). Oxides and carbides are stable at temperatures far above 1,000°C, and the specific flows do not solidify at low temperature.

Several projects implementing this design are being developed internationally (Ho 2016). They differ in terms of the type of the solar receiver and the diameter of the particles employed. PROMES-CNRS laboratory develops the fluidized bed technology with small average diameter particles (about 50 µm; Flamant and Hemati 2010; Benoit et al. 2015). Since particles cannot be pumped, their flow requires mechanical and/or pneumatic means (Zhang et al. 2016).

5.6. Conclusion

The importance of heat transfer fluids for solar thermal power plants should not be underestimated, as their thermophysical and chemical properties dictate, on the one hand, the acceptable fluxes on the absorber tubes and, on the other hand, the limit operating temperatures. The latter constraint determines the efficiency of the thermodynamic cycle. There is no ideal heat transfer fluid, but each choice is a compromise between stability, properties, operability, safety and cost. This multicriteria choice leaves room for innovation.

5.7. References

Benoit, H., Perez Lopez, I., Gauthier, D., Sans, J.L., Flamant, G. (2015). On-sun demonstration of a 750°C heat transfer fluid for concentrating solar systems: Dense particle suspension in tube. *Solar Energy*, 118, 622–633.

Benoit, H., Spreafico, L., Gauthier, D., Flamant, G. (2016). Review of heat transfer fluids in tube-receivers used in concentrating solar thermal systems: Properties and heat transfer coefficients. *Renewable & Sustainable Energy Review*, 55, 298–315.

Cheng, X. and Tak, N. (2006). Investigation on turbulent heat transfer to lead–bismuth eutectic flows in circular tubes for nuclear applications. *Nuclear Engineering and Design*, 236(4), 385–393.

Fink, J.K. and Leibowitz, L. (1995). Thermodynamic and transport properties of sodium liquid and vapour. Reactor Engineering Division, Argonne National Laboratory, Lemont, IL.

Flamant, G. and Hemati, M. (2010). Dispositif collecteur d'énergie solaire (Device for collecting solar energy). French patent FR 1058565, 2010; PCT extension WO2012052661, 2012.

Gnielinski, V. (2013). On heat transfer in tubes. *International Journal of Heat and Mass Transfer*, 63, 134–140.

Grange, B., Ferriere, A., Brillard, D., Vrinat, M., Couturier, R., Pra, F., Fan, Y. (2011). Thermal performances of a high temperature air solar absorber based on compact heat exchange technology. *Journal of Solar Energy Engineering*, 133(3), 031004.

Green, D.W. and Perry, R.H. (2008). Thermodynamic properties. *Perry's Chemical Engineers' Handbook*, 8th edition. McGraw-Hill Professional, New York.

Ho, C. (2016). A review of high-temperature particle receivers for concentrating solar power. *Applied Thermal Engineering*, 109, 958–969.

Kandlikar, S.G. (1997). Further developments in subcooled flow boiling heat transfer. *Engineering Foundation Conference on Convective and Pool Boiling*. IRSEE, Germany, 18–25.

Kang, Q., Flamant, G., Dewill, R., Baeyens, J., Zhang, H.L., Deng, Y.M. (2019). Particles in a circulation loop for solar energy capture and storage. *Particuology*, 43, 149–156.

Lemmon, E.W. and Jacobsen, R.T. (2004). Viscosity and thermal conductivity equations for nitrogen, oxygen, argon, and air. *International Journal of Thermophysics*, 25, 21–69.

Li, Q., Flamant, G., Yuan, X., Neveu, P., Luo, L. (2011). Compact heat exchangers: A review and future applications for a new generation of high temperature solar receivers. *Renewable & Sustainable Energy Review*, 15(9), 4855–4875.

Li, C.J., Li, P., Wang, K., Molina, E.E. (2014). Survey of properties of key single and mixture halide salts for potential application as high temperature heat transfer fluid for concentrated solar power systems. *AIMS Energy*, 2(2), 133–157.

Mehos, M., Turchi, G., Vidal, J., Wagner, M., Ma, Z., Ho, C., Kolb, W., Andraka, C., Kruizenga, A. (2017). Concentrating solar power – Gen3 demonstration roadmap. NREL Technical Report, NREL/TP-5500-67464.

Mey-Cloutier, S., Caliot, C., Kribus, A., Gray, Y., Flamant, G. (2016). Experimental study of ceramic foams used as high temperature volumetric solar absorber. *Solar Energy*, 136(15), 226–235.

NIST (2022). NIST Standard Reference Database 69: NIST Chemistry WebBook. DOI: https://doi.org/10.18434/T4D303.

Pacio, J. and Wetzel, T. (2013). Assessment of liquid metal technology status and research paths for their use as efficient heat transfer fluids in solar central receiver systems. *Solar Energy*, 11, 93–22.

Reinalter, W., Ulmer, S., Heller, P., Rauch, T., Gineste, J.-M., Ferriere, A., Nepveu, F. (2008). Detailed performance analysis of a 10kW dish/stirling system. *Journal of Solar Energy Engineering*, 130(1), 011013.

Siegel, N.P., Bradshaw, R.W., Cordaro, J.B., Kruizenga, A.M. (2011). Thermophysical property measurement of nitrate salt heat transfer fluids. *Proceedings of the ASME 2011 5th International Conference on Energy Sustainability ES2011*, Washington, DC.

Sobolev, V. (2007). Thermophysical properties of lead and lead–bismuth eutectic. *Journal of Nuclear Materials*, 362, 235–247.

Solutia (n.d.). Solutia Technical Bulletin 7239115C (Supersedes 7239115B), Therminol VP-1, Heat Transfer Fluid by Solutia. Available at: https://www.sintelub.com/wp-content/uploads/PDS/37.therminol_vp1.pdf.

SQM International N.V. (n.d.). Available at: https://www.sqm.com/en/producto/ [Accessed 15 December 2020].

Vast (2020). Available at: https://vastsolar.com/ [Accessed 21 December 2020].

Wu, Y.-T., Chen, C., Liu, B., Ma, C.-F. (2012). Investigation on forced convective heat transfer of molten salts in circular tubes. *International Communications in Heat and Mass Transfer*, 39, 1550–1555.

Zavoico, A.B. (2001). Solar power tower design basis document. Sandia National Laboratories, Technical Report SAND2001–2100.

Zhang, H.L., Benoit, H., Gauthier, D., Degreve, J., Baeyens, J., Perez Lopez, I., Hemati, M., Flamant, G. (2016). Particle circulation loops in solar energy capture and storage: Gas–solid flow and heat transfer considerations. *Applied Energy*, 161, 206–224.

6

Numerical Simulations of Flows and Heat Transfers of Solar Receivers

Françoise BATAILLE[1], Adrien TOUTANT[1] and Dorian DUPUY[1,2]

[1] *PROMES-CNRS, UPVD, Perpignan, France*

[2] *CERFACS, Toulouse, France*

6.1. Introduction

Solar receivers are places where turbulent flows are strongly heated in an asymmetric way (irradiated face/hidden face). It is difficult to model these flows, as they combine both dynamic aspects (the flow is turbulent) and thermal aspects (with high-temperature gradients) that are complex and amplified through their interactions. There is a literature on isothermal turbulent flows as well as on anisothermal flows where the temperature is considered as a passive scalar (its behavior does not influence the flow dynamics). An exhaustive review can be found in the books by Bailly and Comte-Bellot (2003) or Lesieur (2012), while keeping in mind that this research is in itself very difficult. On the other hand, scientific literature and studies related to strongly anisothermal turbulent flows are far less abundant.

Concentrating Solar Thermal Energy,
coordinated by Gilles FLAMANT. © ISTE Ltd 2022.

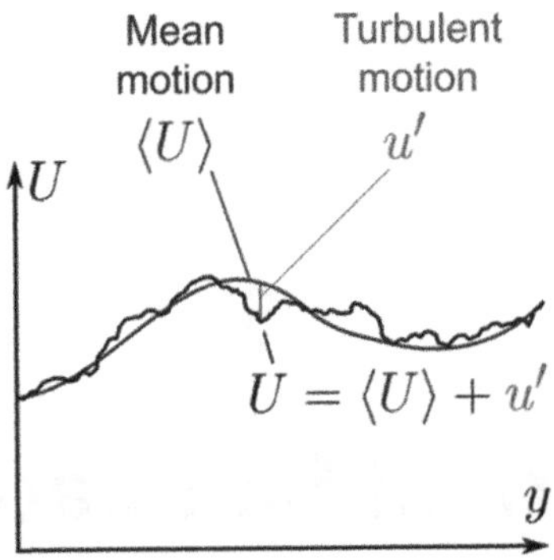

Figure 6.1. *Schematic representation of the profiles of mean and fluctuating fields. For a color version of this figure, see www.iste.co.uk/flamant/energy.zip*

Most flows encountered in nature and engineering applications are turbulent. Turbulence is useful in heat exchangers, as it tends to accelerate the homogenization and mixing processes, intensifying wall heat fluxes. Turbulent flows are chaotic and have, at any point, many eddies at various scales. On the other hand, a laminar flow regime is characterized by a flow with smooth paths and no lateral mixing. Above a critical Reynolds number, the laminar flow is unstable and evolves towards a turbulent state. Given the chaotic nature of turbulent flows, independent experiments or simulations with arbitrarily close initial conditions can lead, after a certain time, to very different results. The mean $\langle\cdot\rangle$ of the results of a large number of independent experiments can be defined and is referred to as Reynolds average, statistical average or ensemble average:

$$\langle G(x,y,z,t)\rangle = \frac{1}{N}\lim_{N\to\infty} G_i(x,y,z,t) \qquad [6.1]$$

where G_i is the value of G during experiment i and N is the number of experiments. The fluctuation of G is the deviation from the mean at a given instant, $G' = G - \langle G\rangle$. These statistical tools can be used to define reproducible quantities for the studies. Figure 6.1 illustrates the nature of the turbulence. A turbulent motion (that may be of significant order of magnitude) is superimposed to a mean motion. The result is an overall change of the flow.

In the case of third-generation solar tower power plants with pressurized gas (Figure 6.2), the expected temperature levels are of the order of 1,000°C and the thermal gradients near the walls are extremely high. These extreme temperature differences modify the turbulent flow and interact with it. The temperature fields are also turbulent and have their own evolution, as represented in Figure 6.3. The physics of the temperature field in a solar receiver is therefore complex and, as we may expect, difficult to grasp.

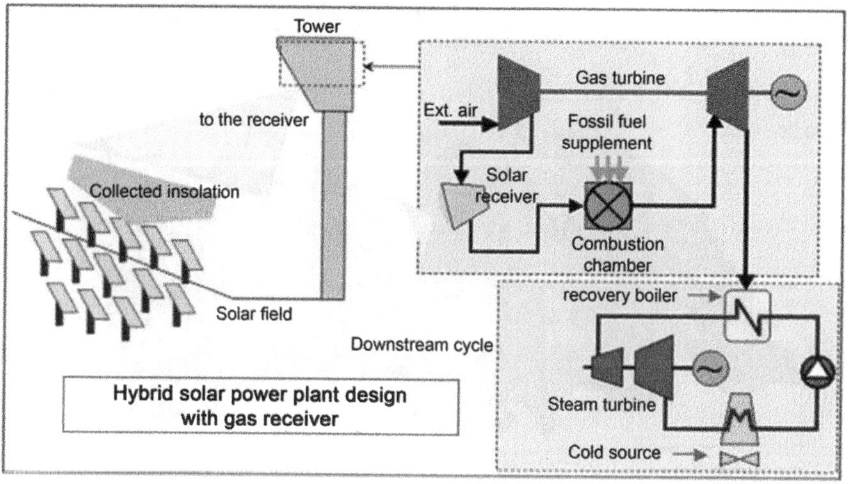

Figure 6.2. *Schematic representation of a hybrid solar power plant with combined cycle employing a solar receiver with pressurized air. The air is compressed by the compressor of the gas turbine. For a color version of this figure, see www.iste.co.uk/flamant/energy.zip*

Prediction of both velocity and temperature fields in the solar receivers is required for their control, in order to increase the internal transfers and their compactness, and to understand the physics underlying these phenomena. A possible approach is that of numerical modeling and simulation, which are at the core of this chapter and the original development pursued for several years at PROMES laboratory. As a first step, section 6.2 presents various modeling approaches starting with complex numerical modeling and simulations, and then simplified models and simulations. Section 6.3 provides detailed presentations of fine models and simulations taking into account the turbulent aspect of the flow and at the same time the specific aspect of extreme thermal transfers present in solar receivers (dissymmetrical heating and high thermal gradients). The fundamental research and results from the study of the coupling between thermal and dynamic aspects will be presented in section 6.4.

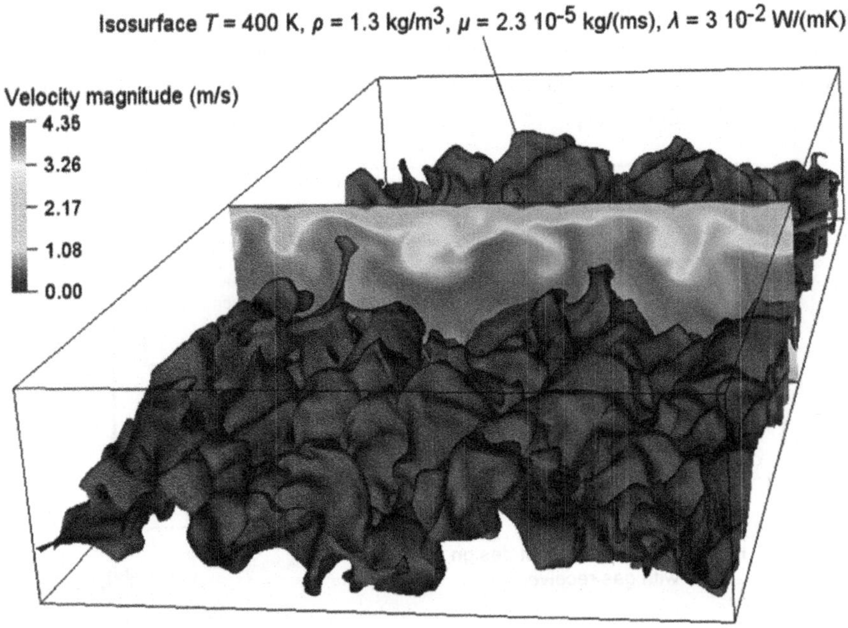

Figure 6.3. *Illustration of dynamic and thermal fields in a solar receiver (isosurface T = 400 K). For a color version of this figure, see www.iste.co.uk/flamant/energy.zip*

6.2. Modeling approaches

The flows inside solar receivers at high temperature are generally turbulent. One of the main difficulties for the study of these solar receivers is the size difference between the geometry of receivers, which is metric, and the size of the smallest scales of turbulence, which is micrometric. To overcome this difficulty, a multiscale approach can be used, differentiating between various levels of studies. Each of these levels is associated with a numerical strategy. Figure 6.4 represents these various numerical strategies: direct numerical simulation (DNS), thermal large-eddy simulation (T-LES), Reynolds-averaged Navier–Stokes (RANS) models and use of empirical correlations. Each of them is briefly presented in the following sections. A crucial point is the relationship between these various approaches. The industrial scale makes it possible to define the conditions of the more detailed studies. These conditions correspond mainly to the transfer fluid employed, the temperature levels, the intensity of heat fluxes and the range of Reynolds numbers. The results of the detailed studies can be used to inform the development of models at more macroscopic scales, in particular with regard to the coupling between thermal and dynamic aspects (see sections 6.3 and 6.4).

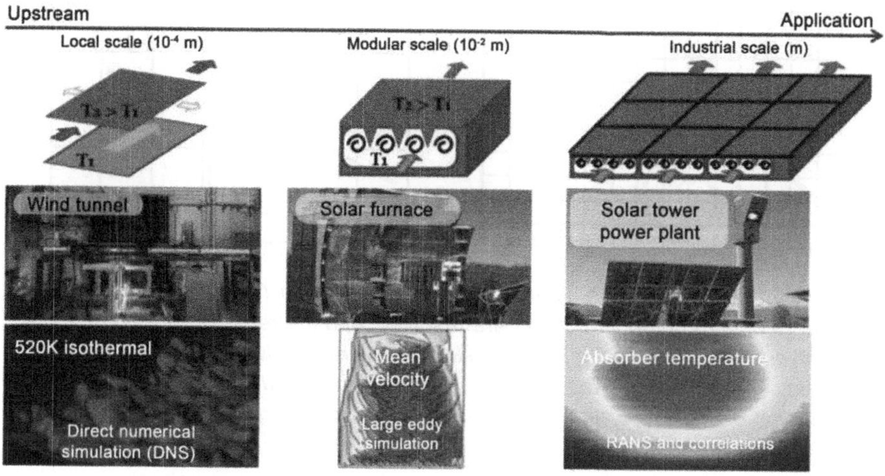

Figure 6.4. *The various modeling approaches for the numerical study of flows and heat transfers in solar receivers. For a color version of this figure, see www.iste.co.uk/flamant/energy.zip*

6.2.1. *Direct numerical simulation*

A DNS uses a grid that is sufficiently fine to capture all the scales of turbulence (Figure 6.5, left), that is, with a cell size smaller than Kolmogorov length scale, the smallest scale of turbulence at a given point of the flow. Hence, the spectrum of turbulence kinetic energy is fully described (Figure 6.6). This approach yields extremely precise and detailed results. To the extent that the turbulence is not modeled (if not through Navier–Stokes equations), these simulations can be considered as "numerical experiments" and their results can be used for the development of more macroscopic models (Toutant and Bataille 2013; Dupuy et al. 2019a, 2019b). The major drawback of DNS is their computing time. They are therefore reserved for simplified geometries, such as flat channel flows of small length (of the order of centimeter) compared to the actual length of the receivers (of the order of meter). Moreover, these simulations require, including under these conditions, the use of supercomputers.

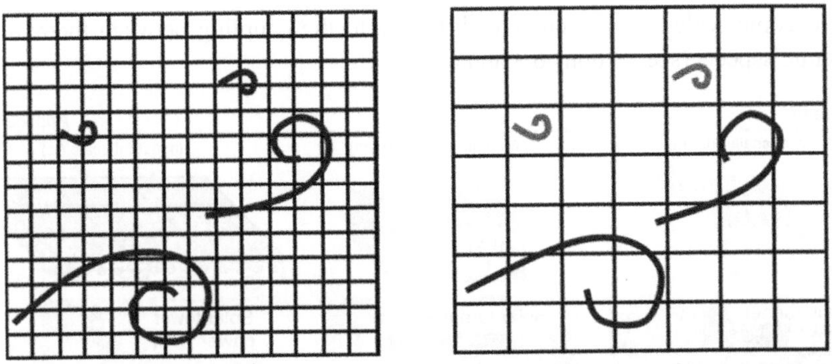

Figure 6.5. *Schematic representation of DNS strategies (left) and LES strategies (right). The blue turbulent structures are resolved, and the red ones are not resolved. In direct numerical simulation, all the structures are resolved (grid on the left). In large-eddy simulation, only the large structures are resolved (grid on the right). For a color version of this figure, see www.iste.co.uk/flamant/energy.zip*

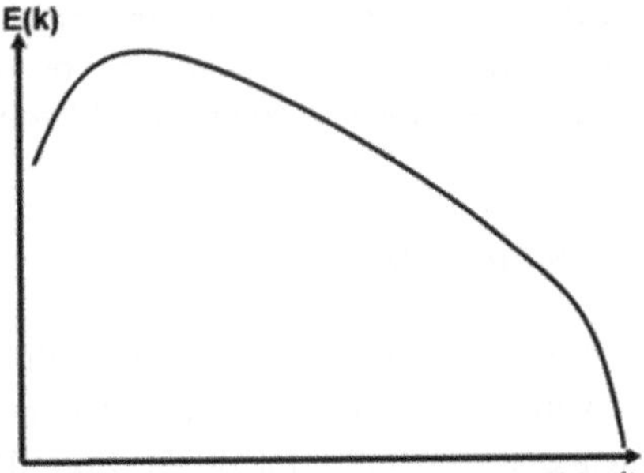

Figure 6.6. *Schematic spectrum of the overall turbulence kinetic energy of the field. On the x-axis: wave number; on the y-axis: spectral density of turbulence kinetic energy*

6.2.2. *Thermal large-eddy simulation*

A thermal large-eddy simulation (T-LES) only resolves the large dynamic and thermal structures of the flow (Figure 6.5, image on the right). It uses a coarser grid than DNS: typically six times fewer grid cells by direction, which means 200 times fewer grid cells overall. The computing cost is therefore significantly lower. The variations of temperature and velocity associated with small scales of length and with small scales of time are not captured. Figure 6.7 illustrates this idea by differentiating between the resolved part of the spectrum of turbulence kinetic energy and the modeled part of this spectrum. The large-eddy simulation can be mathematically represented by the application of a volume filter whose characteristic size is generally related to the size of the grid cells. The application of this filter to the equations of transport of linear momentum or energy brings out terms to be modeled due to nonlinearities. Physically, these terms to be modeled represent the effect of unresolved variations of velocity and temperature on the resolved scales. The strong anisothermal characteristic of the flow gives rise to specific terms to be modeled (e.g. density/velocity correlations) and can modify the behavior of the classical terms to be modeled (Bataille et al. 2005). Therefore, the development of thermal models is required, as discussed in more detail in section 6.3.

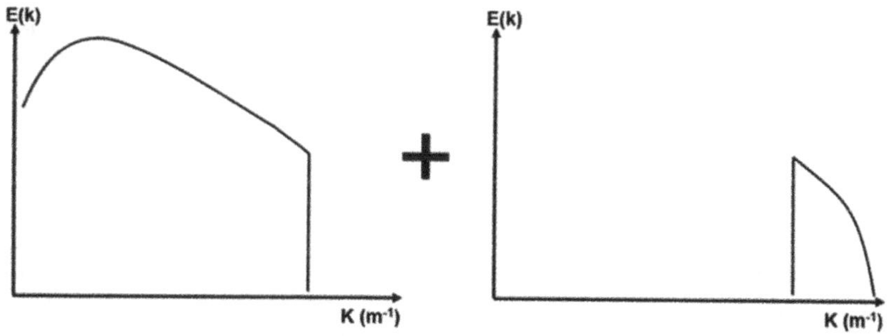

Figure 6.7. *Schematic spectra of LSS decomposition. On the left: resolved turbulence kinetic energy; on the right: modeled turbulence kinetic energy*

6.2.3. *RANS (Reynolds averaged Navier–Stokes equations)*

RANS simulation represents only the statistically mean profiles, that is, those relying on Reynolds mean. The set of turbulent structures is modeled. Continuing with the example of the turbulence kinetic energy spectrum, this means that the whole spectrum is modeled (Figure 6.8). Similar to the large-eddy simulation, the search for equations of averaged quantities highlights terms to be modeled due to nonlinearities. Classically, RANS modeling focuses on the modeling of the Reynolds stress tensor $\mathbf{R} = \langle \boldsymbol{u}' \otimes \boldsymbol{u}' \rangle$. There is a wide variety of approaches to the modeling, generally involving adding one or several differential equations. Further information can be found in Wilcox (1998) or Leschziner (2015). The advantage of the RANS approach is its short computing time. A significant drawback is that it does not allow the detailed study of nonstationary interactions between velocity and temperature (Mathelin et al. 2008). Moreover, the current models are not suited to strongly anisothermal conditions. Our strategy is to use the results of fine numerical simulations (DNS and T-LES) in order to better understand the coupling between dynamic and thermal aspects and adapt the RANS models to flows subjected to high-temperature gradients.

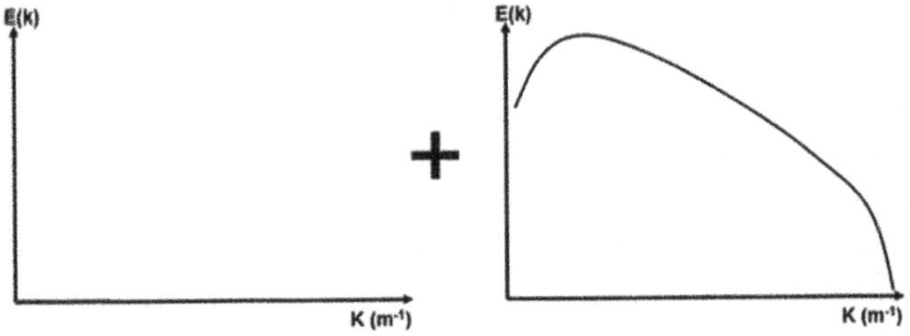

Figure 6.8. *Schematic spectra of RANS decomposition (statistical mean). Left: resolved turbulence kinetic energy; right: modeled turbulence kinetic energy*

6.2.4. *Correlations*

For dimensioning, it is useful to use empirical correlations linking various integral quantities of the flow, that is, quantities that are not associated with a specific spatial position. This is the most macroscopic approach, as it does not offer information on the local evolution of quantities. The integral quantities of interest are generally the pressure losses and the heat transfer between the solid wall of the solar receiver and the pressurized gas. It is important to note that the correlations in the literature are not suited to the strong fluxes and to the asymmetrical heating of solar receivers. For the friction coefficient, for example, asymmetric heating creates a dissymmetry: the friction coefficient at the hot side is higher than that at the cold side. In isothermal configuration, Dean's empirical correlations allow the estimation of this coefficient in a flat channel using the Reynolds number. It is possible to determine a friction coefficient for each wall using this correlation by defining a different Reynolds number at the hot side and at the cold side, but this leads to an underestimation of the dissymmetry. For the heat flux, the only correlation specifically developed for dissymmetric heating conditions under forced convection is given by David et al. (2021) for a smooth flat channel. The use of correlations developed under symmetric heating conditions leads to heat flux errors between 20 and 100%. Similar to experimental studies, DNS, T-LES and RANS simulations can be used to develop correlations suited to the conditions of solar receivers, as presented in section 6.4.1.

6.3. Direct numerical simulation and thermal large-eddy simulation

This section presents the main ideas behind direct numerical simulation (DNS) and thermal large-eddy simulation (T-LES) for strongly anisothermal flows, which include the flows in solar receivers. The approach is illustrated by a simplified model of a solar receiver in order to highlight the physical, numerical and modeling aspects.

6.3.1. *Geometry*

The selected geometry is a channel through which a three-dimensional and fully developed turbulent flow is subjected to a high-temperature gradient (Figure 6.9). This geometry preserves the fundamental and predominant physical phenomena of solar receivers, namely turbulence under conditions of strong fluxes and dissymmetric heating. Moreover, the simplicity of the geometry facilitates the implementation of models and numerical simulations and allows the validation of results by DNS.

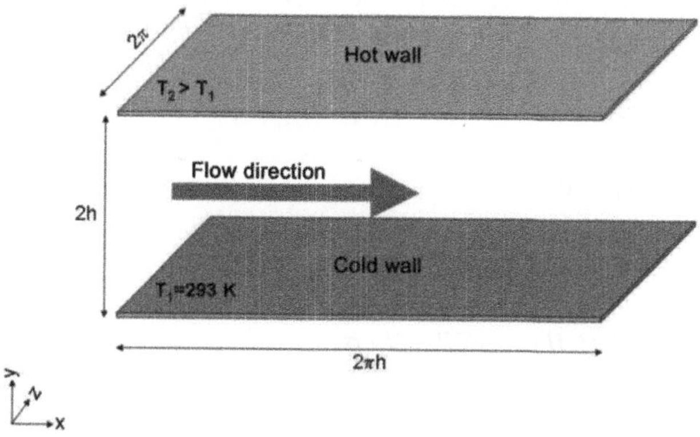

Figure 6.9. *Simplified geometry of a solar receiver. For a color version of this figure, see www.iste.co.uk/flamant/energy.zip*

The channel is periodic in the longitudinal direction (x) and in the transverse direction (z). The channel walls are at constant temperature. The temperature at the cold wall ($y = 0$) is $T_1 = 293$ K. The temperature T_2 at the hot wall ($y = 2h$, where h is the mid-height of the channel) is variable. This creates a temperature gradient in the direction normal to the walls (y). A common practice in wall turbulence is to define the Reynolds number based on the friction velocity $U_\tau = \sqrt{(\mu_\omega \partial_y \langle U_x \rangle_\omega)/\rho_\omega)}$, where $\partial_y \langle U_x \rangle_\omega$ is the wall-normal gradient of the mean wall tangential velocity at the wall, μ_ω is the dynamic viscosity at the wall and ρ_ω is the density at the wall. The Reynolds numbers selected for this study vary from $\mathrm{Re}_\tau = 180$ to $\mathrm{Re}_\tau = 395$, where Re_τ is the arithmetic average of the friction Reynolds numbers $\rho_\omega U_\tau h/\mu_\omega$ at the cold and hot walls. For the sake of simplicity, the examples of this section are all from the case $T_2 = 586$ K and $\mathrm{Re}_\tau = 180$.

6.3.2. *Direct numerical simulation equations*

The flows of solar receivers can be described by the compressible Navier–Stokes equations, which express the laws of classical mechanics and thermodynamics. The incompressible Navier–Stokes equations are not applicable due to the velocity/temperature interactions that undermine the incompressibility assumption. Indeed, the temperature variations (several hundred degrees Celsius) are such that expansion phenomena occur. However, the compressible Navier–Stokes equations can still be simplified, given the low velocity of the fluid compared to the sound velocity in solar receivers. In this section, the DNS relies on the low Mach number equations (Paolucci 1982), which takes advantage of this specificity. Moreover, it is assumed that the fluid behaves as an ideal gas. Under these conditions, the low Mach number equations can be expressed as follows:

– Conservation of mass:

$$\frac{\partial \rho}{\partial t} + \nabla \cdot (\rho \boldsymbol{U}) = 0 \qquad [6.2]$$

– Conservation of linear momentum:

$$\frac{\partial \rho \boldsymbol{U}}{\partial t} + \nabla \cdot (\rho \boldsymbol{U} \otimes \boldsymbol{U}) = -\nabla P + \nabla \cdot \boldsymbol{\sigma} \qquad [6.3]$$

– Conservation of energy:

$$\nabla \cdot \boldsymbol{U} = \frac{\gamma - 1}{\gamma P_0} \nabla \cdot \boldsymbol{Q} - \frac{1}{\gamma P_0}\frac{\partial P_0}{\partial t} \qquad [6.4]$$

– Ideal gas law:

$$P_0 = r\rho T \qquad [6.5]$$

where ρ is the density, T is the temperature, t is the time, $\boldsymbol{U}$ is the fluid velocity, $\boldsymbol{Q}$ is the conductive heat flux, $\boldsymbol{\sigma}$ is the viscous-stress tensor and γ is the heat capacity ratio of the fluid. The specificity of this formulation is the separation of pressure into two terms: the thermodynamic pressure P_0, constant in space, which represents the average pressure in the domain, and the mechanical pressure P, related to the variations of linear momentum and which varies in space. This decomposition eliminates the acoustic phenomena of the flow, which simplifies the numerical resolution of the system of equations.

The air being a Newtonian fluid, the viscous stress and the deformation rate are linearly dependent. According to Stokes' hypothesis, this can be expressed as follows: $\boldsymbol{\sigma} = 2\mu(\nabla\boldsymbol{U} + (\nabla\boldsymbol{U})^T) - (2/3)\mu(\nabla \cdot U)\, \mathbf{I_d}$, where μ is the dynamic viscosity of the fluid. The dynamic viscosity must be modeled as a function of temperature, given the large temperature variations. It is given by Sutherland's law (1893). Fourier's law $\boldsymbol{Q} = -\lambda\nabla T$ is used for the heat flux. Given the definition of the Prandtl number Pr, the variations of thermal conductivity are determined by $\lambda(T) = \mu(T)C_p/\mathrm{Pr}$, where C_p is the isobaric thermal capacity of the fluid. For the simulation of the anisothermal channel, the variations with temperature of the thermal capacity and of the Prandtl number are neglected. The selected values are $\mathrm{Pr} = 0.76$ and $C_p = 1005$ J.kg^{-1}.K^{-1}.

6.3.3. *DNS results*

This section presents the results of the DNS of the anisothermal channel. The low Mach number equations are simulated using a finite difference method on a grid with 384 × 266 × 384 grid cells. The grid is regular in the longitudinal and transverse directions. In the direction normal to the walls, the size of the grid cells decreases near the wall according to a hyperbolic tangent law. This refinement is needed as the scales near the wall are smaller and the variations along the normal to the wall are very significant and rapid. A third-order Runge–Kutta method is used to advance the numerical solution in time. The most expensive step in terms of computing time is the resolution of Poisson's equation, which corresponds to the determination of the pressure P that enforces the constraint established by the energy conservation equation. A multigrid method is used to reduce the computing time required to solve this equation.

Since the spatial and temporal resolution of the simulation is sufficient to capture all turbulent scales, the results serve as a reference to identify the significant physical aspects of the flow. In particular, one of the main effects of the high thermal gradients on the flow is the creation of a dissymmetry between the hot wall and the cold wall. This is illustrated in Figure 6.10 (top), which represents the mean velocity profile as a function of the height of the channel, and Figure 6.10 (bottom), which represents the profile of the standard deviation of velocity. It is clear that the temperature gradient creates a significant dissymmetry between the hot and cold sides, including on dynamic fields.

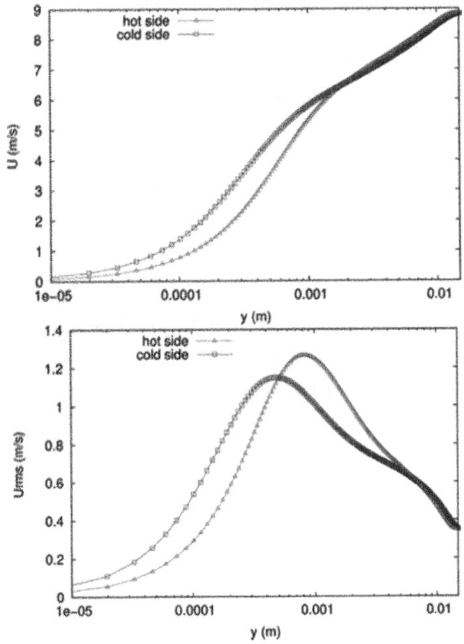

Figure 6.10. *Profiles of longitudinal velocity (top) and of the standard deviation of longitudinal velocity (bottom) at the hot and cold sides of the channel resulting from the direct numerical simulation. For a color version of this figure, see www.iste.co.uk/flamant/energy.zip*

The nature and the origin of this asymmetry, as well as other aspects of the dynamic and thermal coupling, are discussed in further detail in section 6.4. For now, the focus is on the capacity of large-eddy simulation to reproduce this asymmetry, which is characteristic of flow with large thermal gradient present in solar receivers. Indeed, DNS is, as previously explained, extremely expensive in terms of computing time, and cannot be applied to complex configurations.

6.3.4. *Equations of the thermal large-eddy simulation*

The T-LES approach focuses specifically on the large scales of the flow. In order to formalize this point, a spatiotemporal filter $\bar{\cdot}$ can be defined, whose purpose is the elimination of small scales (Sagaut 2006). The large-eddy simulation is within this framework of the simulation of a system of equations whose objective is to describe the evolution of the filtered quantities ($\bar{\rho}$, $\overline{\boldsymbol{U}}$, $\bar{P}$, $\bar{T}$, etc.). The application of the filter $\bar{\cdot}$ to low Mach number equations allows the formal establishment of the equations of the filtered quantities for large-eddy simulation. However, these contain terms that require modeling: the subgrid terms. It is common to write these equations using a density-weighted filter $\widetilde{\cdot} = \overline{\rho\,\cdot}/\bar{\rho}$, known as Favre filter:

– Conservation of mass:

$$\frac{\partial\bar{\rho}}{\partial t} + \nabla\cdot\left(\bar{\rho}\widetilde{\boldsymbol{U}}\right) = 0 \quad [6.6]$$

– Conservation of linear momentum:

$$\frac{\partial\bar{\rho}\widetilde{\boldsymbol{U}}}{\partial t} + \nabla\cdot\left(\bar{\rho}\widetilde{\boldsymbol{U}}\otimes\widetilde{\boldsymbol{U}}\right) = -\nabla\bar{P} + \nabla\cdot\breve{\boldsymbol{\sigma}} + \nabla\cdot(\bar{\rho}\boldsymbol{\tau}) \quad [6.7]$$

– Conservation of energy:

$$\nabla\cdot\widetilde{\boldsymbol{U}} = -\frac{\gamma-1}{\gamma P_0}\left(\nabla\cdot\breve{\boldsymbol{Q}} + \nabla\cdot(\bar{\rho}C_p\boldsymbol{\pi})\right) - \frac{1}{\gamma P_0}\frac{\partial P_0}{\partial t} \quad [6.8]$$

– Ideal gas law:

$$P_0 = r\bar{\rho}\widetilde{T} \quad [6.9]$$

where $\boldsymbol{\tau} = \widetilde{\boldsymbol{U}\otimes\boldsymbol{U}} - \widetilde{\boldsymbol{U}}\otimes\widetilde{\boldsymbol{U}}$ and $\boldsymbol{\pi} = \widetilde{\boldsymbol{U}T} - \widetilde{\boldsymbol{U}}\widetilde{T}$ are subgrid terms, representing the effect of small scales on large scales. For the sake of simplicity, the filter was assumed to commute with the derivatives and to not yield any subgrid terms related to the heat flux or to viscous stresses, that is, $\breve{\boldsymbol{Q}} = -\bar{\lambda}\nabla\widetilde{T}$ and $\breve{\boldsymbol{\sigma}} = 2\bar{\mu}[\nabla\widetilde{\boldsymbol{U}} + \left(\nabla\widetilde{\boldsymbol{U}}\right)^{\mathrm{T}}] - (2/3)\bar{\mu}(\nabla\cdot\widetilde{\boldsymbol{U}})\,\mathbf{I_d}$. The validity of these approximations in the case of an anisothermal channel flow was numerically evaluated.

Various subgrid-scale modeling strategies were proposed in the scientific literature. For the subgrid-scale tensor $\boldsymbol{\tau}$, there are two main approaches: the structural approach and the functional approach. In the structural approach, the effect of the filter is approximated without invoking any physical hypothesis on the nature of the interaction between the subgrid scales and the large scales. This includes, for example, deconvolution approaches aimed at reversing the filtering operation. In the functional approach, the effect of subgrid scales is generally assumed purely dissipative, by analogy with molecular diffusion. This is justified by the fact that the filter removes small scales, which are generally dissipative. It is the subgrid-scale viscosity (or diffusivity) hypothesis:

$$\boldsymbol{\tau} \approx 2\nu_{sm}[\nabla\widetilde{\boldsymbol{U}} + \left(\nabla\widetilde{\boldsymbol{U}}\right)^{\mathrm{T}}] - (2/3)\nu_{sm}(\nabla\cdot\widetilde{\boldsymbol{U}})\,\mathbf{I_d} \quad [6.10]$$

where ν_{sm} is the subgrid-scale viscosity, function of the invariants of the velocity gradient tensor. This approach, which is predominant, includes a variety of models differing only by the expression of the subgrid-scale viscosity. In wall turbulence, the use of WALE, Sigma or AMD models is recommended. This approach is extended to the subgrid-scale vector $\boldsymbol{\pi}$ by the hypothesis of subgrid-scale diffusivity:

$$\pi \approx -\alpha_{sm} \nabla \tilde{T} \qquad [6.11]$$

There are a few specific models for subgrid-scale diffusivity α_{sm}. It is generally determined by assuming a constant subgrid-scale Prandtl number, hence by $\alpha_{sm} = \nu_{sm}/\mathrm{Pr}_{sm}$. It is important to note that the value of the subgrid-scale Prandtl number is a priori not related to the Prandtl number of the fluid.

It should be noted that the dissipative properties of the numerical schemes used are added to the modeled subgrid-scale dissipation. The numerical and modeling aspects are therefore inseparable in large-eddy simulation. The use of higher-order and non-dissipative numerical schemes is recommended to avoid interfering with the subgrid-scale models. For the simulation of the anisothermal channel, a fourth-order centered discretization is used for the convection of linear momentum and a second-order centered scheme is used for the diffusion.

6.3.5. *LES results*

The large-eddy simulation models presented in section 6.3.4 are understudied in this type of turbulent and strongly anisothermal configuration, which characterizes the flows in gas solar receivers. This section addresses the performance evaluation of existing models and the development of new models. Two main approaches can be identified. First of all, an a posteriori performance evaluation of the subgrid-scale models can be performed, by comparing large-eddy simulations implementing the models. As discussed above, this analysis is complex, as it is generally not possible to separate the effect of the model from the numerical method. This stumbling block can be avoided by trying to make an a priori performance evaluation of the models, by comparing the models to the exact terms based on the filtering of experimental fields or of fields from DNS. This is also a sensitive operation, as it is difficult to estimate the effect that discrepancies observed in a priori tests may have on a simulation. A brief presentation of the two approaches in the case of the anisothermal channel is provided here.

For the a priori approach, the results of the DNS are filtered at the resolution of a typical LES grid. Various types of filters can be used. One of the simplest is the box filter, defined in one dimension by:

$$\bar{\varphi}(x) = \frac{1}{\bar{\Delta}}\int_{x-\bar{\Delta}/2}^{x+\bar{\Delta}/2} \varphi(\xi)d\xi \qquad [6.12]$$

and which can be computed in three dimensions by the successive application of the one-dimensional filter in each direction. The exact subgrid terms and the proposed models are then calculated from these filtered fields. Finally, a statistical measure of the agreement between the model and the exact term must be chosen. In Figure 6.11, the energetic contribution of the subgrid term $\boldsymbol{\tau}$ is compared for several models in the literature. It is given by $\rho\boldsymbol{\tau} : [\nabla\widetilde{\boldsymbol{U}} + (\nabla\widetilde{\boldsymbol{U}})^{\mathrm{T}}]$ and represents a kinetic energy transfer between the scales above and below the filter cut-off length. For this measure and in this particular case, the AMD model has a better agreement than the other studied models. These results should be moderated because, while the AMD model is very good in practice, the difference between the models in terms of simulation performances is less prominent.

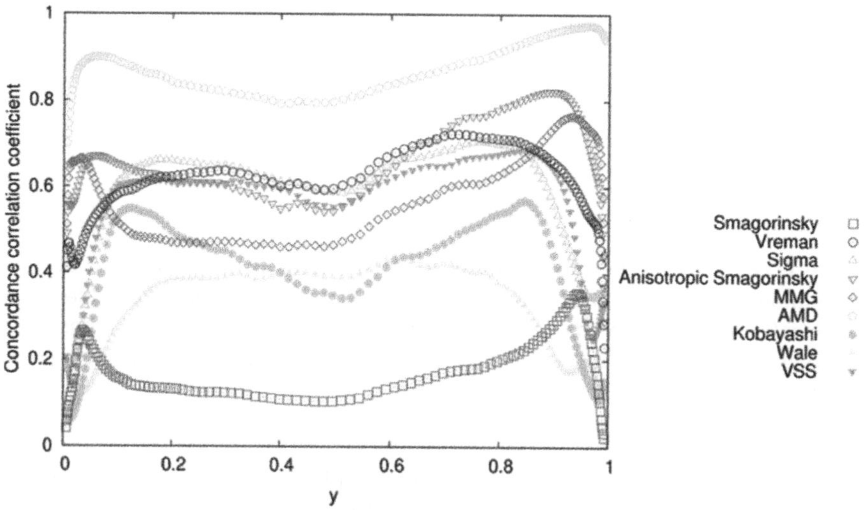

Figure 6.11. *Coefficient of concordance between the exact subgrid-scale dissipation and the subgrid-scale dissipation corresponding to various models of the subgrid-scale tensor τ. For a color version of this figure, see www.iste.co.uk/flamant/energy.zip*

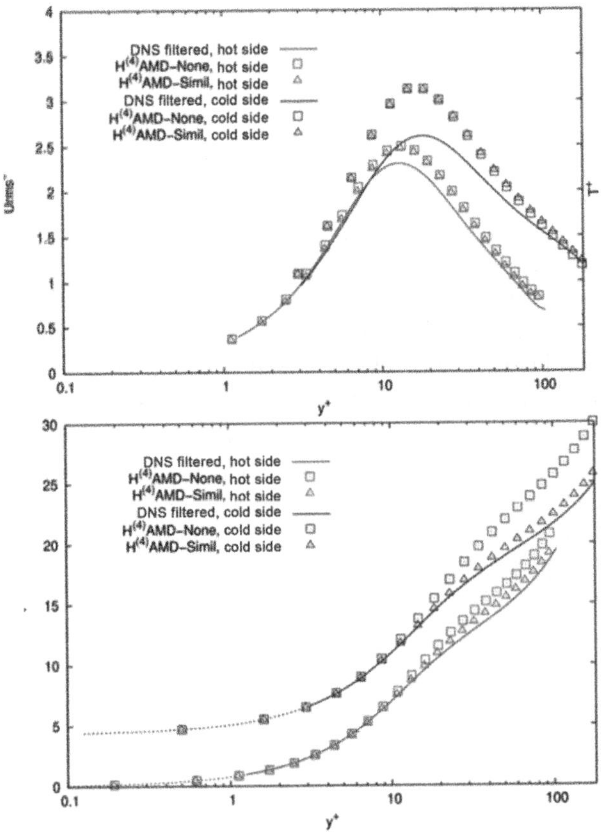

Figure 6.12. *Standard deviation profiles for longitudinal velocity (top) and average temperature at the hot and cold sides of the channel resulting from the direct numerical simulation (filtered) and large-eddy simulations with or without modeling of the subgrid-scale vector* $\boldsymbol{\pi}$*. For a color version of this figure, see www.iste.co.uk/flamant/energy.zip*

For the a posteriori approach, large-eddy simulations are conducted on a grid involving 48 × 50 × 48 grid cells. The subgrid term $\boldsymbol{\tau}$ is modeled by a variant of AMD model, pre-multiplied by a constant tensor. The subgrid-scale vector $\boldsymbol{\pi}$ is considered negligible in a first simulation and modeled by a scale similarity approach in a second simulation. Figure 6.12 (top) shows the profile of the standard deviation of velocity in the two cases, compared to the filtered results of the DNS. It can be observed that the asymmetry between the hot and cold sides is to a certain extent reproduced by large-eddy simulations. The predicted amplitude is, however, overestimated, particularly at the cold side. The model used for the subgrid-scale vector π has little influence on this point. The subgrid-scale vector is, however, essential for predicting the thermal aspects of the flow. This is illustrated in Figure 6.12 (on the right), which represents the average temperature profile, and in Figure 6.13, which represents the conductive heat flux. It can be noted that the prediction of the statistics related to temperature is improved by the use of a scale similarity model for π, similar to the wall flux prediction. Due to this property, the use of a model for this subgrid term is essential for precise predictions in the strongly anisothermal flows of solar receivers.

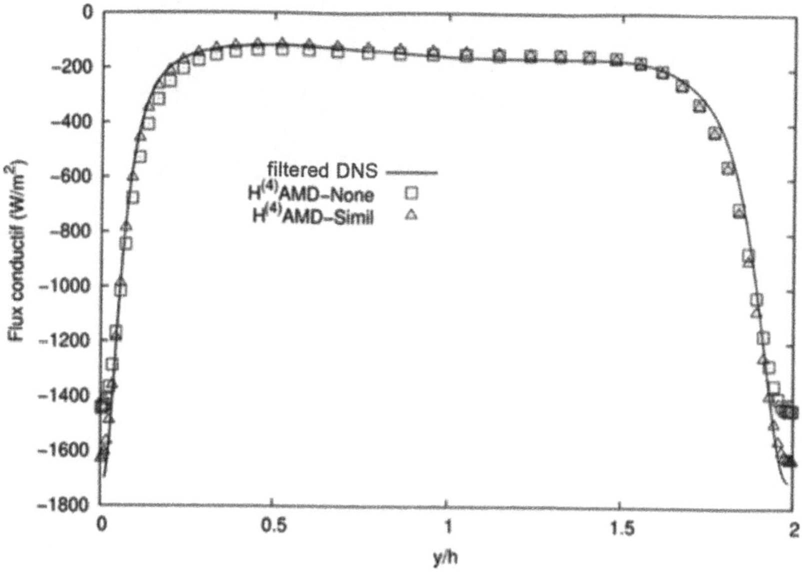

Figure 6.13. *Profiles of the conductive heat flux at the hot and cold sides of the channel resulting from the direct numerical simulation (filtered) and large-eddy simulations with or without modeling of the subgrid-scale vector* π

6.4. Dynamic and thermal couplings – physical approach

The strong fluxes, dissymmetric heating and high temperatures characteristic of solar receivers give rise to couplings between dynamic and thermal phenomena. Figure 6.14 illustrates these interactions between velocity and temperature: the turbulent structures (represented by iso-surfaces of the Q-criterion) follow the deformations of the isotherm (in red on the figure). These interactions modify the behavior of the flow. This section analyzes the modifications of the flow reflected by the integral quantities of the flow, and then by the evolution of variables in the spatial and spectral domains.

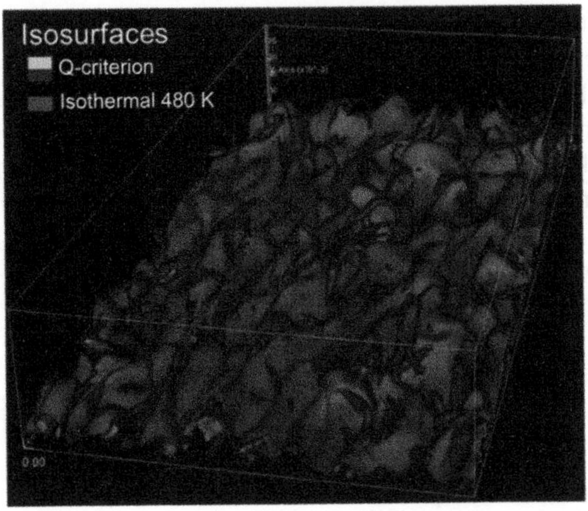

Figure 6.14. *Illustration of turbulence/temperature interactions. Pressurized gas, bi-periodic channel with dissymmetric heating. For a color version of this figure, see www.iste.co.uk/flamant/energy.zip*

6.4.1. *Analysis of integral quantities*

The wall heat transfer is an integral quantity that is particularly important for solar receivers and difficult to predict with the correlations in the literature. This is why the PROMES laboratory developed a correlation suited to asymmetric-heating configurations (David et al. 2021). For this purpose, DNS and T-LES simulations were used. It was thus possible to precisely calculate the gas/wall heat transfer for various Reynolds and Prandtl numbers and for various temperature levels. A shape is assumed for the correlation, similar to the correlation of Battista and Perkins (1970) within the limit of symmetric heating, but including a new term that takes into account the dissymmetric heating. A least-squares regression is then used to determine the coefficients of the expression postulated by the correlation. The resulting correlation is:

$$Nu_w = 0{,}024 Re_b^{0{,}8} Pr_b^{0{,}4} \left(\frac{T_w}{T_b}\right)^{-0{,}9} \left(\frac{T_w}{T_w - T_b}\right)^{1{,}4\left(1-\frac{T_w}{T_m}\right)\frac{T_b}{T_w}} \quad [6.13]$$

where T_w, T_b and T_m are the wall temperature, the average temperature of the fluid and the average temperature of the walls, respectively. The correlation covers the following ranges: wall/fluid thermal flux between 4 and 578 kW/m^2, Reynolds number defined with the mean streamwise linear momentum from 10,000 to 180,000 and wall temperature between 293 and 1,300 K. The error between the

Nusselt numbers obtained by the proposed correlation and by reference simulations is always below 10%, as illustrated in Figure 6.15.

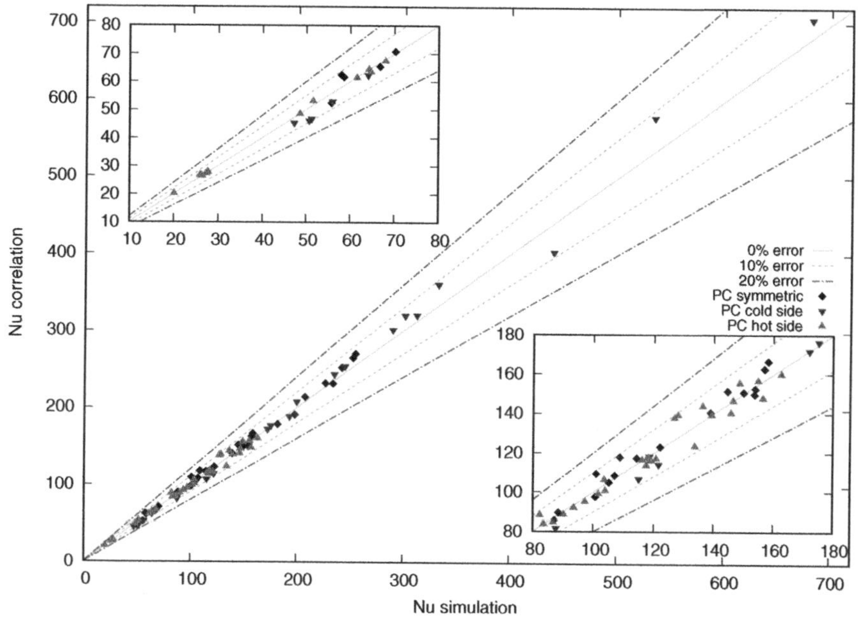

Figure 6.15. *Comparison of Nusselt numbers obtained by the proposed correlation and by reference simulations. For a color version of this figure, see www.iste.co.uk/flamant/energy.zip*

6.4.2. *Analysis in the spatial domain*

This section describes the behavior of average quantities and fluctuating quantities as a function of the average friction Reynolds number Re_τ and the temperature ratio between the two walls T_2/T_1, using the anisothermal channel as an example.

6.4.2.1. *Average quantities*

Figure 6.16 shows the evolution of the profiles of velocity and average temperature with increasing temperature ratio between the wall subjected to solar radiation and the back wall. The asymmetric heating creates a dissymmetry of the velocity profiles. This dissymmetry increases with the temperature ratio. For a temperature ratio of 5, the flow at the hot side is close to a laminar flow. This is essentially explained by the increase in the viscosity of the pressurized gas with temperature (Serra et al. 2011). Figure 6.17 (top) shows that the slope of the logarithmic part of the velocity profile is modified by the temperature gradient. It has been shown that this modification is related to the existence of a nonzero average velocity in the direction perpendicular to the walls. This velocity is zero in the isothermal case and results from expansion phenomena. Indeed, the conservation of mass imposes a displacement from the hot side (low density) to the cold side (high density). An analogy can be drawn between the expansion-related velocity in the anisothermal channel and the injection velocity of a flow with effusion. In both cases, the scaling of Van Driest (1951) can be used to correct the average velocity profiles and therefore to interpret the effect of the temperature gradient. With this nondimensionalization, the profiles at the hot and cold sides are superimposed to the isothermal logarithmic law (Figure 6.18).

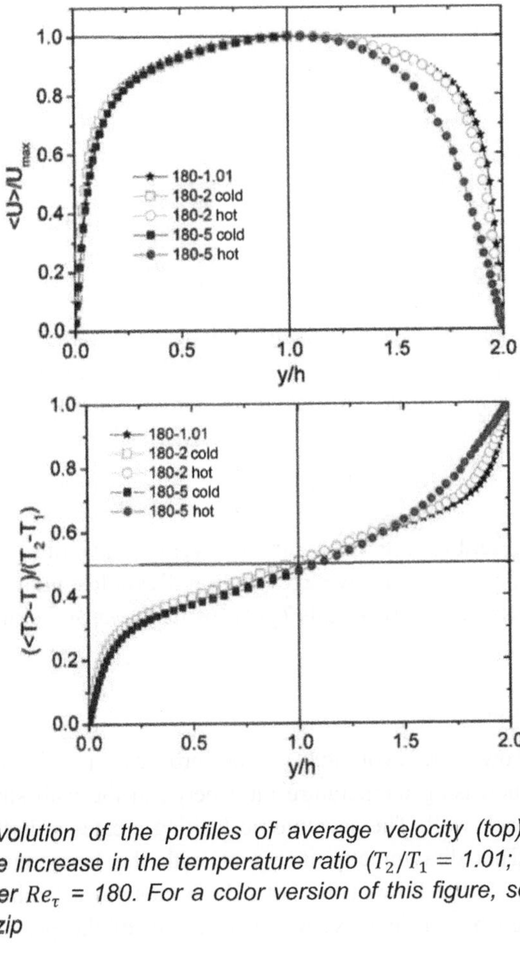

Figure 6.16. *Evolution of the profiles of average velocity (top) and temperature (bottom) with the increase in the temperature ratio (T_2/T_1 = 1.01; 2 or 5) and friction Reynolds number Re_τ = 180. For a color version of this figure, see www.iste.co.uk/flamant/energy.zip*

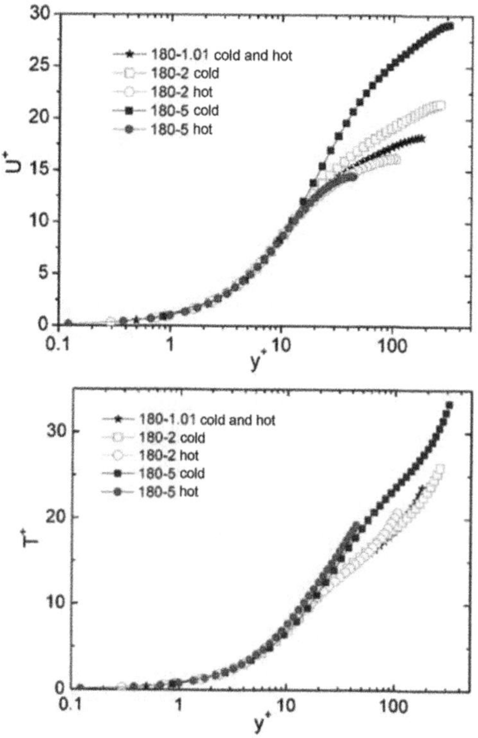

Figure 6.17. *Comparison of the profiles of velocity (top) and temperature (bottom) at the hot and cold sides for various temperature ratios. The velocity and temperature are scaled by the friction velocity and temperature. For a color version of this figure, see www.iste.co.uk/flamant/energy.zip*

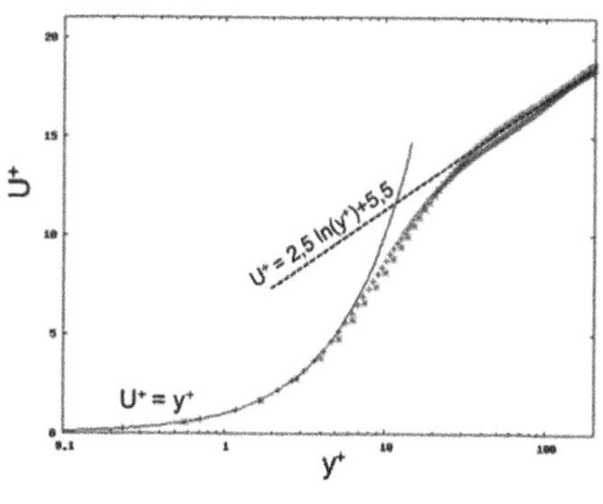

Figure 6.18. *Longitudinal average velocity with the scaling of Van Driest. For a color version of this figure, see www.iste.co.uk/flamant/energy.zip*

6.4.2.2. *Fluctuating quantities*

Concerning the study of fluctuating quantities, let us first of all fix the average friction Reynolds number in order to study the effect of increasing temperature ratio between the two walls. Figure 6.19 (top) shows the standard deviation of velocity for an average friction Reynolds number $\mathrm{Re}_\tau = 180$. The main effect of the temperature gradient is to create a dissymmetry between the hot and cold sides. It can be noted that the standard deviation at the hot side is lower than that obtained through isothermal simulation, while the one at the cold side is higher. This dissymmetry increases with the temperature ratio. This is mainly due to a local Reynolds number effect. Indeed, the gas viscosity increases with temperature, and consequently the local Reynolds number is higher at the cold side than at the hot side. This is a drawback from the perspective of the solar receiver application, as the heat transfer between the heat transfer fluid and the wall is directly related to the turbulent intensity. It is therefore important to develop internal receiver geometries that increase the turbulence level and thus intensify heat transfers. Figure 6.20 shows the longitudinal velocity/wall-normal velocity and longitudinal velocity/temperature correlations. The same observations and analyses than those for the standard deviation of velocity can be made. Figure 6.19 (bottom) shows the profiles of standard deviation of temperature. The dissymmetry between the hot and cold sides is again obvious. However, the standard deviation of temperature at the hot side increases with the temperature ratio (while the standard deviation of velocity at the hot side decreases). It is in particular higher than the standard deviation of temperature in quasi-isothermal situation (temperature ratio equal to 1.01). This is explained by the competition between two phenomena. Similar to velocity, the dissymmetry between the sides is mainly due to a local Reynolds number effect. This is the first phenomenon. The second phenomenon is a mechanical increase of the standard deviation with the increase of the temperature ratio that partly compensates for the decrease at the hot side related to the local Reynolds number effect.

For a better separation of these two phenomena, it is interesting to conduct two isothermal simulations: a first one whose friction Reynolds number is the Reynolds number at the hot side and a second one with the Reynolds number at the cold side (Serra et al. 2012). Figure 6.21 shows the standard deviation of velocity for these simulations. The anisothermal simulation with a temperature ratio $T_2/T_1 = 2$ and an average friction Reynolds number $Re_\tau = 180$ has a friction Reynolds number at the cold side of $Re_{\tau,f} = 262$ and a friction Reynolds number at the hot side of $\mathrm{Re}_{\tau,c} = 106$. Two isothermal simulations are thus conducted: one at $Re_\tau = 262$ and the other at $Re_\tau = 106$. Figure 6.21 (left) shows as expected the effect of the Reynolds number: the standard deviation at $Re_\tau = 106$ is below that at $Re_\tau = 262$. On the other hand, the dissymmetry between the standard deviation at these two Reynolds numbers is lower than that obtained for the anisothermal simulation: the standard deviation at the hot side is lower than that of the isothermal simulation at the same Reynolds number and the standard deviation at the cold side is higher than that of the isothermal simulation. Everything is happening as if there was a turbulence kinetic energy transfer from the hot side to the cold side. This transfer is related to the wall-normal velocity (oriented from the hot side to the cold side). For a more significant temperature ratio $T_2/T_1 = 5$, the results are different (Figure 6.21, right). The Reynolds number at the hot side $Re_{\tau,c} = 44$ corresponds to a laminar flow. As expected, the standard deviation of velocity of the isothermal simulation is zero. On the other hand, the standard deviation of velocity under anisothermal configuration is not zero. Indeed, the turbulence at the cold side is sufficient to induce a transition of the flow at the hot side to a turbulent state despite the very low local Reynolds number.

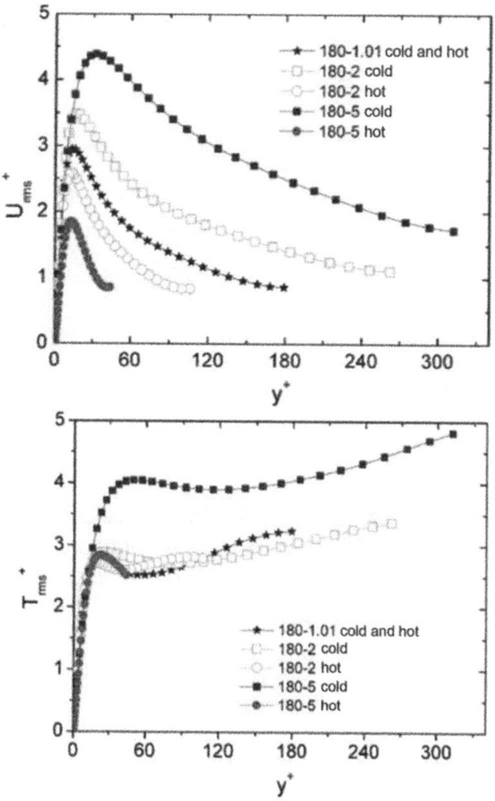

Figure 6.19. *Standard deviation of velocity (top) and temperature (bottom) for various temperature ratios (1.01; 2 and 5) and a friction Reynolds number Re_τ= 180. For a color version of this figure, see www.iste.co.uk/flamant/energy.zip*

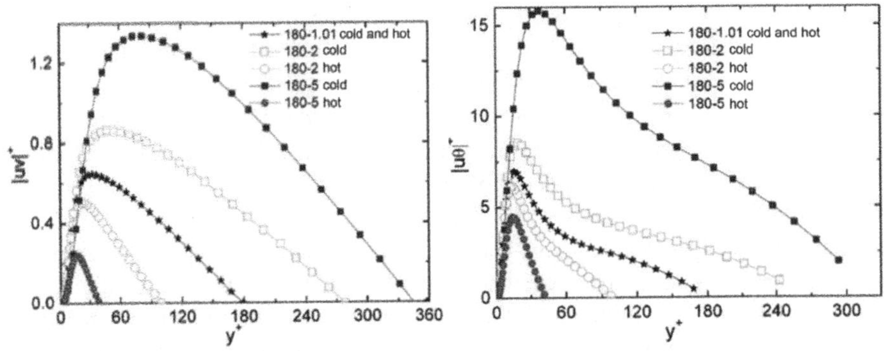

Figure 6.20. *Correlations between longitudinal velocity and velocity normal to the walls on the left and between longitudinal velocity and temperature on the right for various temperature ratios. For a color version of this figure, see www.iste.co.uk/flamant/energy.zip*

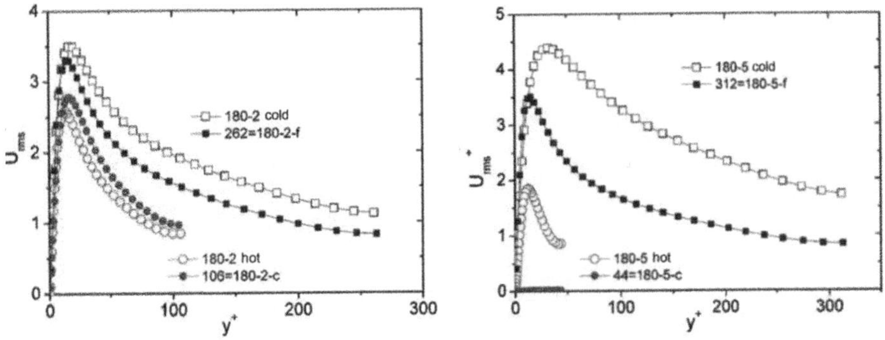

Figure 6.21. *Comparison of the standard deviation of velocity under isothermal and strongly anisothermal configurations. On the left: temperature ratio of 2; on the right: temperature ratio of 5. For a color version of this figure, see www.iste.co.uk/flamant/energy.zip*

6.4.3. *Analysis in the spectral domain*

The analysis of the effect of dissymmetry induced by the temperature gradient can be extended to the spectral domain, as a function of the size of turbulent structures. As previously described, turbulent flows include eddy of various sizes covering a wide range of scales. Generally speaking, the large scales have the highest energy, while the small scales are the most dissipative. The interscale energy transfer from large structures to small-scale dissipative structures is referred to as the energy cascade. This process is well documented for incompressible isotropic homogeneous turbulence (Batchelor 1953; Sagaut and Cambon 2008). For a high Reynolds number, there is a clear separation between the production and dissipation zones (Figure 6.22, top). In the intermediate zone without production or dissipation, known as inertial region, a spectral flux carries energy to the low-scale structures. The spectral density of turbulence kinetic energy follows in the inertial region a -5/3 slope predicted by the works of Kolmogorov (1941), as shown in Figure 6.22 (bottom).

In the real flows found in solar receivers, this canonical description is enhanced by several phenomena. The geometry of the receiver and the presence of walls add a spatial dimension to the energy transfers and can lead to an overlapping of productive and dissipative scales. Finally, these phenomena are modulated by the dissymmetric heating of the receiver. Once again, the anisothermal channel is an interesting case study that highlights these effects (Aulery et al. 2017). Four fundamental transfers can be identified in this case: production (interaction with the mean motion), interscale transfer, spatial transfer and interaction with the internal energy (whose irreversible, predominant component corresponds to dissipation). Production occurs around $y^* = 12$ and $k^* = 0.07$ (Figure 6.23). Interscale transport redistributes the energy generated among scales, mainly towards small scales, as well as towards large scales (Figure 6.24). Moreover, the energy is spatially transferred, mainly towards the wall and with small-scale variations (Figure 6.25). There, it is dissipated by the interaction with internal energy (Figure 6.26).

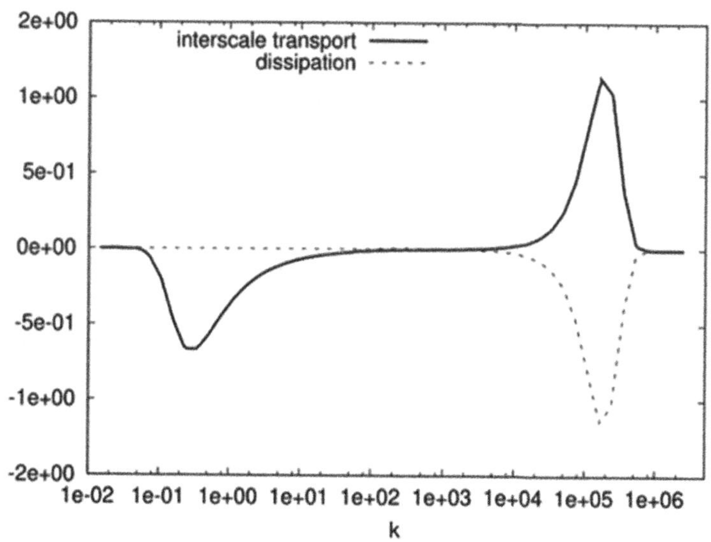

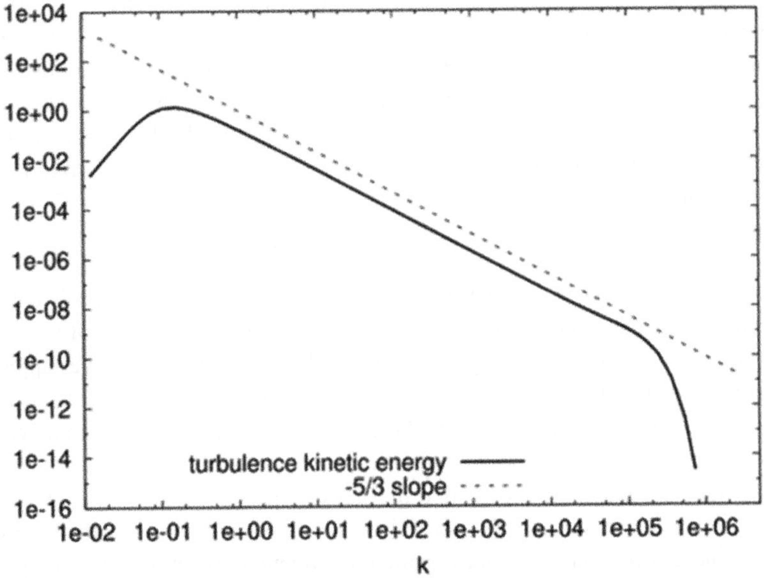

Figure 6.22. *Typical spectra of interscale transport and dissipation (top), premultiplied by the wave number, and of turbulence kinetic energy (bottom), for an incompressible isotropic homogeneous turbulence. The scales are arbitrary. Source: data from W. Bros spectra provided in Sagaut and Cambon (2008)*

Each of these transfers is influenced by the compressibility of the flow, the variations or fluctuations of the properties of the fluid and/or the existence of an average wall-normal velocity. In an anisothermal channel, these phenomena create an asymmetry between hot and cold sides. A large part of this asymmetry can be explained using semi-empirical scalings. Near the wall, a semi-local scaling proposed by Huang et al. (1995) is used, relying on a combination of the semi-local friction velocity $U_\tau^* = \sqrt{(\mu_\omega \partial_y \langle U_x \rangle_\omega)/\langle \rho \rangle}$ and the mean local kinematic viscosity $\langle \nu \rangle$. Towards the center of the channel, a semi-local scaling based on U_τ^* and the half-height of channel h is more relevant (Dupuy et al. 2019c). These scalings account for an important part of the effect of variations of the fluid properties but do not capture all the effects of the asymmetric heating. The spectral energy transfers have a lower amplitude at the cold side than at the hot side, but they occur at smaller wave numbers, closer to the wall and over a wider range of scales at $\mathrm{Re}_\tau^* = 395$ than at $\mathrm{Re}_\tau^* = 180$ (Figures 6.23–6.26). Moreover, the largest spatial amplitude of the terms at the cold side seems to result from the inclusion of a larger number of scales in the spectral domain.

One phenomenon not taken into account is the low Reynolds number effects corresponding to the variations of local friction Reynolds number $U_\tau^* h/\langle\nu\rangle$ in the channel. This effect can be identified to the variations of the scaled transfers with the friction Reynolds number of the channel in the isothermal case. While Reynolds number effects are always present, the impact is particularly prominent for a low Reynolds number. The semi-local scalings are generally expected to be less effective for low Reynolds number. This can be verified in Figures 6.23–6.26 where, for each of the energy transfers, the asymmetry between hot and cold sides is lower at $Re_\tau = 395$ than at $Re_\tau = 180$.

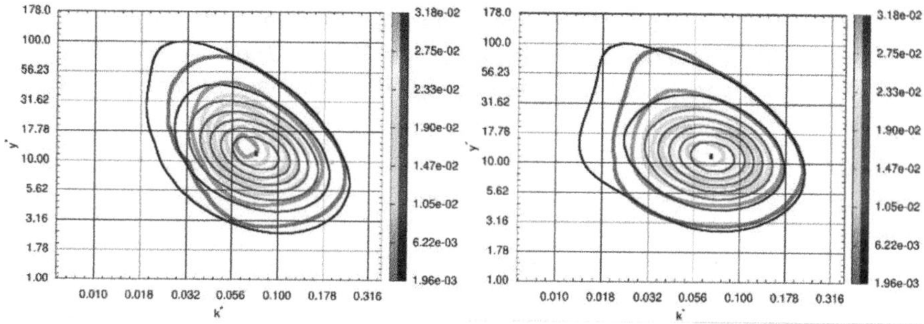

Figure 6.23. *Spectrum of the production with semi-local scaling at Re_τ = 180 (thick lines, light color) and Re_τ = 395 (fine lines, dark color). For a color version of this figure, see www.iste.co.uk/flamant/energy.zip*

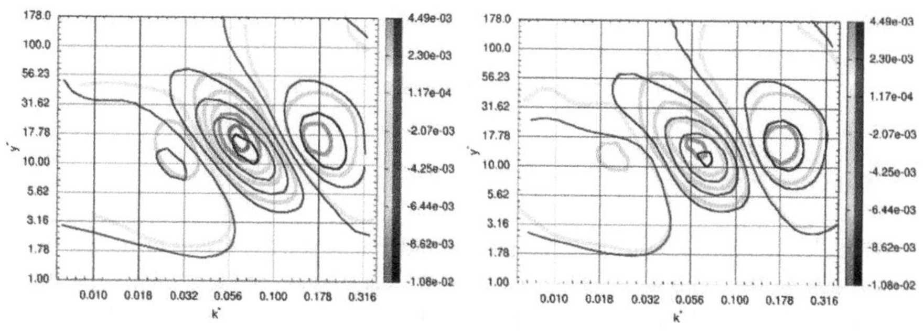

Figure 6.24. *Spectrum of the interscale transfer with semi-local scaling at Re_τ = 180 (thick lines, light color) and Re_τ = 395 (fine lines, dark color). For a color version of this figure, see www.iste.co.uk/flamant/energy.zip*

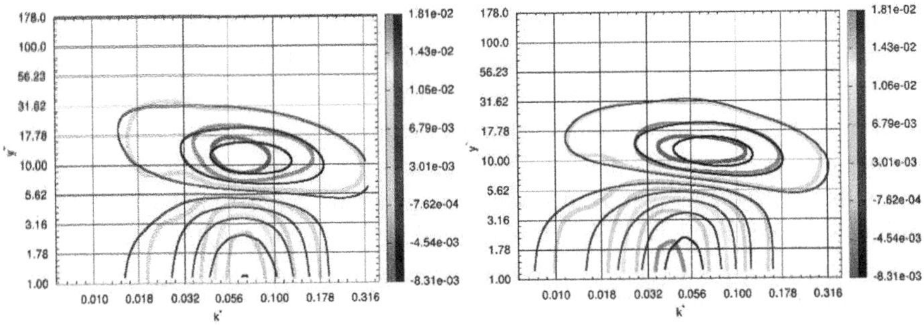

Figure 6.25. *Spectrum of the spatial transfer with semi-local scaling at* Re_τ *= 180 (thick lines, light color) and* Re_τ *= 395 (fine lines, dark color). For a color version of this figure, see www.iste.co.uk/flamant/energy.zip*

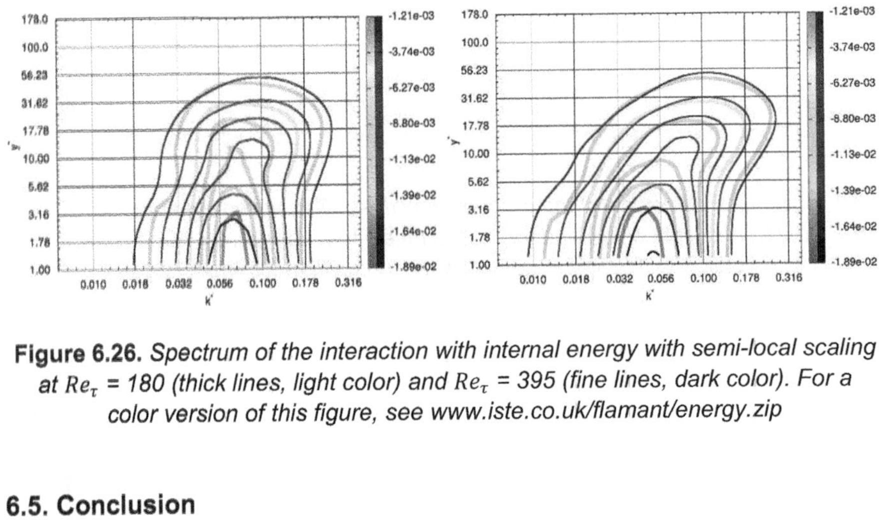

Figure 6.26. *Spectrum of the interaction with internal energy with semi-local scaling at* Re_τ *= 180 (thick lines, light color) and* Re_τ *= 395 (fine lines, dark color). For a color version of this figure, see www.iste.co.uk/flamant/energy.zip*

6.5. Conclusion

This chapter focused on the necessity to understand anisothermal turbulent flows inside gas solar receivers. This is not an easy task, and it requires many developments considering the complexity of the physics involved in these solar receivers. Several approaches are possible, from the most detailed (DNS) to the more macroscopic (correlations), passing through large-eddy simulation, which is a very good compromise between computing cost and information detail. These various studies can be used to determine profiles of velocity and temperature (average and fluctuating), to calculate heat fluxes, and also to understand how energy is distributed. A significant number of work remains to be done to improve the understanding and modeling of the high-temperature gradients encountered in solar receivers. Besides the traditional up-scaling approaches to model development, another promising path is the exploitation of high-fidelity simulations by artificial-intelligence tools, such as machine learning.

6.6. References

Aulery, F., Dupuy, D., Toutant, A., Bataille, F., Zhou, Y. (2017). Spectral analysis of turbulence in anisothermal channel flows. *Computers and Fluids*, 151, 115–131.

Bailly, C. and Comte-Bellot, G. (2003). *Turbulence. Sciences et techniques de l'ingénieur*. CNRS Éditions, Paris.

Bataille, F., Rubinstein, R., Hussaini, Y. (2005). Eddy viscosity and diffusivity modeling. *Physics Letters A*, 346, 168–173.

Batchelor, G.K. (1953). *The Theory of Homogeneous Turbulence*. Cambridge University Press, Cambridge.

Battista, E. and Perkins, H. C. (1970). Turbulent heat and momentum transfer in a square duct with moderate property variations. *International Journal of Heat and Mass Transfer*, 13(6), 1063–1065.

David, M., Toutant, A., Bataille, F. (2021). Numerical development of heat transfer correlation in asymmetrically heated turbulent channel flow. *International Journal of Heat and Mass Transfer*, 164, 120599.

Dupuy, D., Toutant, A., Bataille, F. (2019a). A posteriori tests of subgrid-scale models in strongly anisothermal turbulent flows. *Physics of Fluids*, 31(6), 065113.

Dupuy, D., Toutant, A., Bataille, F. (2019b). A priori tests of subgrid-scale models in an anisothermal turbulent channel flow at low mach number. *International Journal of Thermal Sciences*, 145(November), 105999.

Dupuy, D., Toutant, A., Bataille, F. (2019c). Effect of the Reynolds number on turbulence kinetic energy exchanges in flows with highly variable fluid properties. *Physics of Fluids*, 31(1), 015104.

Huang, P.G., Coleman, G.N., Bradshaw, P. (1995). Compressible turbulent channel flows: DNS results and modelling. *Journal of Fluid Mechanics*, 305, 185–218.

Kolmogorov, A.N. (1941). Локальная структура турбулентности в несжимаемой жидкости при очень больших числах Рейнольдса. *Dokl. Akad. Nauk SSSR*, 30(4), 299–303.

Kolmogorov, A.N. (1991). The local structure of turbulence in incompressible viscous fluid for very large Reynolds numbers. *Proceedings of the Royal Society A*, 434, 9–13.

Leschziner, M. (2015). *Statistical Turbulence Modelling for Fluid Dynamics-Demystified: An Introductory Text for Graduate Engineering Students*. World Scientific, London.

Lesieur, M. (2012). *Turbulence in Fluids*. Springer Science & Business Media, Cham.

Mathelin, L., Bataille, F., Zhou, Y. (2008). Theoretical investigation of some thermal effects in turbulence modelling. *Theorical and Computational Fluid Dynamics*, 22, 471–483.

Paolucci, S. (1982). On the filtering of sound from the navier–stokes equations. Technical Report SAND82-8257, Sandia National Labs, Livermore, CA.

Sagaut, P. (2006). *Large Eddy Simulation for Incompressible Flows: An Introduction*. Springer Science & Business Media, Paris.

Sagaut, P. and Cambon, C. (2008). *Homogeneous Turbulence Dynamics*. Cambridge University Press, Cambridge.

Serra, S., Toutant, A., Bataille, F. (2011). Thermal large eddy simulation in a very simplified geometry of a solar receiver. *Journal of Heat Transfer Engineering*, 33, 505–524.

Serra, S., Toutant, A., Bataille, F., Zhou, Y. (2012). High temperature gradient effect on a turbulent channel flow using thermal large-eddy simulation in physical and spectral spaces. *Journal of Turbulence*, 13, 1–25.

Sutherland, W. (1893). The viscosity of gases and molecular force. *The London, Edinburgh, and Dublin Philosophical Magazine and Journal of Science*, 36(223), 507–531.

Toutant, A. and Bataille, F. (2013). Turbulence statistics in a fully developed channel flow submitted to a high temperature gradient. *International Journal of Thermal Sciences*, 74, 104–118.

Van Driest, E.R. (1951). Turbulent boundary layer in compressible fluids. *Journal of the Aeronautical Sciences*, 18(3), 145–160.

Wilcox, D.C. (1998). *Turbulence Modeling for CFD*. DCW Industries, La Canada, CA.

7

Materials for Concentrated Solar Power

Audrey SOUM-GLAUDE[1] and Antoine GROSJEAN[2]
[1] *PROMES-CNRS, Font-Romeu, France*
[2] *PROMES-CNRS, UPVD, Perpignan, France*

7.1. Introduction

All solar concentrating technologies rely on the implementation of components having various optical functions. Solar radiation is collected and concentrated by systems of highly reflective mirrors. The role of the solar receiver is to absorb the radiation concentrated by the mirrors and to convert it into thermal energy or heat that is transferred to a heat transfer fluid. In some cases, a transparent window is placed in front of the receiver in order to protect it from chemical and mechanical degradations and limit thermal losses (see Chapter 4).

These optical functions are most often performed by thin coatings of one or several layers deposited on a substrate. These coatings add a surface optical functionality, while taking advantage of the substrate characteristics (rigidity, thermal conductivity, etc.). Several dozens or hundreds of nanometers of materials in thin films are in fact sufficient to modify the surface optical properties and obtain the intended function. This serves to limit the costs of the components, having access to low cost substrate materials and using only small quantities of high optical performance materials.

Concentrating Solar Thermal Energy,
coordinated by Gilles FLAMANT. © ISTE Ltd 2022.

These economic considerations are explained by the size of the solar field needed for a concentrated solar power plant. For example, the solar field of the Andasol power plant in Spain has 510,000 m^2 of parabolic trough mirrors and 88 km of receiver tubes, representing the highest investment item (about 30% of the total cost of the power plant).

Moreover, the operating conditions of concentrated solar power technologies are a challenge for the optical components that are exposed to direct or concentrated solar radiation, oxidizing and corrosive environments, erosive particles, high temperatures, etc., over lengthy periods of time of the order of 25 years. The resistance and durability of the materials and components under these demanding conditions are therefore critical for limiting the costs related to their maintenance and replacement.

In this context, this chapter on the materials for concentrated solar power systems focuses on optical components and is structured in four sections. The first section describes the general optics elements, a prerequisite for understanding the chapter. The next three sections provide details on the three types of optical surfaces that are used in solar fields: reflective, transparent and absorbing surfaces. Each of these sections defines the performance indicators associated with the considered optical function, and describes the existing technologies and those being developed in order to reach these performances. Particular attention is also paid to the operating conditions specific to each type of optical surface, the material aging phenomena that these conditions may trigger, and the solutions being considered in order to limit the aging-related degradations.

7.2. Optical properties of materials

7.2.1. *Spectral properties*

During a radiation–matter interaction, the incident radiation can be either reflected at the surface of the material, or transmitted through the material, or absorbed by the material (Figure 7.1).

The capacity of a material to reflect the incident radiation is quantified by its reflectance R, its capacity to transmit this radiation is quantified by its transmittance Tr, and its capacity to absorb is quantified by its absorptance A. These quantities are spectral and directional, meaning that they depend on the wavelength λ and on the incidence angle θ of the incident radiation, and also on the temperature T of the material. They are referred to as spectral, and directional, reflectance, transmittance and absorptance. These three quantities are related by the law of conservation of energy (equation [7.1], in which R is the hemispherical directional reflectance):

$$R(\lambda,\theta,T)+Tr(\lambda,\theta,T)+A(\lambda,\theta,T)=1 \qquad [7.1]$$

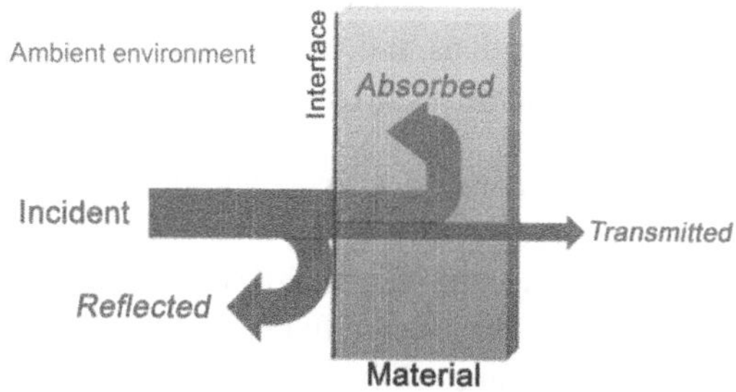

Figure 7.1. *Interaction between an incident radiation and a material. For a color version of this figure, see www.iste.co.uk/flamant/energy.zip*

By themselves, they illustrate the optical behavior of a material with respect to an incident radiation of given wavelength and direction. In practice, only the spectral reflectance and transmittance can be readily measured, as they refer to rays reflected and transmitted outside the material. As an internal property of the material, spectral absorptance is difficult to measure and is simply deduced from equation [7.1].

The spectral reflectance and transmittance of a surface are classically measured by spectrophotometry. Typically, the emission of a light source is separated in wavelength by a monochromator, and the intensity of the beam reflected or transmitted by the material is compared to that of the beam incident on the material. These quantities can also be simulated using spectral complex refractive indices $n(\lambda) + i \cdot k(\lambda)$ of the material, thanks to Fresnel coefficients and the transfer matrix method (Furman, Tikhonravov 1992).

Solar radiation varies strongly in intensity with the wavelength. It is therefore essential to know the spectral response of a material in order to deduce from it its actual solar performance.

7.2.2. *Solar performance*

The solar performance of a material, denoted here by η_S, is defined as the ratio of the solar flux density (W/m^2) available after interaction with the material, denoted by J_u, to the incident solar flux density, denoted by J_i.

The solar spectrum yields the spectral irradiance $E_{sol}(\lambda)$ in W/m^2/nm, as a function of the wavelength λ in nm. The incident solar flux density J_i (W/m²) is therefore deduced from the integration of this spectrum over its whole spectral range, typically from λ_1 = 280 nm to λ_2 = 2.5 µm. ASTM G173-03 Direct + Circumsolar (DC) AM1.5 is the standard solar spectrum considered for the concentrating solar power applications. Indeed, only the irradiance coming from the solar disk is useful for these systems, as diffuse radiation cannot be concentrated.

Similarly, the useful flux density J_u is deduced from the integration over wavelength of the spectral irradiance resulting after interaction with the material. The latter is the product of the spectral function $f(\lambda)$ of the material and the incident solar spectral irradiance $E_{sol}(\lambda)$. The spectral function corresponds to one of the three spectral optical properties, reflectance $R(\lambda)$, transmittance $Tr(\lambda)$ or absorptance $A(\lambda)$, depending on the considered type of component.

The J_u/J_i ratio determines the solar performance η_S, namely the part of flux density that is actually reflected, transmitted or absorbed by the surface, with respect to the incident solar flux density. Equation [7.2] yields the general expression of solar performance to be used in the rest of the chapter:

$$\eta_S = \frac{J_u}{J_i} = \frac{\int_{\lambda_1}^{\lambda_2} f(\lambda) \cdot E_{sol}(\lambda) \cdot d\lambda}{\int_{\lambda_1}^{\lambda_2} E_{sol}(\lambda) \cdot d\lambda} \qquad [7.2]$$

7.3. Reflective components: solar mirrors

7.3.1. *Optical performance indicator: solar reflectance*

The optical function of solar mirrors is to reflect as much as possible the incident solar radiation, in order to collect and concentrate the latter towards the receiver. Their main performance indicator is therefore solar reflectance R_S, resulting from expression [7.2] and defined as follows (Fernández-García et al. 2018), with $\Delta\lambda = 5$ nm:

$$R_S = \frac{\sum_{280\,nm}^{2500\,nm} R(\lambda) \cdot E_{sol}(\lambda) \cdot \Delta\lambda}{\sum_{280\,nm}^{2500\,nm} E_{sol}(\lambda) \cdot \Delta\lambda} \qquad [7.3]$$

Moreover, a solar mirror is performing only if its spectral reflectance is mostly specular (reflection along a single known direction, given by the Snell–Descartes law) and not diffuse (reflection in all directions). A specular reflection actually allows the precise positioning of the mirror so that the reflected solar flux is directed towards the receiver. The solar performance of a mirror therefore relies on its total (specular + diffuse) solar reflectance, calculated from its hemispherical spectral reflectance $R(\lambda)$, and also from its specular solar reflectance, calculated from its specular spectral reflectance.

The part of specular reflection of a surface is directly related to its roughness, and depends on the incident wavelength (Figure 7.2). In order to have a high specular solar reflectance, the roughness of the mirror surface must be very low, typically below 5 nm.

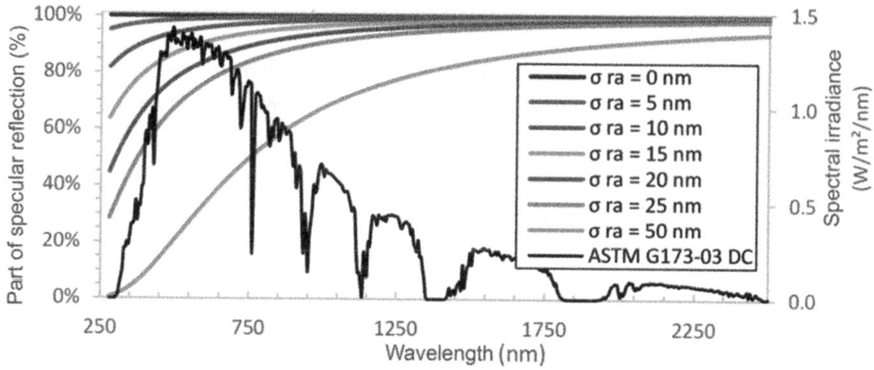

Figure 7.2. *Part of the specular reflection in the total reflection as a function of the incident wavelength (normal incidence) for various roughness values σ_{ra}. For a color version of this figure, see www.iste.co.uk/flamant/energy.zip*

7.3.2. *Materials and structures of solar mirrors*

The most reflective materials are metals. In particular, silver and aluminum are highly reflective all over the solar range (Figure 7.3). Theoretically speaking, they can reach a solar reflectance of 95.5% and 92.2%, respectively (pure metal without roughness). These two materials are therefore naturally part of the composition of most solar mirrors.

Reflectance being a surface property, it is not necessary to use a bulk silver or aluminum mirror, due to its lack of rigidity and stability under bad weather conditions and to its high cost in the case of silver. A thin metallic film of several dozen nanometers deposited on a glass substrate is indeed sufficient to achieve the same reflectance as the bulk metal (Figure 7.4).

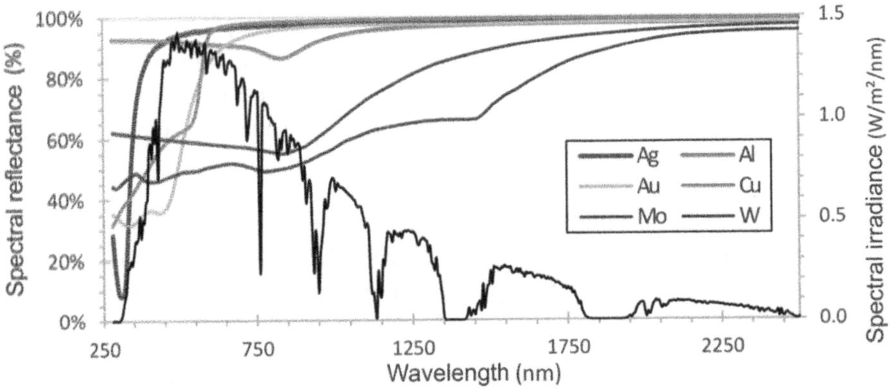

Figure 7.3. *Theoretical spectral reflectance of pure metals in the solar range. For a color version of this figure, see www.iste.co.uk/flamant/energy.zip*

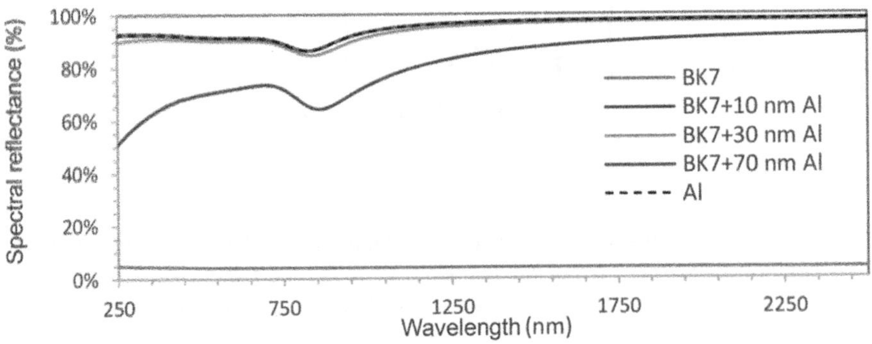

Figure 7.4. *Simulated spectral reflectance of pure Al and Al thin films of various thicknesses deposited on a BK7 glass slide. For a color version of this figure, see www.iste.co.uk/flamant/energy.zip*

7.3.2.1. *Silver-based mirrors*

Silver-based mirrors are the most common in concentrated solar power plants. Their structure is illustrated in Figure 7.5. A stack of thin films is deposited on a glass substrate, which is then turned over, so that the glass faces the solar radiation and protects the underlying thin films. This leads to optical losses, and also enhances the durability of the mirror. The first layer encountered by the radiation is a silver layer, the most reflective metal. A copper layer is behind the silver layer in order to reflect or absorb the UV radiation that is not reflected by silver, the latter becoming partially transparent below 400 nm (Figure 7.3). It also serves to protect silver against corrosion. Finally, paints are applied on the backside in order to protect the reflective layers, particularly against scratches.

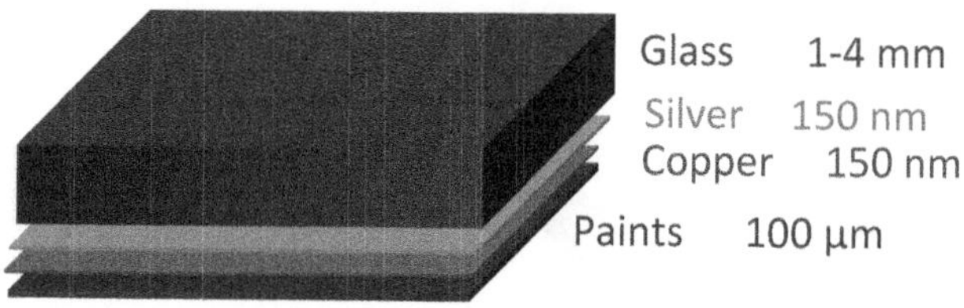

Figure 7.5. *Architecture of a silver-based solar mirror. For a color version of this figure, see www.iste.co.uk/flamant/energy.zip*

A variety of silver-based commercial solar mirrors (e.g. Rioglass, Flabeg) are installed on concentrating systems, reaching solar reflectances of the order of 93%. The main drawback of these mirrors is that they are heavy and fragile, because of the thick glass slide, which serves as substrate. They are also sensitive to breakage, particularly during mirror cleaning phases. Given their weight, they also represent an additional investment cost related to their transport and installation when a power plant is built. Finally, although small quantities of matter are necessary for manufacturing a mirror, silver remains a strategic metal for many other domains (optics, PV, photography, medical, jewelry, etc.) and its reserves are limited, which is an impediment to large-scale deployment of concentrated solar power. Hence, despite their excellent optical performances, alternative solutions to silver-based mirrors are sought after that are lighter, more flexible and employ less expensive and less critical materials.

7.3.2.2. *Aluminum-based mirrors*

Aluminum, which is the second most reflective material after silver (Figure 7.3), is proposed as an alternative to silver for the manufacturing of solar mirrors. Commercial solutions exist (e.g. Alanod, Almeco). A typical structure of an aluminum-based mirror is presented in Figure 7.6. Solar reflection is provided by the intermediate layer of pure aluminum. Other transparent layers serve essentially to chemically and physically protect this layer, which is very sensitive to oxidation. The substrate is a thin aluminum foil, making the mirror flexible, light, versatile and low cost. This mirror can then be fixed on a more rigid substrate during on-site installation, limiting transport costs. The major drawback of this type of mirror compared to silver-based mirrors is a lower solar performance, about 90%. It is also more fragile and less durable to this day. It is therefore fit for solar applications of moderate concentration, in which cost is considered more important than performance.

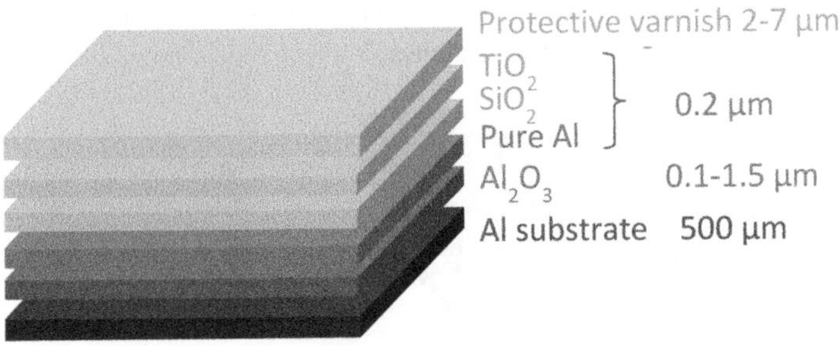

Figure 7.6. *Architecture of an aluminum-based mirror. For a color version of this figure, see www.iste.co.uk/flamant/energy.zip*

7.3.3. *Aging and durability of solar mirrors*

Concentrated solar power technologies, and particularly electric power plants, are generally installed in very sunny locations, for obvious productivity-related reasons. These are most often deserts, sometimes located at high altitude with brighter sunshine (e.g. Chile), and sometimes on coastal regions (e.g. Middle East). These environments are particularly aggressive for solar mirrors, due to the presence of erosive particles (sand, dust) often accelerated by winds and/or corrosive species (water vapor, atmospheric pollutants, salt sprays), as well as large temperature differences between day and night. Lengthy exposure to solar radiation, particularly ultraviolet (high energy) radiation, is obviously also causing mirror aging.

These difficult conditions often generate corrosion of the reflective metal layers (cut edge or pit corrosion), scratches, increase in surface roughness, cracks, etc. (Sutter et al. 2015). Moreover, these damages lead to the degradation of the optical performances of mirrors, particularly their specularity, and also their embrittlement which shortens their service life. These phenomena are well known (García-Segura et al. 2016, 2019; Sutter et al. 2018), so that nowadays there are standards for testing mirror resistance under representative weather conditions, as summarized in Table 7.1. These tests are used to certify mirrors. Moreover, regular cleaning of mirrors, besides maintaining high optical efficiency, is also an opportunity to remove the particles and impurities that may serve as precursors for localized degradation phenomena, such as pit corrosion. Finally, some manufacturers such as GlassPoint propose systems fully enclosed in glasshouses, in order to protect the mirrors from bad weather, particularly from soiling and the erosive effect of sand storms.

Type of test	Conditions
Damp heat	2,000 h 85°C 85% HR *(IEC 62108, Test 10.7)*
Salt spray	3,000,h 35°C 100% HR, NaCl (50 g/L, pH 6.5–7.2) *(ISO 9227)*
UV + humidity	2,000,h cyclic (4 h 60°C UV 340 nm + 4 h 50°C 100% RH) *(ISO 11507)*
Thermal cycling	150 cycles (-40°C – 65°C) + 2,000 h damp heat *(IEC 62108, Test 10.8)*

Table 7.1. *Standard tests for mirror aging*

7.4. Transparent components: protective glass

7.4.1. *Optical performance indicator: solar transmittance*

Transparent surfaces are present, in the form of sheets of glass (windows or tubes), in many concentrated solar technologies. Their main role is thermal insulation and/or physical protection of other elements, such as absorber components, or even sometimes reflective mirrors. Therefore, the impact they have on the optical performances of the components they protect must be as limited as possible. In other words, their transmittance must be high throughout the whole solar range. Hence, these windows are made of glass, a transparent, rigid and low-cost material. Nevertheless, their presence necessarily generates optical losses by absorption in the glass and by reflection at the two interfaces with the ambient environment. Indeed, the glass being mostly transparent, a large part of the incident radiation travels through it and a parasitic reflection occurs not only on its frontside, but also on its backside.

Hence, the performance criterion for these transparent surfaces is their solar transmittance T_S "for two sides". This quantity results from the integration over wavelength, according to equation [7.2], $f(\lambda)$ being the spectral transmittance for the two sides, $T_{2f}(\lambda)$. This takes into account the multiple reflections on the front and back sides, as well as the internal absorption of the glass $A(\lambda)$, such as illustrated in Figure 7.7. This yields equation [7.4] for $T_{2f}(\lambda)$ (normal incidence). It is important to note that if glass absorption is zero, the expression is reduced to the first term:

$$T_{2f}(\lambda) = [1 - R(\lambda)]^2 \cdot \left[\left(1 - A(\lambda)\right) + R(\lambda)^2 \cdot \left(1 - A(\lambda)\right)^3\right] \quad [7.4]$$

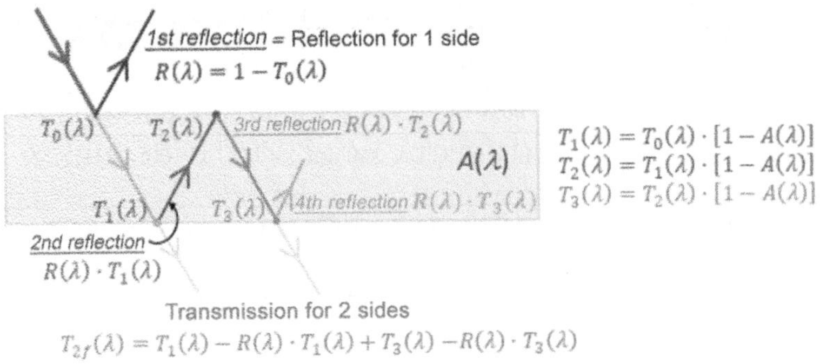

Figure 7.7. *Reflection at the two interfaces of a glass and transmission for two sides. For a color version of this figure, see www.iste.co.uk/flamant/energy.zip*

7.4.2. Materials and structures of protective glass

In order to have a high solar transmittance of these components, the optical losses by absorption and reflection should be reduced to the minimum. This is possible by adjusting the selection of the component materials and by employing surface functional thin films.

7.4.2.1. *Limitation of optical losses by absorption: glass selection*

Glass absorption is most often illustrated by its "internal transmittance" τ_i whose complement to unity it is. It is related to its (dimensionless) extinction coefficient k or to its absorption coefficient α (typically in cm^{-1}), and to its thickness d_{glass}, according to equation [7.5] for a given wavelength. The solar absorption A_S of the glass is obtained from equation [7.2], with $f(\lambda) = A(\lambda)$:

$$A(\lambda) = 1 - \tau_i(\lambda) = 1 - e^{-\alpha(\lambda) \cdot d_{glass}}\ ;\ \alpha(\lambda) = \frac{4\pi \cdot k(\lambda)}{\lambda} \qquad [7.5]$$

To limit this absorption, the extinction coefficient k of the glass must be small or zero over the wide range of wavelengths corresponding to the solar spectrum, and it should be thin while maintaining its mechanical strength. Thus, in practice, the solar absorptance of protective glass is not zero, as their thickness is of several millimeters, but it not significant. Among the many existing types of glass, those with a low iron content are less absorbing and are used particularly for silver-based mirrors (see section 7.3.2.1). The windows subjected to concentrated solar radiation are made from borosilicate crown glass, such as BK7. Their optical properties are adapted, with a low extinction coefficient k, and a refractive index n that is low and does not vary much with wavelength (small dispersion) (Grosjean et al. 2018).

Besides being transparent, they have a small thermal expansion coefficient and good resistance to thermal shocks. For example, a BK7 glass with a thickness of 1 mm and a refractive index n = 1.516 at 610 nm has a solar absorptance A_s = 0.27% over the range 280–2,500 nm.

7.4.2.2. *Limitation of optical losses by reflection: antireflective coatings*

The losses by reflection at the interface between the ambient environment – most often air or vacuum – and a material are due to their difference in refractive index n. On a naked glass, this phenomenon generates non-negligible optical losses. With n ranging between 1.5 and 1.9 for glasses, the losses by reflection range between 4 and 10% at each change of medium (air–glass or glass–air). For a glass through which solar radiation travels, this corresponds to 8–20% losses due to the double reflection, at the air–glass interface (frontside) and then at the glass–air interface (backside).

In order to limit these losses by reflection, thin films known as antireflective coatings (AR) are deposited on each side of the glass. Their principle relies on the creation of destructive interferences between the rays reflected at the AR/air (R_1) and glass/AR (R_2) interfaces (Figure 7.8), so that the resulting total reflection is canceled for a given wavelength.

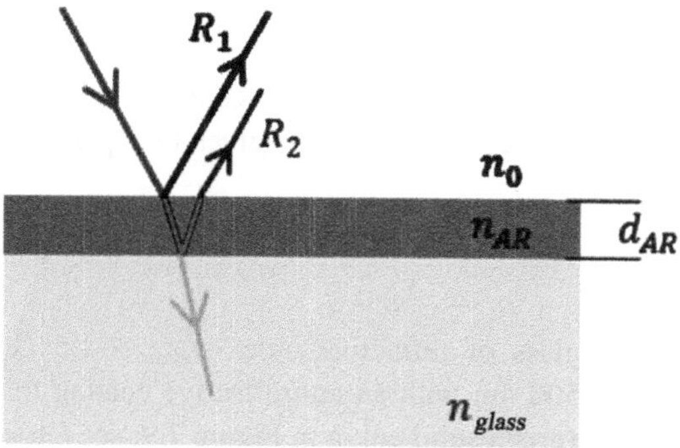

Figure 7.8. *Reflection at the interfaces of a glass coated with antireflective layer (AR). For a color version of this figure, see www.iste.co.uk/flamant/energy.zip*

This phenomenon appears when the reflected rays are out of phase: their path difference Δ is then as given by equation [7.6]:

$$\Delta = R_1 - R_2 = \lambda/2 + m\lambda \quad (m \in \mathbb{Z}^{0+}) \qquad [7.6]$$

Light travels twice through the antireflective coating, of thickness d_{AR}, to generate an outgoing reflected ray R_2, so that $R_1 - R_2 = 2d_{AR}$. Therefore, in order to cancel the reflection at wavelength λ, the thickness of the antireflective coating must be similar to that in equation [7.7]. The antireflective coatings are therefore quarter-wave plates:

$$d_{AR} = \lambda/4 + m\lambda/2 \quad (m \in \mathbb{Z}^{0+}) \qquad [7.7]$$

Total reflection resulting from the glass + AR set is classically deduced from Fresnel coefficients for the reflection at each interface encountered by the radiation (Furman and Tikhonravov 1992). These coefficients deriving from the Maxwell equations express the ratio of the amplitude of the wave reflected or transmitted at the interface between two media, to that of the incident wave. They depend on the complex refractive indices of the two media and the incidence angle of the radiation. Their mathematical development shows that in order to obtain an antireflective effect, the refractive index of the antireflective coating, n_{AR}, must have a particular relation with that of the glass, n_{glass}, and that of the ambient environment, n_0. This relation is illustrated by equation [7.8] in the simple case of a glass and an antireflective material, both non-absorbent ($k_{glass} = k_{AR} = 0$) and under normal incidence:

$$n_{AR} = \sqrt{n_0 \cdot n_{glass}} \qquad [7.8]$$

The resulting total reflection is then fully canceled at wavelength λ_0 if the thickness of the antireflective coating is the one given by equation [7.9]:

$$d_{AR} = \frac{\lambda_0}{4 \cdot n_{AR}} \qquad [7.9]$$

For example, for a glass of refractive index $n_{glass} = 1.5$ ($k_{glass} = 0$), a zero reflection is obtained at 500 nm with an antireflective coating of index $n_{AR} = 1.23$ and thickness of 102 nm (green solid curve in Figure 7.9; see Grosjean 2018). If n_{AR} deviates from the ideal (red curves), a reflection minimum is still obtained, but reflection is not zero. A variation of the thickness of the antireflective coating (green dotted curve) simply shifts this reflection minimum towards another wavelength.

In practice, there are no natural materials with refractive indices sufficiently low as to meet the condition [7.8]. The materials getting the closest to this ideal value are the fluorides, for example MgF_2 ($n = 1.38$ at 600 nm). Hence, in order to improve transmittance, porous materials are often employed: as the air has a very low index ($n = 1$), its incorporation in a dense material gives a porous material whose effective index is lower than that of the dense material. These materials must also be transparent in the solar spectrum, and resistant to weather conditions. The dielectric materials meet these needs. For example, porous SiO_2 is widely used in solar applications: a SiO_2 layer with 50% porosity and thickness of 125 nm on each side of a (1 mm) BK7 glass cancels the reflection at 610 nm and guarantees a high solar transmittance, T_S = 97.2% (Grosjean et al. 2018). This system can be deposited on large surfaces by sol–gel methods (dip coating) (San Vicente et al. 2011), based on precursors such as TetraEthyl OrthoSilicate (TEOS, $SiC_8H_{20}O_4$) (Xin et al. 2013), leading directly to a porous coating.

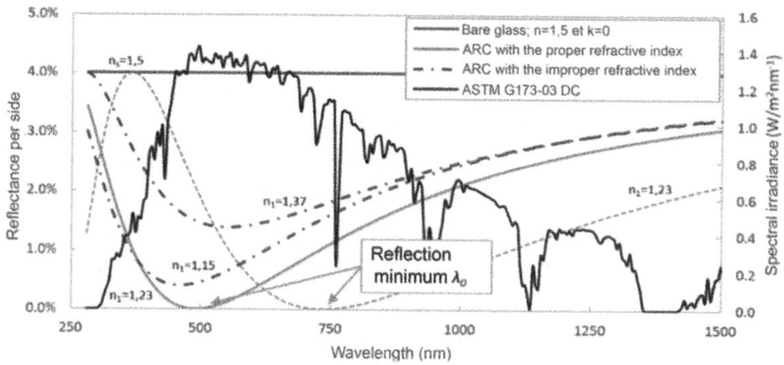

Figure 7.9. *Reflection on a glass with ideal or non-ideal antireflective coating. For a color version of this figure, see www.iste.co.uk/flamant/energy.zip*

To further improve T_S by several tenths, it is possible to stack several antireflective coatings, in order to extend the reflection minimum over a wider spectral range. For example, with a porous SiO_2 two-layer antireflective coating (porosities of 29% and 70%, thicknesses of 113 and 133 nm, respectively), the solar transmittance of a BK7 glass slide (1 mm) reaches 98.8% (+1.6 points). A compromise between complexity, cost and performance gain must however be considered.

7.4.2.3. *Complete components*

In practice, protective glass is often associated during the fabrication of absorber components that they protect and are sold together. In particular, the receivers of parabolic trough technologies are commercially available from various manufacturers, such as Rioglass, RoyalTech or Archimede Solar Energy. They are composed of an absorber tube (see section 7.5) inserted in a borosilicate glass tube with an antireflective coating, typically of porous SiO_2. Their solar transmittance ranges between 96% and 97%, depending on the manufacturer. Fresnel technologies also involve protective borosilicate glass with antireflective coating, but in this case, there is no standard receiver on the market; each operator develops its own receiver.

7.4.3. *Aging and durability of antireflective glasses*

The glasses and their antireflective coatings must resist to working conditions that are specific to concentrated solar power applications. They are by nature little affected by the concentrated solar radiation reflected by the mirrors, as they are largely transparent to it. On the contrary, the weather conditions, particularly erosive particles, water vapor and pollutants, can significantly degrade the porous antireflective coatings. For example, water vapor contamination causes an alteration of their optical performance: the refractive index of water (n = 1.32) being higher than that of the air (n = 1), the refractive index of the H_2O–material mixture is higher than that of the air–material mixture and the efficiency of the antireflective coating significantly decreases. For porous SiO_2, one of the current solutions is to withdraw the active hydroxyl (-OH) groups of the surface resulting from residual silanol (Si-OH) groups and replacing them by blocking trimethylsilyl groups ($Si\text{-}(CH_3)_3$). This additional surface treatment is conducted by dipping in a hexamethyldisilane (HMDS) bath (San Vicente et al. 2011). But HMDS is a flammable and toxic product. The bath must be sufficiently large to take in components of several meters, requiring a large quantity of product, which must be renewed regularly and retreated. Therefore, the associated environmental and financial cost is not negligible. Alternative solutions to this chemical treatment are expected, but they are still lacking.

7.5. Absorbing components: solar absorbers

7.5.1. *Optical performance indicators for solar absorbers*

7.5.1.1. *Solar absorptance*

In concentrating solar power technologies, the concentrated solar radiation is converted into heat by a solar absorber. The main optical function of the absorber is therefore, logically, to best absorb solar radiation. This capacity is illustrated by its solar absorptance α_S, defined by equation [7.10], deriving from expressions [7.1] and [7.2]. α_S is deduced from the spectral reflectance $R(\lambda)$, considering the absorber opaque ($Tr(\lambda) = 0$), and represents the ratio of the flux density absorbed by the material in W/m^2 (numerator) and the incident solar flux density in W/m^2 (denominator). α_S is therefore a dimensionless quantity:

$$\alpha_S = \frac{\int_{\lambda_1}^{\lambda_2} [1 - R(\lambda)] \cdot E_{sol}(\lambda) \cdot d\lambda}{\int_{\lambda_1}^{\lambda_2} E_{sol}(\lambda) \cdot d\lambda} \quad [7.10]$$

Most often, α_S is calculated over the range λ_1 = 0.28 µm – λ_2 = 2.5 µm, representing over 99% of the solar spectrum. The spectral reflectance of a material can be measured within this range with a single UV-Vis-NIR spectrophotometer, which is more convenient and less expensive for the developers of solar absorbers. The remaining 1% are emitted between 2.5 and 4 µm. The measurement of reflectance within this range requires an additional infrared spectrophotometer, but for more precision, λ_2 is sometimes considered equal to 4 µm.

7.5.1.2. *Thermal emittance*

The absorption of solar radiation heats the absorber, which transfers its heat to the heat transfer fluid flowing through it. Any heat that is not transferred to the fluid represents a thermal loss to the system. Most of these losses are radiative, as any body with a temperature above absolute zero emits radiation naturally: on the one hand, losses by conduction and most of all by convection are minimized by the very structure of the solar receiver (particularly the protective glass, under which the absorber is moreover sometimes under vacuum); on the other hand, conductive and convective losses depend linearly on temperature, while radiative losses depend on the fourth power of temperature, according to the Stefan–Boltzmann law on black body irradiance (equation [7.11], where σ = 5.67·10^{-8} W/m^2/K^4). This irradiance derives from the integration of the Planck law giving the spectral irradiance of the black body $M_\lambda^0(\lambda, T)$ at temperature T (equation [7.12] where h and k_B are the Planck and Boltzmann constants, respectively, and c is the speed of light):

$$M^0(T) = \sigma T^4 = \int_0^\infty M_\lambda^0(\lambda, T) \cdot d\lambda \quad [7.11]$$

$$M_\lambda^0(\lambda, T) = \frac{2\pi hc^2}{\lambda^5} \cdot \frac{1}{exp\left(\frac{hc}{\lambda k_B T}\right) - 1} \quad [7.12]$$

The secondary function of the solar absorber is to minimize these radiative losses to ensure a better efficiency of the solar-to-heat conversion. Its radiative performance is illustrated by its thermal emittance $\varepsilon(T)$ at temperature T (equation [7.13], resulting from expressions [7.1] and [7.2]). It is also deduced from the spectral reflectance $R(\lambda)$ considering the absorber opaque ($Tr(\lambda) = 0$). The spectral emittance of the material $E(\lambda)$ derives from the Kirchhoff law of radiation (equation [7.14]). According to the latter, at thermal equilibrium, if an incident radiation of a given wavelength λ and in a given direction θ is absorbed by a material, in order to maintain its temperature T constant, the material re-emits all this radiation at the same wavelength and in the same direction:

$$\varepsilon(T) = \frac{\int_{\lambda_1}^{\lambda_2} [1 - R(\lambda)] \cdot M_\lambda^0(\lambda, T) \cdot d\lambda}{\int_{\lambda_1}^{\lambda_2} M_\lambda^0(\lambda, T) \cdot d\lambda} \quad [7.13]$$

$$A(\lambda, \theta, T) = E(\lambda, \theta, T) = 1 - R(\lambda, \theta, T) \quad [7.14]$$

The thermal emittance represents the ratio of the flux density emitted by the material at temperature T (W/m^2) to the flux density emitted by an ideal black body at the same temperature (W/m^2). This is also a dimensionless quantity. The black body emission at the operating temperatures of concentrated solar power systems (e.g. 200–800°C) is in the near and mid-infrared (Figure 7.10). Hence, values of $\lambda_1 = 0.28$ µm and $\lambda_2 = 25 - 30$ µm are generally chosen, corresponding to the upper limit of most of infrared spectrophotometers measuring $R(\lambda)$.

A high performance absorber would have the lowest possible thermal emittance $\varepsilon(T)$, so that its total irradiance $M(T) = \varepsilon(T) \cdot \sigma T^4$, representing the radiative losses of the material, is also low.

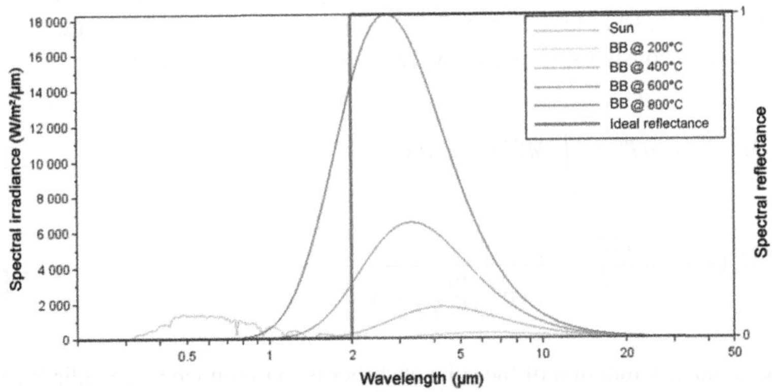

Figure 7.10. *Spectral irradiances of Sun and black body (BB) at various temperatures; ideal spectral reflectance of a solar selective absorber. For a color version of this figure, see www.iste.co.uk/flamant/energy.zip*

7.5.1.3. *Spectral selectivity*

In order to combine high solar absorptance and low thermal emittance, a solar absorber must ideally be spectrally selective, meaning that its optical behavior changes drastically depending on the spectral range (Figure 7.10). It thus has a spectral reflectance $R(\lambda)$ that is the lowest possible in the solar range, to maximize $[1 - R(\lambda)]$ in equation [7.10], and the highest possible in the infrared range, to minimize $[1 - R(\lambda)]$ in equation [7.13]. As the temperature of the absorber increases, the irradiance of the black body shifts towards the solar range (Wien displacement law). The two previous requirements are therefore sometimes incompatible, and a choice must be made between high absorptance and low emittance, using (through the absorber structure) the position (wavelength) of the change in optical behavior, known as *cut-off*.

7.5.1.4. *Heliothermal efficiency*

The compromise between high absorptance and low emittance must be guided by the influence of these two parameters on the solar-to-heat conversion efficiency, or heliothermal efficiency η_h. The latter depends on:

– the solar flux density (W/m^2) incident on the absorber: $\Phi_{conc} = C \cdot I$, where C is the solar concentration ratio and I is the direct solar irradiance at the ground ($\approx$ 1,000 W/m^2);

– the solar flux density (W/m^2) actually absorbed by the absorber, depending on its solar absorptance: $\Phi_{abs} = \alpha_s \cdot \Phi_{conc} = \alpha_s \cdot C \cdot I$;

– the flux density emitted by the hot absorber, depending on its temperature and on its thermal emittance (radiative thermal losses): $\Phi_{emit}(T) = \varepsilon(T) \cdot \sigma T^4$, in first approximation. For more precision, the radiative exchange with the ambient environment at temperature T_0, or even with the sky, can also be taken into consideration. In this case, $\Phi_{emit}(T) = \varepsilon(T) \cdot \sigma(T^4 - T_0^4)$.

The heliothermal efficiency is then given by equation [7.15]. It corresponds to the ratio of (absorbed flux – emitted/lost flux) and the concentrated solar flux. It depends on temperature T. Thermal losses by conduction and convection (depending on the first power of T) are considered negligible compared to radiative thermal losses (in T^4) and are not often taken into account in the calculation of heliothermal efficiency:

$$\eta_h(T) = \frac{\Phi_{abs} - \Phi_{emit}(T)}{\Phi_{conc}} = \alpha_s - \frac{\varepsilon(T) \cdot \sigma(T^4 - T_0^4)}{C \cdot I} \qquad [7.15]$$

As an example, these various flux densities are illustrated in Figure 7.11, at $C = 50$, $T = 500°C$, $I = 894\ W/m^2$ ($\Phi_{conc} = 45\ kW/m^2$), $T_0 = 20°C$, for:

– a solar absorber without spectral selectivity (non-selective, Pyromark 2500 painting): $\alpha_s = 0.97$, $\varepsilon(500°C) = 0.75$, $\Phi_{abs} = 43\ kW/m^2$, $\Phi_{emit} = 15\ kW/m^2$, $\eta_h(500°C) = 0.63$;

– a modeled absorber with spectral selectivity (selective): Fe / W / 25%W – Al_2O_3 / 22%W – Al_2O_3 / Al_2O_3 (1 mm / 160 nm / 60 nm / 50 nm / 75 nm), $\alpha_s = 0.94$, $\varepsilon(500°C) = 0.13$, $\Phi_{abs} = 41\ kW/m^2$, $\Phi_{emit} = 3\ kW/m^2$, $\eta_h(500°C) = 0.87$. In this case, the emitted flux is significantly reduced. Consequently, the heliothermal efficiency is higher (+0.24 points).

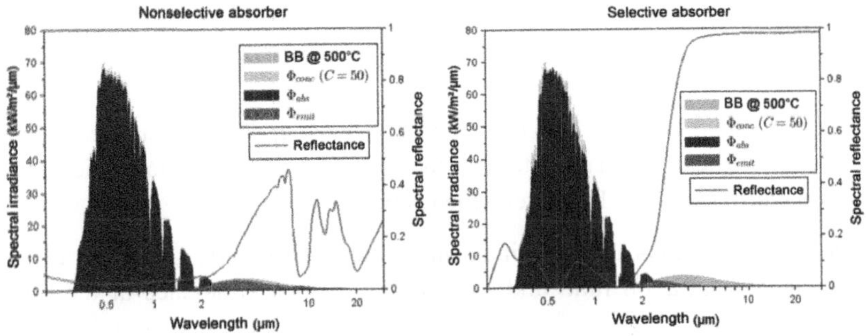

Figure 7.11. *Spectral reflectance (green) and flux densities concentrated (yellow), absorbed (blue) and emitted at 500°C (red), for two types of absorbers. For a color version of this figure, see www.iste.co.uk/flamant/energy.zip*

The analysis of expression [7.15] reveals the following:

– radiative losses $\Phi_{emit}(T)$ increase with T, so that $\eta_h(T)$ decreases when T increases (Figure 7.12). Working at higher concentration C increases the solar contribution and decreases the impact of radiative losses on the efficiency; hence, for $C = 500$ in the example in Figure 7.11 (other conditions maintained), the heliothermal efficiencies of the two absorbers become equal (0.93);

– $\eta_h(T)$ is significantly more sensitive to α_s than to $\varepsilon(T)$: $\partial\eta_h/\partial\alpha_s = 1$, hence if α_s increases by 10% (absolute scale), η_h also increases by 10%; $\partial\eta_h/\partial\varepsilon = -\sigma T^4/CI$, hence for [$T = 400°C$, $C = 100$, $I = 1{,}000\ W/m^2$], $\partial\eta_h/\partial\varepsilon = -0.116$ and if $\varepsilon(T)$ decreases by 10% (absolute scale), η_h increases only by 1.2%. If C is increased to 200, η_h increases only by 0.6%. It is therefore often more advantageous, if a compromise must be reached between the two, to favor a high α_s to the detriment of a low $\varepsilon(T)$, particularly at high concentration.

These considerations are illustrated in Figure 7.12, showing the conversion efficiencies depending on temperature, at two solar concentration ratios ($C = 50$, $C = 500$)[1], for a standard non-selective absorber (Pyromark 2500, $\alpha_S = 0.96$, $\varepsilon = 0.85$; see Ho et al. 2013) and for a standard selective absorber ($\alpha_S = 0.96$, $\varepsilon(T) = 0.062 + 2 \cdot 10^{-7} \cdot T^2 \approx 0.06 - 0.26$; see Burkholder and Kutscher 2009). It is clear that in all cases, the selectivity of the absorber has a positive impact on the heliothermal efficiency, slowing down its decrease with temperature (reduction of radiative losses). It is therefore always more advantageous to use a selective absorber, to the extent that it is possible. Spectral selectivity is all the more beneficial at low concentration, starting at low temperatures, as solar contribution is less, and radiative losses are compensated less efficiently.

In a concentrated solar power plant, the solar-to-heat conversion stage is followed by a heat-to-mechanical work conversion stage (most often in order to drive a steam turbine and generate electricity). This second stage is governed by a thermodynamic efficiency, for example, that of Chambadal–Novikov ($\eta_{thermo} = 1 - \sqrt{T_0/T}$) (green curve in Figure 7.12), which increases with temperature. Therefore, the global efficiency of a power plant (blue curve in Figure 7.12), a product of the heliothermal and thermodynamic conversion efficiencies[2], passes through a maximum at an optimal temperature. A selective absorber can be used not only to increase the value of the overall efficiency at a given temperature, but also to increase the value of the optimal temperature, and therefore the maximal reachable global efficiency, particularly at low solar concentration.

These theoretical examples highlight the difficulty of determining the optimal optical function of solar absorbers. It depends on the effect of numerous factors, related to the operating conditions of the absorber: temperatures of the heat transfer fluid, of the absorber surface and of the ambient environment, solar concentration level, etc. This optical function is however essential for the application, as it has a very strong impact on the overall performance of the installation, and must therefore be optimized with discernment.

---

1 It is important to note that certain values in Figure 7.12 are forbidden, as in practice it is not possible to obtain high temperatures ($T > 500°C$) at low concentration ($C = 50$), or to maintain low temperatures ($T < 300°C$) at high concentration ($C = 500$). Nevertheless, Figure 7.12 illustrates the actual tendencies and particularly the interest of selective absorbers.

2 These two efficiencies are also multiplied by a “reflection efficiency” of the concentrating system, taking into account the solar reflectance of the mirror, the losses due to errors of pointing towards the receiver, etc. In the present example, this efficiency is considered equal to 1.

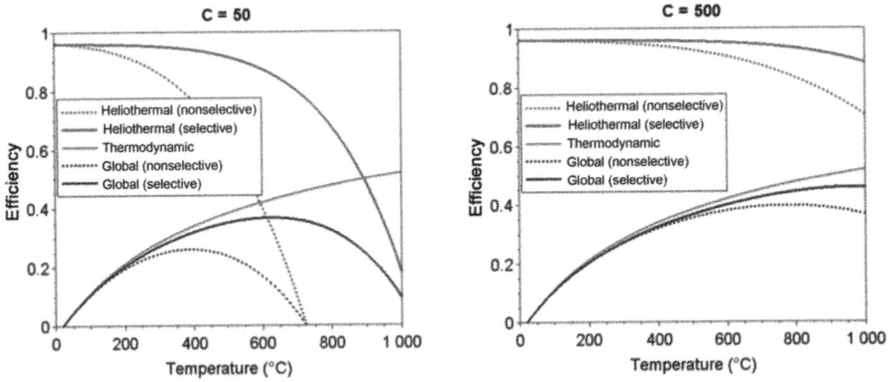

Figure 7.12. *Efficiencies of solar-to-heat (heliothermal) conversion, heat-to-work (thermodynamic) conversion and solar-to-work (global) conversion depending on the operating temperature, for two concentration ratios C. For a color version of this figure, see www.iste.co.uk/flamant/energy.zip*

7.5.2. Materials and structures of solar absorbers

7.5.2.1. *Selective absorbers*

No material by itself is able to meet the demanding and adaptive criteria required for the optical function of solar selective absorbers. In practice, they are made from several materials with complementary optical behaviors, associated in complex architectures, which must be finely optimized in order to have a good optical performance. Classically, the absorbers are composed of metallic receivers from stainless steel or other high-temperature alloys (such as Inconel), through which the heat transfer fluid flows. These metallic substrates ensure the mechanical functionalities (rigidity, creep strength, etc.) and thermal functionalities (conduction and thermal transfer to the fluid). They are coated with a stack of thin films providing optical functionalities, referred to as selective coating (Figure 7.13).

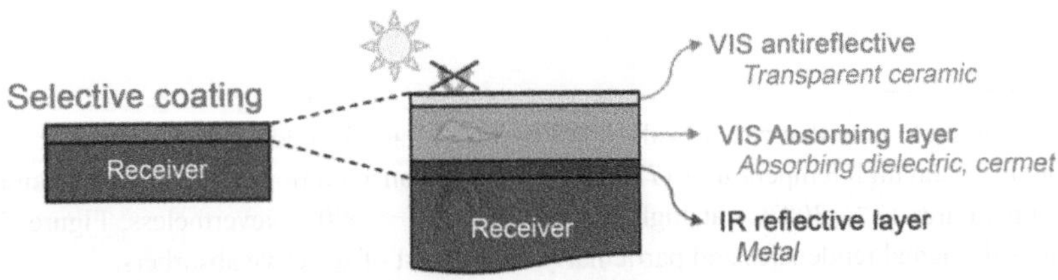

Figure 7.13. *Architecture of a solar selective absorber coating. For a color version of this figure, see www.iste.co.uk/flamant/energy.zip*

Metals are naturally reflective; therefore, they are not emissive, in the domain of infrared thermal emission (Figure 7.3). The low thermal emittance of the absorbers can thus be ensured either by the metallic substrate itself, or most often by a sublayer of pure metal, more reflective than the one in the substrate. The latter blocks the infrared radiation emitted by the hot receiver and sends it back to it, generating a greenhouse effect and significantly reducing radiative losses.

As for the solar absorption, it is ensured by an absorbing system, coated with an adapted antireflective layer (section 7.4). This absorbing system is composed of several layers, generally associating a metal and a dielectric or a semiconductor. There are various types; the most commonly used being the following:

– Metal-ceramic composites or cermets. A ceramic matrix ensures chemical and thermal stability. Metallic inclusions are used to adapt the complex refractive index of the composite and make it absorbing. For a better absorption, two cermets with different composition are used (HMVF and LMVF, standing for high and low metal volume fraction, respectively). This way, the refractive index decreases continuously between the substrate (metal, high n) and the ambient environment (air/vacuum, n = 1), creating a gradual accommodation and trapping the incident radiation (Figure 7.14), which ensures its absorption.

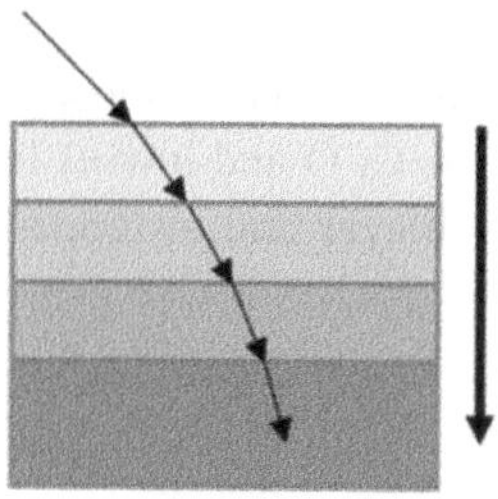

Figure 7.14. *Accommodation of refractive indices and trapping of incident radiation. For a color version of this figure, see www.iste.co.uk/flamant/energy.zip*

The commonly used cermets include, for example, Al-AlN, Cr-Cr_2O_3, Mo-Al_2O_3, Mo-SiO_2, W-Al_2O_3 and W-AlN (Kennedy 2002; Selvakumar and Barshilia 2012; Xu et al. 2020). These cermets are inserted between an infrared reflective sublayer, often composed of the same metal as the cermet inclusion, and an upper antireflective layer, often composed of a transparent dielectric that is the matrix of the cermet (Figure 7.15). The selective absorber in Figure 7.11 is an example. The dielectric is often an oxide or a metallic nitride, sometimes resulting from the same metal as the one in the cermet inclusions. The manufacturing of the absorber by classical vacuum plasma techniques is simplified by these practical choices (simple and reactive cathode sputtering of the same metallic targets), and the associated costs are reduced.

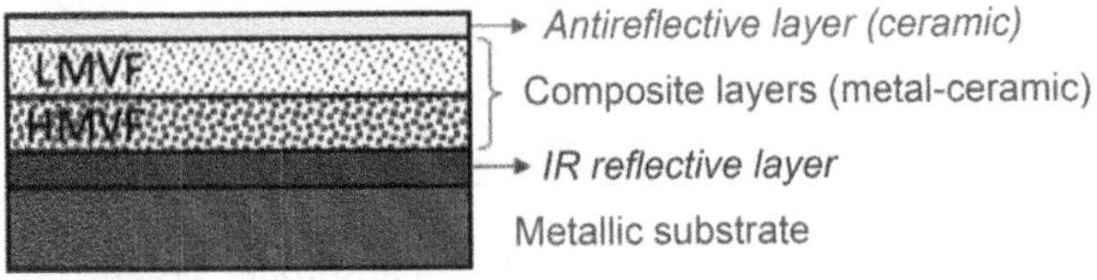

Figure 7.15. *Classical architecture of a composite absorber. For a color version of this figure, see www.iste.co.uk/flamant/energy.zip*

– Tandem absorbers, associating two layers of materials with different optical behavior: one with a behavior tending towards that of a metal, the other one towards that of a semiconductor. The oxides, nitrides, carbides, oxynitrides, oxycarbides, silicides, etc. of transition metals are materials of choice for tandem absorbers. There are many such examples (Xu et al. 2020): $TiAlN_x/TiAlN_y$, TiAlN/TiAlON, TiAlN/TiAlSiN, TiAlSiN/TiAlSiON, TiAlN/AlON, TiAlC/TiAlCN/TiAlSiCN, HfMoN/HfON, NbTiON/NbTiON, AlSiN/AlSiON, NbAlN/NbAlON, CrMoN/CrON, etc. Once more, a gradient of refractive index is created (Figure 7.14) by varying the composition of the two layers to obtain either a metallic behavior (low rate of N, O, C and high n) for the layer closest to the substrate, or a semiconductor behavior (high rate of N, C, O and low n) for the upper layer of the tandem. Similarly, an IR reflective sublayer and an upper antireflective layer are often used (Figure 7.16).

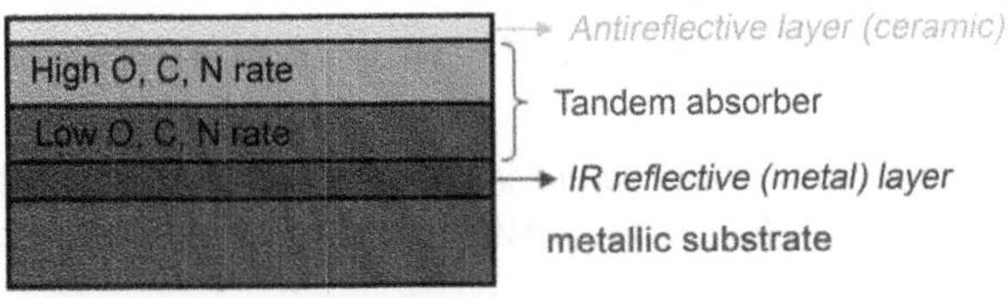

Figure 7.16. *Classical architecture of a tandem absorber. For a color version of this figure, see www.iste.co.uk/flamant/energy.zip*

In practice, cermet-based absorbers are the most frequently encountered, particularly for parabolic trough technologies, where the receiver is maintained under vacuum inside a glass tube. The selective coating is thus protected against oxidation, a major cause of its degradation. It is the case of receivers commercialized by Rioglass, RoyalTech, Archimede Solar Energy, etc. Tandem absorbers are still in the research and development stage, and are expected to be candidates of choice for future applications, when the absorber operates at high temperature in air, such as low-cost Fresnel systems.

There are also other configurations of selective solar absorbers, but they are less frequently used, due to their more complex design, such as the dielectric-metal-dielectric (DMD) multilayer absorbers, whose principle relies on the generation of destructive interferences between the reflections at the interfaces between layers. The best known example is the "AMA" (Bogaerts 1983). This historic absorber was an $Al_2O_3/Mo/Al_2O_3$ sequence. Since then, many studies have used various metals (Ag, Al, Cu, Cr, Mo, Ni, Ti, W, Pt) and dielectrics (Al_2O_3, AlN, SiO_2, MgF_2, ZnS, Si_3N_4, SiCH) (Selvakumar and Barshilia 2012).

In all cases, the optical performance is very sensitive to the "variables" of the architecture of the selective absorber, which are the thicknesses and compositions of the layers of the stack. For example, for the cermets (Figure 7.15), solar absorptance increases with its thickness and its metal volumetric concentration, similar to the emittance, as the cermet also absorbs the infrared radiation coming from the Sun. The thickness and the porosity of the antireflective layer must be readapted for each new configuration of the cermet. Hence, a preliminary stage of optical simulation and optimization is essential in order to optimize these variables and determine the design with the highest performances. From the complex refractive indices of the constituent materials, the Fresnel coefficients can be calculated: they give access to reflections and transmissions at the interfaces. Then, the transfer matrix method gives the resulting reflection and transmission for the complete stack (Furman and Tikhonravov 1992), and by extension its performance criteria (solar absorptance, thermal emittance, heliothermal efficiency). An optimization algorithm (Levenberg–Marquardt, genetic, etc.) can then be applied to test ranges of variables in order to find their optimum. Often the optimization is simply intended to maximize solar absorptance and minimize thermal emittance. For a given solar application, when the operating conditions are known, it is however more relevant to aim for maximal heliothermal efficiency. This contributes to finding the best compromise between the two parameters α_S and $\varepsilon(T)$ and to giving the algorithm more freedom to find the optimal design, notably by varying the *cut-off* position (Figure 7.10) and by allowing higher emittance values, when this does not affect the efficiency.

7.5.2.2. *Solar absorbers for high concentrations and temperatures*

For the applications at higher concentrations and temperatures ($T > 600°C$), when the absorber must operate under air, such as solar towers, the previous solutions are difficult to consider, as they reach their limit of thermal and chemical stability. Moreover, spectral selectivity is less useful in this case, for the already mentioned reasons. The focus is then on solar absorptance, to the detriment of the emittance, and also on the resistance to high temperatures. The metallic substrates of the receivers are also subjected to high temperatures under air and corrosive fluids: for example, nickel-based superalloys, such as Inconels, are commonly used for this type of application. Concerning the optical function, high-temperature absorbing paints are used, particularly Pyromark 2500, whose optical behavior is illustrated in Figure 7.11 (left). Their solar absorptance and also thermal emittance are very high ($\alpha_s = 0.97$, $\varepsilon(500°C) = 0.75$) as they also strongly absorb in the infrared range. The resulting radiative losses are high (e.g. $\Phi_{emit}(500°C) = 15$ kW/m^2), but negligible in comparison with concentrated and absorbed solar flux densities, which are very high at high concentration (e.g. $\Phi_{abs} = 430$ kW/m^2 at $C = 500$). This matt black paint, based on silicone, is stable up to 1100°C under air. It is applied by spray on thicknesses of the order of 10–15 μm and requires a thermal pretreatment for stabilization, and regular application (every three to five years approximately) as it undergoes unsticking due to usage, which increases maintenance and solar electricity generation costs (Boubault et al. 2016).

Higher performance alternatives, in terms of spectral selectivity (but tolerating a higher emittance than at low concentration) and durability at high temperature, are the object of active research:

– In the form of thin films on metallic substrates: spinels (e.g. $CuCr_2O_4/CuFeMnO_4$; see Kim et al. 2016), carbon nanotubes in ceramic matrices (Abendroth et al. 2015), naturally selective refractive ceramics (e.g. carbides and borides of transition metals such as Zr, Ti, Ta, Hf; see Gao et al. 2019), etc.

– In bulk form, the receiver itself is no longer made of metal, but of a selective refractory ceramic (the above list; see also Sciti et al. 2013; Charpentier and Caliot 2017) or not (e.g. SiC), or an association of both (e.g. SiC-TiC; see Aréna et al. 2020); or in surface/2D form such as plates or tubes (Leray et al. 2017), or in volume/3D form such as foams (Mey-Cloutier et al. 2016) or honeycombs (Ali et al. 2020). Moreover, such receivers are more chemically and mechanically resistant to new generation heat transfer fluids that are more erosive (particles) or corrosive (chlorides, liquid sodium). On the contrary, they are more sensitive to fracture.

7.5.3. *Aging and durability of solar absorbers*

All the absorber materials must be able to operate, ideally for several decades, under extreme conditions: concentrated solar radiation, high temperatures, short period and high amplitude thermal cycling (cloudy spells, day/night cycles), contact with corrosive fluids (air, water vapor, atmospheric pollutants, molten salts, etc.), etc. These conditions induce many thermal and/or thermomechanical phenomena: oxidation, corrosion, atomic diffusion, densification, cracking, fatigue, creep, etc.

These complex and strongly coupled processes may damage the optical properties, and therefore must be studied prior to any concrete application.

Unlike solar mirrors, there is no standard protocol for the certification of absorbers. Therefore, all the developers of absorbers test the durability of their materials under different and relatively conservative conditions, typically for several dozen to hundreds of hours in an electric furnace at temperatures close to the operating conditions under consideration (Xu et al. 2020). Few are those who push the limit of their tests up to several thousand hours and/or apply intense thermal cycling. Some, such as IEA Task X, adapt the aging protocols from low-temperature thermal solar power (without concentration; see Carlsson et al. 2004). Others apply accelerated aging protocols based on the hypothesis that the optical degradations of absorbers are only due to thermally induced phenomena following an Arrhenius law: materials are tested at several temperatures beyond the considered operating temperature ($T_{acc} = T_{op}$ + 100–200°C), a degradation activation energy is determined and a prediction of the service life may even be proposed (Raccurt et al. 2017). Finally, some try to get closer to the actual operating conditions, by testing the absorber materials under concentrated solar flux which is either simulated (Dan et al. 2019) or actual (Boubault et al. 2014; Reoyo-Prats et al. 2019), notably to apply rapid and high amplitude thermal cycling, which is not possible with classical furnaces.

The problem of absorber durability, both for substrate materials and for selective and non-selective coatings, is absolutely essential and becomes increasingly central as concentrating solar power technologies tend towards high temperatures. It is and will be however difficult to harmonize knowledge and aging protocols for solar absorbers, due to the wide diversity of proposed materials and structures, and to the limited technical capacities available to developers, notably for tests under concentrated solar flux. The sharing and availability of unique dedicated installations within national projects (e.g. Equipex SOCRATE and DURASOL in France) and international projects (e.g. EU SFERA projects) represent a first step towards this harmonization.

Most of the solutions proposed for improving the durability of absorbers rely on the preselection of high temperature materials known in other applications (aeronautics, turbines, etc.) and their incremental microstructural adaptation, which consumes time, materials and energy. The demands of the application are so high and the aging phenomena involved so complex that it is, however, difficult to this day to consider other reliable strategies for the development of solar absorbers.

7.6. Conclusion

Concentrated solar power involves a vast number of materials with complementary properties: mechanical properties (e.g. stiffness, resistance to scratches), chemical properties (e.g. resistance to oxidation/corrosion) or optical properties (reflection, transmission, absorption). Most often, a solid substrate material of reasonable price is used (e.g. glass, steel), its surface being functionalized, generally by stacks of thin films of other materials, in order to modify the optical behavior. Indeed, the latter is determined by the first few hundred nanometers encountered by the incident radiation. This results in a mirror that is very efficient in reflecting solar radiation using a thin film of silver or aluminum deposited on a glass substrate; reducing the natural reflection of glass by using a porous SiO_2 thin film in order to improve its transmission in the solar range; making a naturally reflective and highly absorbant metallic substrate in the solar range, thanks to a complex stack of thin films that accommodate and trap solar radiation, and also low emissive in order to limit radiative thermal losses, a case that is known as spectral selectivity.

These thin films must be finely optimized in order to get the intended performance, the latter being very sensitive to thicknesses and compositions of the component materials, either bulk or in thin films. An optical design and simulation stage is necessary. This is often the case in optical applications, the difference here being that materials are additionally subjected to highly demanding operating conditions, in terms of temperatures, received energy fluxes, erosive and corrosive environments, etc. The performance of the materials for concentrated solar power is therefore assessed not only based on their optical properties, which must be perfectly controlled and adapted to the application, but also and most of all by maintaining these properties in the long term and under these potentially damaging conditions. Durability is therefore an absolutely essential point for the materials for concentrated solar power applications, and it is sometimes preferable to choose a component architecture with lower initial optical performance, but with better chances to maintain this performance over time.

7.7. References

Abendroth, T., Althues, H., Mäder, G., Härtel, P., Kaskel, S., Beyer, E. (2015). Selective absorption of carbon nanotube thin films for solar energy applications. *Solar Energy Materials and Solar Cells*, 143, 553–556.

Ali, M., Rady, M., Attia, M.A.A., Ewais, E.M.M. (2020). Consistent coupled optical and thermal analysis of volumetric solar receivers with honeycomb absorbers. *Renewable Energy*, 145, 1849–1861.

Aréna, H., Coulibaly, M., Soum-Glaude, A., Jonchère, A., Arrachart, G., Mesbah, A., Pradeilles, N., Vandenhende, M., Maître, A., Deschanels, X. (2020). Effect of TiC incorporation on the optical properties and oxidation resistance of SiC ceramics. *Solar Energy Materials and Solar Cells*, 213, 110536.

Bogaerts, W.F. (1983). Materials for photothermal solar energy conversion. *Materials Science*, 18, 2847–2875.

Boubault, A., Claudet, B., Faugeroux, O., Olalde, G. (2014). Aging of solar absorber materials under highly concentrated solar fluxes. *Solar Energy Materials and Solar Cells*, 123, 211–219.

Boubault, A., Ho, C.K., Hall, A., Lambert, T.N., Ambrosini, A. (2016). Levelized cost of energy (LCOE) metric to characterize solar absorber coatings for the CSP industry. *Renewable Energy*, 85, 472–483.

Burkholder, F. and Kutscher, C. (2009). Heat loss testing of Schott's 2008 PTR70 parabolic trough receiver. Report No. NREL/TP--550-45633, 1369635 [Online]. Available at: http://www.osti.gov/servlets/purl/1369635/ [Accessed 25 December 2020].

Carlsson, B., Möller, K., Köhl, M., Heck, M., Brunold, S., Frei, U., Marechal, J.-C., Jorgensen, G. (2004). The applicability of accelerated life testing for assessment of service life of solar thermal components. *Solar Energy Materials and Solar Cells*, 84(1), 255–274.

Charpentier, L. and Caliot, C. (2017). The impact of the oxidation on the optical properties of TaC. *Solar Energy Materials and Solar Cells*, 171, 16–23.

Dan, A., Soum-Glaude, A., Carling-Plaza, A., Ho, C.K., Chattopadhyay, K., Barshilia, H.C., Basu, B. (2019). Temperature- and angle-dependent emissivity and thermal shock resistance of the W/WAlN/WAlON/Al2O3-Based Spectrally Selective Absorber. *ACS Applied Energy Materials*, 2(8), 5557–5567.

Fernández-García, A., Sutter, F., Montecchi, M., Sallaberry, F., Heimsath, A., Heras, C., Le Baron, E., Soum-Glaude, A. (2018). Parameters and method to evaluate the reflectance properties of reflector materials for concentrating solar power technology. *Solar PACES Official Reflectance Guideline Version 3.0* [Online]. Available at: https://www.solarpaces.org/csp-research-tasks/task-annexes-iea/task-iii-solar-technology-and-advanced-applications/reflectance-measurement-guidelines/ [Accessed 10 September 2018].

Furman, S.A. and Tikhonravov, A.V. (1992). *Basics of Optics of Multilayer Systems*. Editions Frontières, Gif-sur-Yvette.

Gao, X.-H., Qiu, X.-L., Li, X.-T., Theiss, W., Chen, B.-H., Guo, H.-X., Zhou, T.-H., Liu, G. (2019). Structure, thermal stability and optical simulation of ZrB2 based spectrally selective solar absorber coatings. *Solar Energy Materials and Solar Cells*, 193, 178–183.

García-Segura, A., Fernández-García, A., Ariza, M.J., Sutter, F., Valenzuela, L. (2016). Durability studies of solar reflectors: A review. *Renewable and Sustainable Energy Reviews*, 62, 453–467.

García-Segura, A., Fernández-García, A., Ariza, M.J., Sutter, F., Diamantino, T.C., Martínez-Arcos, L., Reche-Navarro, T.J., Valenzuela, L. (2019). Influence of gaseous pollutants and their synergistic effects on the aging of reflector materials for concentrating solar thermal technologies. *Solar Energy Materials and Solar Cells*, 200, 109955.

Grosjean, A. (2018). Etude, modélisation et optimisation de surfaces fonctionnelles pour les collecteurs solaires thermiques à concentration. PhD Thesis, Université de Perpignan, Perpignan.

Grosjean, A., Soum-Glaude, A., Neveu, P., Thomas, L. (2018). Comprehensive simulation and optimization of porous SiO_2 antireflective coating to improve glass solar transmittance for solar energy applications. *Solar Energy Materials and Solar Cells*, 182, 166–177.

Ho, C.K., Mahoney, A.R., Ambrosini, A., Bencomo, M., Hall, A., Lambert, T.N. (2013). Characterization of Pyromark 2500 for high-temperature solar receivers. *ASME 2012 10th International Conference on Fuel Cell Science, Engineering and Technology. American Society of Mechanical Engineers Digital Collection*, 509–518.

Kennedy, C.E. (2002). Review of Mid- to high-temperature solar selective absorber materials. Report No. NREL/TP-520-31267, National Renewable Energy Laboratory, Golden, CO.

Kim, T.K., Van Saders, B., Caldwell, E., Shin, S., Liu, Z., Jin, S., Chen, R. (2016). Copper-alloyed spinel black oxides and tandem-structured solar absorbing layers for high-temperature concentrating solar power systems. *Solar Energy*, 132, 257–266.

Leray, C., Ferriere, A., Toutant, A., Olalde, G., Peroy, J.-Y., Chéreau, P., Ferrato, M. (2017). Design and proof of concept of an innovative very high temperature ceramic solar absorber. *International Conference on Concentrating Solar Power and Chemical Energy Systems (SolarPACES 2016)*, Abu Dhabi, 030032.

Mey-Cloutier, S., Caliot, C., Kribus, A., Gray, Y., Flamant, G. (2016). Experimental study of ceramic foams used as high temperature volumetric solar absorber. *Solar Energy*, 136, 226–235.

Raccurt, O., Matino, F., Disdier, A., Braillon, J., Stollo, A., Bourdon, D., Maccari, A. (2017). In air durability study of solar selective coating for parabolic trough technology. *AIP Conference Proceedings*, 1850(1), 130010.

Reoyo-Prats, R., Carling Plaza, A., Faugeroux, O., Claudet, B., Soum-Glaude, A., Hildebrandt, C., Binyamin, Y., Agüero, A., Meißner, T. (2019). Accelerated aging of absorber coatings for CSP receivers under real high solar flux – Evolution of their optical properties. *Solar Energy Materials and Solar Cells*, 193, 92–100.

San Vicente, G., Bayón, R., Germán, N., Morales, A. (2011). Surface modification of porous antireflective coatings for solar glass covers. *Solar Energy*, 85(4), 676–680.

Sciti, D., Silvestroni, L., Mercatelli, L., Sans, J.-L., Sani, E. (2013). Suitability of ultra-refractory diboride ceramics as absorbers for solar energy applications. *Solar Energy Materials and Solar Cells*, 109, 8–16.

Selvakumar, N. and Barshilia, H.C. (2012). Review of physical vapor deposited (PVD) spectrally selective coatings for mid- and high-temperature solar thermal applications. *Solar Energy Materials and Solar Cells*, 98, 1–23.

Sutter, F., Fernández-García, A., Heller, P., Anderson, K., Wilson, G., Schmücker, M., Marvig, P. (2015). Durability testing of silvered-glass mirrors. *International Conference on Concentrating Solar Power and Chemical Energy Systems, SolarPACES 2014*, 69, 1568–1577.

Sutter, F., Wette, J., Wiesinger, F., Fernández-García, A., Ziegler, S., Dasbach, R. (2018). Lifetime prediction of aluminum solar mirrors by correlating accelerated aging and outdoor exposure experiments. *Solar Energy*, 174, 149–163.

Xin, C., Peng, C., Xu, Y., Wu, J. (2013). A novel route to prepare weather resistant, durable antireflective films for solar glass. *Solar Energy*, 93, 121–126.

Xu, K., Du, M., Hao, L., Mi, J., Yu, Q., Li, S. (2020). A review of high-temperature selective absorbing coatings for solar thermal applications. *Journal of Materiomics*, 6(1), 167–182.

8

Thermal Energy Storage

Aubin TOUZO[1], Quentin FALCOZ[1] and Gilles FLAMANT[2]

[1] *PROMES-CNRS, UPVD, Perpignan, France*

[2] *PROMES-CNRS, Font-Romeu, France*

8.1. Introduction

8.1.1. *Advantages related to thermal energy storage*

Although free and abundant, solar energy has a major drawback: it is intermittent. A solar power plant cannot generate during the night, or during cloudy spells, in the absence of a storage solution.

The integration of thermal energy storage in a concentrated solar power plant has three main advantages:

– It compensates for the daily sunshine fluctuations, mainly due to cloudy spells. During cloudy spells, the power received by the transfer fluid varies, and this affects the power block supply. The scattered cloudy spells can be compensated by the thermal inertia of the transfer fluid or of various pieces of equipment (receivers, cladding, etc.), but as soon as the sunshine drops for several tens of minutes, the power block must be turned off. The power block generally involves a steam turbine or a gas turbine. This equipment is very sensitive, and the successive turn-ons and turn-offs generate significant maintenance expenses. Moreover, turbines are designed for a rated operating point. Storage makes operation at this rated point possible and thus maximizes the generated power output.

– It shifts the production phase. The surplus energy generated during the day can be stored and then released later on. This leads to an increase in the capacity factor

Concentrating Solar Thermal Energy,
coordinated by Gilles FLAMANT.

of the solar power plant, which is the ratio of the number of hours of production at rated power and the number of hours in a year. Due to the thermal energy storage, the annual capacity factor of the solar power plant increases from 25% to 43% for the solar power plants with parabolic trough concentrators (with 7 h of storage capacity on average), and up to 55% for solar towers (10 h of storage capacity) (Kuravi et al. 2013). The thermal energy storage compensates for the phase shift between sunshine and electricity consumption. Therefore, a solar power plant with storage may be able to generate electricity up to 20 h per day. Generally, the choice is to release, and therefore generate, electric power during consumption peaks. Storage thus contributes to improving the economic indicators of the power plant, the price of electricity being higher during consumption peaks, generally between 18 h and 22 h, when sunshine declines and is absent.

– Participates in balancing the grid and facilitates the integration of renewable energies. A further significant part of these intermittent energies in the energy mix is susceptible to destabilize the grid. Thanks to storage, concentrated solar power can become a technology of choice to help the operators of power grids in matching generation and consumption. Energy is then released only in case of need and in addition to the energy generated by photovoltaic panels and wind turbines.

8.1.2. *An overview of thermal energy storage*

Thermal energy storage should have the following strengths:

– low cost;

– high-energy density;

– low thermal losses;

– good compatibility between the materials used and the transfer fluid;

– low environmental impact.

The significant advantage presented by storage is proven by the fact that 74% of the power plants under construction had one in 2017 (Pelay et al. 2017). It appears as a key element in compensating for the drawback of solar energy intermittence, while offering flexibility in power plant management. Hence, it is important to use a technically and economically attractive technology. Moreover, nowadays most of the calls for projects in the CSP field require building power plants with storage.

8.1.3. *Integration of storage in the solar power plant dimensioning*

Storage sizing must be considered in the design of the solar power plant. The factor used is referred to as the solar multiple (SM), which corresponds to the ratio of the thermal power of the solar field and the thermal power absorbed by the power block to generate the nominal power of the power plant.

A ratio of 1 means that when the collected solar thermal power corresponds to the nominal power, it can supply the power block to generate the nominal electrical power. This type of dimensioning leads to an operation in harmony with the Sun, with periods during which the solar resource (DNI, see Chapter 2) is below the nominal value. Under these conditions, the power block operates under partial charge during extended periods of time, which reduces its efficiency.

Dimensioning a power plant with a SM above 1 means that the amount of energy collected by the solar field is larger than what the power block can absorb. The surplus energy is stored (yellow in Figure 8.1), and the power block operates at its nominal point for a longer period, which maximizes its efficiency. The storage can then supply the power block at the moment intended by the power plant operator.

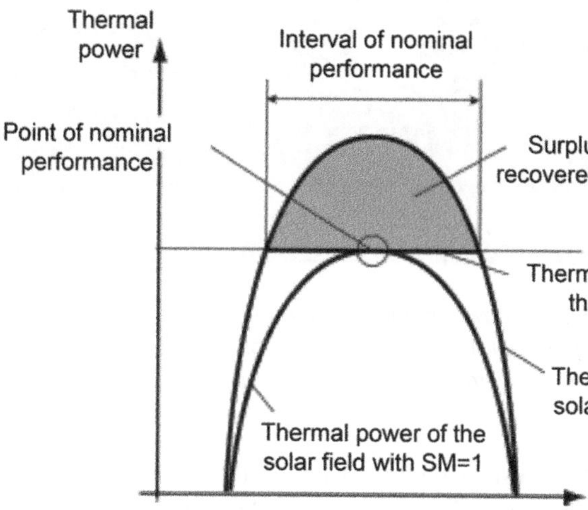

Figure 8.1. *Illustration of SM factor (Odru et al. Forthcoming)*

When dimensioning a solar power plant, the SM must therefore be chosen depending on the intended storage capacity. The expected values range between 1.5 and 2.5. By integrating the difference between the solar power and the power absorbed by the electrical block at the rated point, it is possible to calculate the storage capacity of the power plant.

8.2. Two-tank molten salt storage

8.2.1. *Examples of existing power plants*

The vast majority of storage in existing commercial solar power plants operates with sensible heat. Among them, one of the reference power plants is in Andasol, near Seville, Spain. This parabolic trough power plant has a nominal power of 50 MWe.

Figure 8.2. *Andasol power plant; the storage tanks are visible in the upper right. For a color version of this figure, see www.iste.co.uk/flamant/energy.zip*

The power collected by the solar field increases in the morning, until it reaches the rated thermal power for supplying the turbine. Beyond this rated power, the thermal energy collected by the solar field is transferred to the storage. At the end of the day, when the power collected by the solar field is not sufficient, the storage discharges, allowing the solar power plant to continue its production in the hours corresponding to the peak of power consumption, between 18 and 22 h.

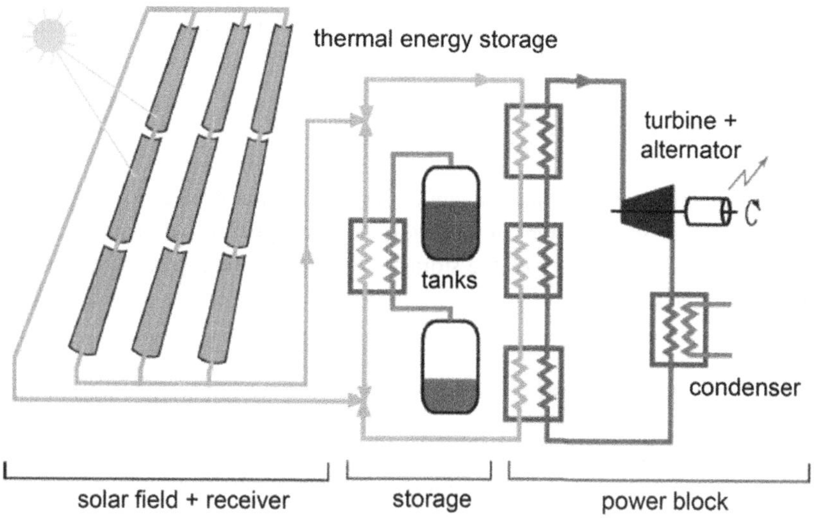

Figure 8.3. *Simplified diagram of the components of Andasol power plant (Odru et al. Forthcoming). For a color version of this figure, see www.iste.co.uk/flamant/energy.zip*

Its storage system comprises two tanks filled with 28,500 tons of molten salts, whose composition is 60% $NaNO_3$ and 40% KNO_3 by weight. It has a capacity of 1,010 MWh_{th}, and enables 7.5 h of storage with full power generation. The SM for this power plant is 1.6.

The heat of molten salts is stored in the form of sensible heat. The solar field and the power block are represented in Figure 8.3. The molten salts are first stored at a temperature of about 292°C in a tank, referred to as low temperature. During the thermal storage charging phase, molten salts are pumped from the low temperature tank through an exchanger, which absorbs the heat of the oil coming from the solar field at 393°C. The salts are heated at 385°C and are stored in the high-temperature tank. When the solar field no longer brings enough power to supply the steam, the storage system starts to discharge and the process is reversed. The molten salt is pumped from the hot tank, releases its heat to the heat transfer fluid (synthetic oil) through the exchanger, and is stored in the low temperature tank.

8.2.2. *Operating principle*

Storage by sensible heat involves bringing an object to a higher temperature when heat is available, so that this object can release this heat when needed. The equation for determining the amount of heat stored when passing from state 1 to state 2 is related to the enthalpy of the object:

$$Q = \Delta H = \int_{T_1}^{T_2} mc_p dT = \int_{T_1}^{T_2} V_{stock}\rho c_p dT \qquad [8.1]$$

where m is the mass of the storage material (solid, liquid or gas), c_p its heat capacity, ρ its density and V_{stock} is the storage volume. The product of density and heat capacity is the energy density of the storage. The higher the latter, the more heat is stored for a given volume.

8.2.3. *Materials employed*

In general, the heat transfer fluids used in solar receivers are oils, molten salt or water/steam (see Chapter 5). The oils employed in parabolic trough power plants have good energy density, but their cost is high, and this impedes their use in large quantities as storage element. Molten salts are therefore preferred, as they have good thermal performances at lower cost. This choice involves the integration of an oil-molten salt heat exchanger in the thermal system. Consequently, three different fluids are used in a commercial parabolic trough power plant, oil (heat transfer), solar salt (storage) and steam (working fluid). This drawback does not exist in molten salt tower power plants, as it is employed as heat transfer fluid and storage (see Chapter 1).

8.2.4. *Economic advantage*

The reference economic indicator is LCOE (levelized cost of electricity), which is used to compare power generation technologies. Considering this single criterion, concentrated solar power with storage is not attractive compared to other renewable technologies whose cost is provided without storage (between 25 and 55 USD/MWh for photovoltaics and wind power, and between 70 and 110 USD/MWh for concentrated solar power) (IRENA 2016; Dowling et al. 2017; USEIA 2020). But this comparison is not correct, as a key advantage of concentrated solar power is the capacity to generate electricity as needed thanks to the storage, and this service is not taken into account in the calculation of LCOE (Dowling 2017). A plus-value indicator is used in order to estimate the service related to storage. The latter is particularly interesting in the case of strong integration of renewable energy in the energy mix, with a plus-value generated by the storage that can reach 80% more than for conventional renewable energies (Denholm and Hummon 2013). A comparison of the cost of (thermal and electrochemical) storage technologies is given in section 8.8.

8.2.5. *Drawbacks of molten salt storage systems*

Despite being considered commercial benchmarks, molten salts have many drawbacks:

– Safety: oil is a fuel, and molten salts based on nitrates are oxidizers, besides being extremely corrosive, which involves fire hazard. A fire started on the oil circuit of the Andasol 1 power plant in 2009. Although the incident had no major consequence, as it did not affect the elements where oil and salt are present together, it is a reminder that safety problems should be considered.

– Usage conflict: molten salts are obtained from natural nitrates, mainly extracted from Chile. But these nitrate salts are used for manufacturing fertilizers. A usage conflict with the food sector, coupled with limited extraction, may slow down massive development of concentrated solar power (Pihl 2012).

– Cost: although molten salts-based heat storage is the most cost-effective option, the cost of the storage system remains relatively high in the total cost of a parabolic trough power plant, from 15 to 20%. Nevertheless, this relative cost is significantly reduced in tower power plants, as the specific energy stored per unit weight is multiplied by three (increase in the temperature difference between hot and cold stocks).

– Limited high temperature: in order to have higher efficiencies, there is a tendency to increase the supply temperature of the power block, in particular for tower power plants. For example, the implementation of supercritical cycles (water and carbon dioxide), that operate at a temperature of about 750°C is considered instead of the subcritical steam turbines in use. The maximal temperature of use of molten salts being of the order of 600°C, new storage systems must emerge to address this problem.

In order to compensate for these drawbacks, several paths are considered. Thermocline storage is one of the most promising ones. Other types of storage, such as the storage of latent or thermochemical energy, are also considered as development paths.

8.3. Thermocline storage

Thermocline storage could be used to address problems of usage conflicts and environmental impacts, notably by using recycled materials. It also leads to a reduction of the global cost of storage, by using a single tank instead of the two tanks used for molten salt-based storage. The principle is simple: the storage material is used as filler material through which the heat transfer fluid flows.

Consequently, thermocline storage can operate with gas or liquid as heat transfer fluid, and therefore allows the storage of heat up to 1,000°C.

8.3.1. *Examples of solar power plants with thermocline storage*

Solar One is a solar tower power plant in California (7 MWe), which was operational between 1982 and 1987. Its thermocline storage, framed in red in Figure 8.4, had a diameter of 18.2 m, a height of 13.3 m, representing a storage capacity of 170 MWh (Faas 1986). The energy was stored by sensible heat thanks to a porous bed made from granite and sand, through which oil, the heat transfer fluid of the power plant, flows. The storage temperature fluctuated between 224°C, its low temperature, and 304°C, its high temperature.

Figure 8.4. *Solar One solar power plant (storage framed in red). For a color version of this figure, see www.iste.co.uk/flamant/energy.zip*

Another example of thermocline storage is that installed in the solar power plant of Ait-Baha. This pilot solar power plant was put into service in 2014 (Good et al. 2014). It operates with air as heat transfer fluid, heated at 650°C by parabolic trough concentrators. Heat is stored in a rock bed of 3.9 MWh at 650°C.

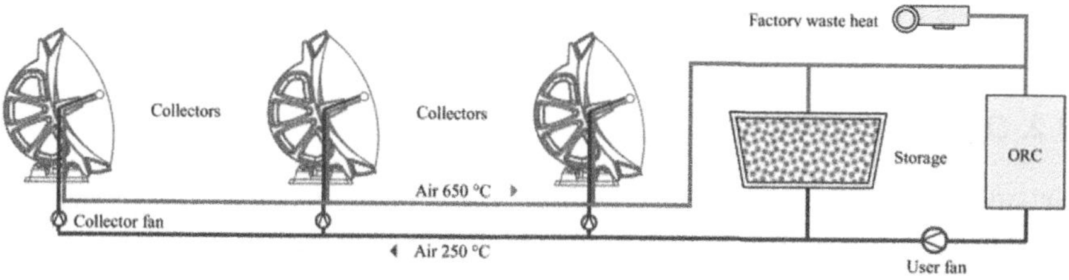

Figure 8.5. *Process diagram of the power plant in Ait-Baha (Good et al. 2014). For a color version of this figure, see www.iste.co.uk/flamant/energy.zip*

Figure 8.6. *Eco-Stock thermocline storage used to store 2 MWh heat on an industrial site (Eco-Tech Ceram). For a color version of this figure, see www.iste.co.uk/flamant/energy.zip*

Eco-Stock is a horizontal thermocline storage developed by Eco-Tech Ceram, an industrial ecology engineering company providing solutions for energy use in order to improve the energy efficiency of manufacturers. The main objective of Eco-Stock is to use the industrial heat (Figure 8.6), but experimental tests with thermal energy sources characteristic for concentrated solar power plants yielded promising results for the integration of this type of storage. The capacity of an Eco-Stock is about 2 MWh at 600°C.

8.3.2. *Operating principle*

Unlike the system with two molten salt tanks, the thermocline storage has only one tank, containing both hot and cold fluid. When charging starts, the storage system is at its low temperature. The high-temperature heat transfer fluid is injected in the storage. The heat transfer fluid releases its heat to the porous bed, which heats gradually. The storage is then separated into three parts: a part heated by the fluid, a cold part and a thermal gradient zone, known as thermocline. During charging, the thermocline moves in the storage from the hot side, where the fluid is injected, to the cold side, where fluid exits the storage. When thermocline reaches the storage exit, the charging stops. During the discharge, the direction of the fluid flow in the storage is reversed. The fluid is injected on the cold side, is charged with heat in contact with the porous bed, and exits on the hot side. The thermocline moves in the opposite direction and the system gets discharged. The thermocline is then extracted from the tank. This phenomenon is illustrated in Figure 8.7.

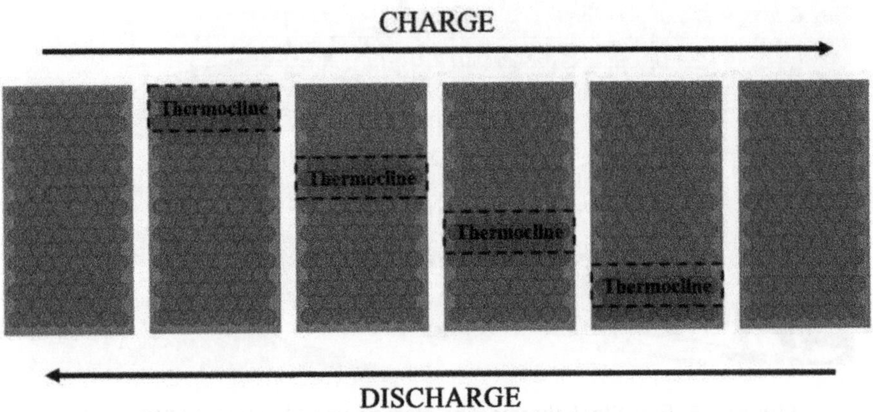

Figure 8.7. *Diagram of thermocline development. For a color version of this figure, see www.iste.co.uk/flamant/energy.zip*

A thermocline is the result of the non-ideality of heat transfers between fluid and solid. In the case of an ideal thermal energy transfer, the thermocline would be an infinitely thin layer of transition between cold and hot states.

If the fluid is liquid, such as oil or molten salts, the porous bed offers the possibility of making without a large part of the liquid, which is generally expensive, and replaces it with a less expensive solid. Oil and molten salts play only a role of heat transfer fluid, and not that of storage. The storage is integrated in the power plant, and the heat transfer fluid flows directly through it. It plays the role of exchanger and storage at the same time.

8.3.3. *Modeling*

The reference model is Schumann model (1929). He was among the first to study the flow of a fluid in a bed of particles, and notably the temperature profile in this bed. The formulated hypotheses are the following:

– solids have infinite thermal diffusivity; therefore, there is no thermal gradient in the particles;

– conduction between particles is considered negligible with respect to convection;

– heat transfer between fluid and solid at each point of the system is proportional to the difference of temperature between fluid and solid;

– variations of fluid and solid volume due to temperature variations are considered negligible;

– properties are constant and independent of temperature.

He therefore establishes two equations of the energy, one for the solid (index s), and the other one for the fluid (index f):

$$\varepsilon\left(\rho c_p\right)_f\left(\frac{\partial T_f}{\partial t}+u\frac{\partial T_f}{\partial x}\right)=h\,a_c\left(T_s-T_f\right) \qquad [8.2]$$

$$(1-\varepsilon)\left(\rho c_p\right)_s\frac{\partial T_s}{\partial t}=h\,a_c\left(T_f-T_s\right) \qquad [8.3]$$

where ε is the bed porosity, ρ is the heat capacity, u is the velocity of the fluid, h is the convection heat transfer coefficient, a_c is the specific surface (exchange surface per bed volume). The first equation is the equation of the fluid, which comprises a heat accumulation term, a left side transport term and a right side convective transfer term. The second equation is that of the solid, which comprises an accumulation term and a convective term. The two equations are coupled by the convective term.

8.3.4. *Integration challenges in a solar power plant*

The main drawback of thermocline storage is the control of the gradient zone inside the storage. During the charging phase, the fluid exits the storage at the low temperature of the tank, and returns to the solar field.

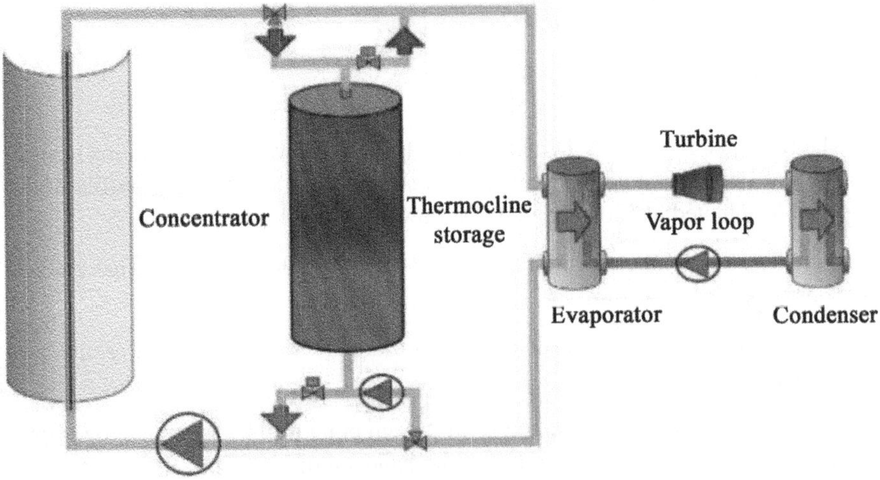

Figure 8.8. *Storage integration in a solar power plant, adapted from Fasquelle et al. (2018). For a color version of this figure, see www.iste.co.uk/flamant/energy.zip*

When the thermocline is extracted, the temperature of the fluid exiting the storage gradually increases.

The fluid entering the solar field may then overheat, which may lead to oil degradation in the case of low overheating, or even generate fires in the most serious cases. A threshold charging temperature must therefore be set, corresponding to the maximal output temperature, which defines the charging stop. The threshold charging temperature is defined as follows:

$$T_{thresh-ch} = T_H - \kappa_c(T_H - T_B) \quad [8.4]$$

where $0 < \kappa_c < 1$

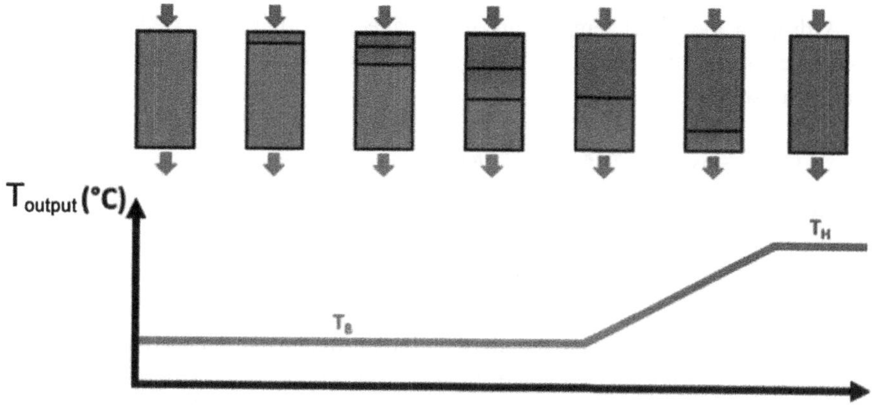

Figure 8.9. *Extraction of thermocline during charging (Fasquelle 2017). For a color version of this figure, see www.iste.co.uk/flamant/energy.zip*

Thermocline control is also challenging during the discharge. At the beginning of the discharge, the storage exit temperature is the high temperature. Then, the temperature drops when the storage material at the outlet gets cooler, represented in Figure 8.10.

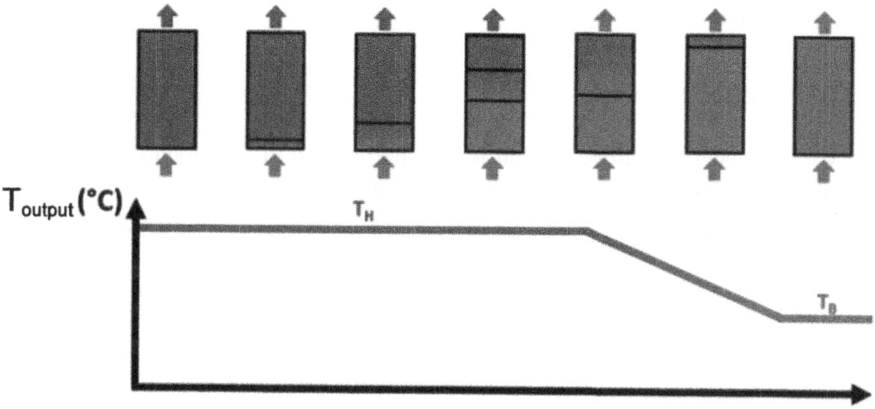

Figure 8.10. *Thermocline extraction during discharge (Fasquelle 2017). For a color version of this figure, see www.iste.co.uk/flamant/energy.zip*

A drop in the outlet temperature may leave the turbine underfed. A threshold discharge temperature should therefore be set, with a definition similar to that of the charge threshold:

$$T_{thresh-dis} = T_B + \kappa_d(T_H - T_B) \quad [8.5]$$

with $0 < \kappa_d < 1$

The thermocline storage must therefore be controlled with charge and discharge threshold temperatures. These are generally defined at $\kappa = 0.2$, but threshold temperatures can be adjusted depending on the upstream and downstream processes, for example the solar collector and the steam generator in the case of a parabolic trough solar power plant (Fasquelle et al. 2018).

The choice of charge and discharge threshold temperatures influences storage performances. Imposing a low charge threshold temperature involves a less charged storage, as the material near the outlet is at a low temperature. Its storage capacity is lower. During the discharge, imposing a high threshold temperature involves that all the heat below this temperature cannot be discharged, and must be preserved in the storage. The charge and discharge threshold temperatures therefore reduce the useful power provided by the storage and the useful storage capacity with respect to the tank volume.

8.3.5. *Storage materials*

Various materials can be chosen for the porous bed of thermocline storage. An adapted material is characterized by the following:

– low cost, essential for the economic sustainability of the storage;

– high-energy density (product ρc_p);

– chemical compatibility with the transfer fluid;

– thermal and chemical stability at high temperature;

– reduced environmental footprint.

Two geometries of materials are considered: nonstructured and structured.

Nonstructured or bulk materials can take various forms: pebbles, spheres, Raschig rings, cylinder, etc. Nonstructured materials are small in size, and offer a large transfer surface allowing an optimal exchange between the heat transfer fluid and the storage solid. These small size elements can also be used to obtain a low

Biot number (below 0.1), guaranteeing temperature uniformity in the solid and therefore maximal stored heat. Table 8.1 presents several solids considered as potential storage materials.

Material	ρ (kg.m^{-3})	c_p (J.kg^{-1}.K^{-1})	ρc_p (kJ. m^{-3}.K^{-1})	T_{max} (°C)	Price
Basalt	3,000	920	2,760	900	+
Gabbro	2,911	643	1,872	-	+
HT Concrete	2,750	916	2,519	600	+
HT Ceramic	3,500	866	3,031	–	+++
Zirconia	5,999	597	3,581	2,300	+++
Refractory bricks	3,000	1,150	3,450	1,200	++
Alumina spheres	3,953	1,157	4,574	1,800	++
Vitrified asbestos	3,120	1,034	3,226	1,200	+
Slags	3,770	912	3,438	1,000	+

Table 8.1. *Characteristics of several storage materials (Esence 2017)*

Alumina spheres, with several centimeters in diameter, were used as reference material in the first prototypes of thermocline storage, as their properties are known and the geometry of the bed of spheres can be easily apprehended. Alumina has high-energy density (4,574 kJ.m^{-3}.K^{-1}) and a maximal theoretical use temperature of 1,800°C. However, in large quantity, this material is relatively expensive and its environmental impact is non-negligible.

In order to drastically cut costs, natural rocks such as basalt or granite are used by Solar One and Ait-Baha solar power plants. The characteristic diameter of these rocks is generally of several centimeters. Natural rocks have also been used in a thermocline storage of 120 MWh located in Hamburg and built by Siemens Gamesa. This storage is not integrated in a solar power plant, but it was developed in order to store renewable energy in thermal form.

In order to reduce environmental impact, storage materials resulting from industrial waste were developed. For example, vitrified asbestos and metallurgical slags can be used as storage materials. These recycled materials can be easily obtained in small amounts in laboratory, but the passage to industrial manufacturing is difficult, as the demand for storage materials is low at the moment.

Nonstructured materials (pebbles, spheres, etc.) have two drawbacks. Small size solid objects are used in order to maintain high thermal performances, but they are susceptible to strong pressure drops, as it is difficult for the flow to make its way in the porous bed. The energy expenses related to the flow of the transfer fluid are consequently increased. A compromise should therefore be found between thermal performances and pressure drops. The second drawback is related to thermo-mechanical constraints. During the charge, the temperature of the solid bed increases and it expands, then it regains its initial size during discharge. The sequence of successive charges and discharges can then modify the layout of the particles in the bed. The porous bed is compressed, which generates strong constraints on the walls, potentially leading to its breakage. This phenomenon is referred to as *thermal ratcheting*. Specific studies of thermo-mechanical constraints must be conducted in parallel with thermal dimensioning.

Structured geometries may have various forms, such as perforated bricks or honeycomb bricks, as shown in Figure 8.11.

Figure 8.11. *Examples of structured materials. For a color version of this figure, see www.iste.co.uk/flamant/energy.zip*

Structured geometries can be used when thermal ratcheting is problematic. It is also possible to find them in high thermal capacity installations, in order to reduce the generated pressure drops, and therefore the cost of fluid flow. A honeycomb-shaped material is, for example, used in the solar power plant in Julich (1.5 MWe, 9 MWh_{th} storage capacity). On the contrary, they offer a smaller transfer surface, and therefore less convective transfer between the fluid and the solid. Moreover, the characteristic thickness of the material is greater, which affects the hypothesis of a thermally thin body, and may therefore reduce the quality of the thermal transfer. Structured materials can also have higher costs, related to the specific shape of the material.

8.3.6. *Life cycle analysis*

Life cycle analysis is a tool that can be used to consider the environmental impact of a product or service during its life cycle: extraction of materials, manufacturing, installation, use and disposal or recycling. For example, it can be used to assess the greenhouse gas emissions related to electricity generation, expressed as tons of CO_2 equivalent released. Hence, even though a source is referred to as renewable, its life cycle has nevertheless generated an environmental impact. It is however always lower than that of conventional power generation (coal, gas, fuel oil, etc.) in terms of CO_2 equivalent.

The pollutions generated by renewable energies, including CSP, are low during their operation. Pollution is essentially generated upstream operation, during the extraction, the manufacturing and the installation of the power plant. Storage accounts for approximately 40% of the CO_2 emissions related to the manufacturing of the power plant, most of which is dedicated to the production of storage material (Heath 2009). Compared to a system with two molten salt tanks, thermocline storage would allow a 54% reduction of CO_2 emissions, which is a significant reduction of the environmental impact of the solar power plant.

8.3.7. *Economic considerations*

The control of thermocline must be further improved so that this type of storage replaces molten salt systems, and thermocline storage plays a more important role in the market of storage applied to CSP. It would allow a reduction of costs by using only one tank instead of two tanks required for molten salts. The storage cost reduction may reach 33% (Angelini et al. 2013), the cost of a commercial storage using molten salts ranging between 40 and 90 USD/kWh (Jacob et al. 2016). However, an 11% over-dimensioning of the thermocline storage is required in order to obtain equivalent performances (Fasquelle 2017). The integration of thermocline storage in a solar power plant seems economically viable and promising.

8.4. Processes with steam accumulator

8.4.1. *Existing power plants*

PS10 is a concentrated solar power plant located near Seville, Spain. This solar power plant generates 11 MWe power. It is a tower power plant that directly generates low temperature saturated steam (250°C, 40 bar). This power plant does not employ oils or molten salts. A diagram of this type of power plant is presented in Figure 8.12.

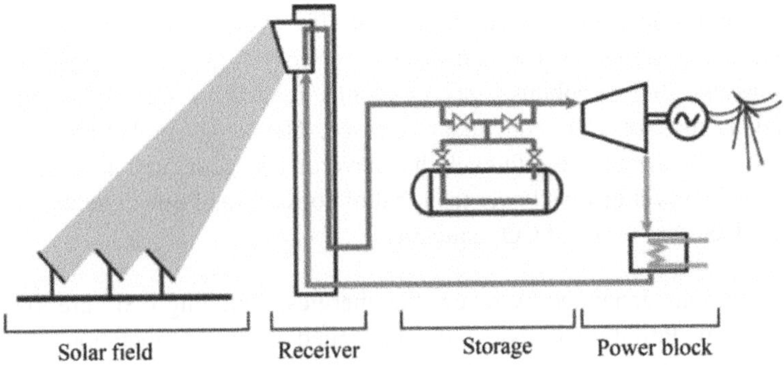

Figure 8.12. *Diagram of a solar power plant with direct steam generation (Odru et al. Forthcoming). For a color version of this figure, see www.iste.co.uk/flamant/energy.zip*

Figure 8.13. *Four steam accumulators of the PS10 power plant (Abengoa Solar). For a color version of this figure, see www.iste.co.uk/flamant/energy.zip*

This type of power plant is very sensitive to cloudy periods, even for short periods, of the order of several dozens of minutes. Indeed, sunshine absence would force the turn-off and then turn-on of the turbine at each cloudy spell, which strongly increases the maintenance costs and damages the turbine. For a continuous supply of the turbine, a short duration storage solution was chosen. The role of thermal storage differs from the molten salt-based storage solutions, since the main purpose is not to postpone production, but rather to allow operating continuity during the short periods without sunshine. This is referred to as protective storage. The heat transfer fluid being the steam, a steam accumulator was chosen as storage solution (see Figure 8.13).

The storage capacity of PS10 power plant is 20 MWh, which ensures a 50-minute supply of the turbine at 50% of partial charge. The storage contains four tanks of 5 MWh, which can be used independently, depending on the duration of the cloudy period.

The eLLO solar power plant built by SUNCNIM, installed at Llo in the Eastern Pyrenees (department no. 66) was inaugurated in 2019. It is a solar power plant with linear Fresnel concentrator (150,000 m^2 mirrors) covering 35 hectares. This power plant uses direct steam generation and can generate up to 9 MW of power. Steam can be stored in one of the nine tanks of 120 m^3 (see Figure 8.14), which ensure 4 h full-power operation. The steam is stored at 70 bar and 286°C.

Figure 8.14. *Image of the storage tanks of eLLO power plant (SUNCNIM). For a color version of this figure, see www.iste.co.uk/flamant/energy.zip*

8.4.2. *Operating principle*

At the beginning of the charge, the steam accumulator is filled with a water and steam mixture at a relatively low temperature. The high-temperature steam from the receiver(s) is injected in the liquid water of the accumulator (see Figure 8.15). Steam releases its heat to the liquid water, which heats up. This energy contribution shifts the liquid/steam point of equilibrium to a higher temperature. The continuous inflow of steam is accompanied by an increase in water pressure and water level in the accumulator. When the maximal charge pressure is reached, the steam inflow is cut off.

During the discharge, steam is extracted from the storage, which leads to a drop in pressure in the storage. Pressure is then below the saturation value, and liquid water evaporates, which drops the water level in the tank. The tank temperature decreases following the saturation curve.

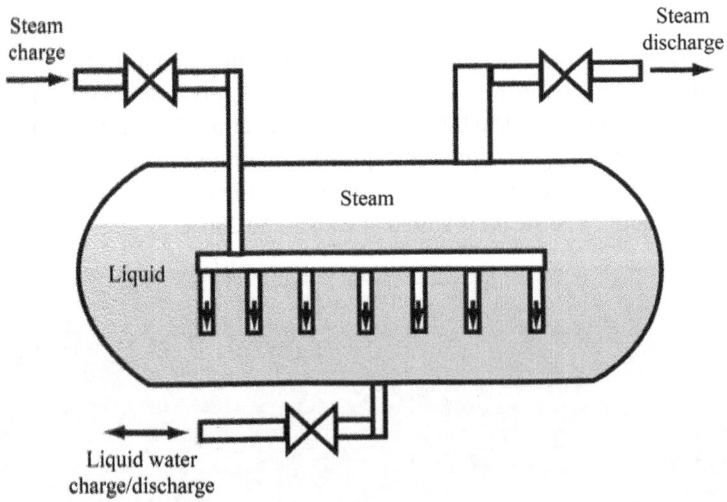

Figure 8.15. *Diagram of a steam accumulator (Odru et al. Forthcoming)*

The models that can be used to obtain energy from an accumulator are quite complex. A relatively simple model is proposed by Steinmann and Eck (2006).

8.4.3. *Drawbacks*

Steam storage has several drawbacks. Its energy density is low, of the order of 20 to 30 kWh/m^3 (from 30 to 100 kWh/m^3 for solar salt) which does not make this type of storage attractive for large storage capacities. Storage tanks must operate under high pressure, which makes them expensive and require additional safety standards. This cost should be compared to the maintenance costs related to an intermittent operation of the turbine to determine whether the use of this type of storage is economically viable.

8.5. Solar power plant with particle receiver

The THEMIS solar power plant in Targasonne is a research prototype operated by PROMES-CNRS. It offers the possibility of conducting experiments and research works on architectures of future power plants, using solar receivers, transfer fluids or innovative storage systems. A possible option is the use of a dense suspension of fluidized particles as transfer fluid. In this case, the particles play also the role of storage material. The diagram of this solar power plant with fluidized bed is shown in Figure 8.16.

The particles are stored in a low temperature tank (1). This tank supplies a distributor (feeder) under the solar receiver. A suspension of air-fluidized particles is injected in the solar receiver with vertical tubes (2). The particles are driven by the overpressure imposed in the distributor and the air up in the receiver and are heated in contact with the wall of the tubes (3), they then fall due to gravity in a storage tank that accumulates high-temperature particles (4). These particles release their heat to the air supplying the turbine inside an exchanger (5) before being mechanically transported in the cold tank (6).

The interest of this solar power plant design is to allow high-temperature (750–800°C) operation and to employ a very low-cost storage material (approximately five times less expensive than solar salt).

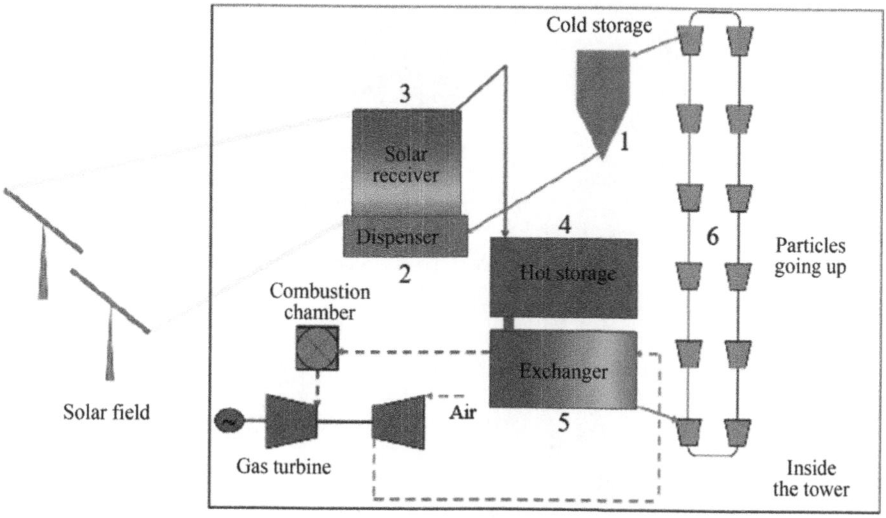

Figure 8.16. *Diagram of a solar power plant with fluidized bed (adapted from Grange et al. 2020). For a color version of this figure, see www.iste.co.uk/flamant/energy.zip*

Thermal energy storage is perfectly integrated here in the operation of the thermal power plant. This type of power plant is promising (Behar et al. 2020), but research works still need to be conducted on three aspects: solar receiver, particle exchanger and particle transportation.

8.6. Research and development of latent heat processes

Latent heat-based storage has a high-energy density and can absorb and deliver heat at constant temperature. Materials that can change their state (solid/liquid or liquid/gas) at high temperature have been developed. The phase change material (PCM) can be integrated in an exchanger or encapsulated.

8.6.1. *PCM exchanger*

A PCM can be integrated in an exchanger, involving the heat transfer fluid on the one side, and the PCM on the other side. CEA developed a storage of this type for solar power plants with direct evaporation, supplied with steam at 350°C (145 bar). This employs 6,350 kg of $NaNO_3$.

Figure 8.17. *Pilot PCM storage by CEA-Liten (Garcia et al. 2015). For a color version of this figure, see www.iste.co.uk/flamant/energy.zip*

Steam releases its heat to the material, which liquefies. Since PCMs have low conductivity, aluminum is added to the PCM to ensure good thermal energy transfer.

8.6.2. *PCM encapsulation*

In order to have a large surface of exchange with the heat transfer fluid, the PCM is enclosed in a capsule, made from an inert material with good thermal conductivity. PROMES and RAPSODEE laboratories have, for example, encapsulated $NaNO_3$ ($T_{melting}$ = 308°C) in steel tubes, in order to integrate it in the Micro-Sol-R power plant in PROMES-Odeillo that already had a thermocline storage. Adding a PCM layer at the higher level of the lining material can compensate for the drop in fluid temperature at the outlet at the end of discharge (see Figure 8.10). The efficiency of the discharging system is therefore improved.

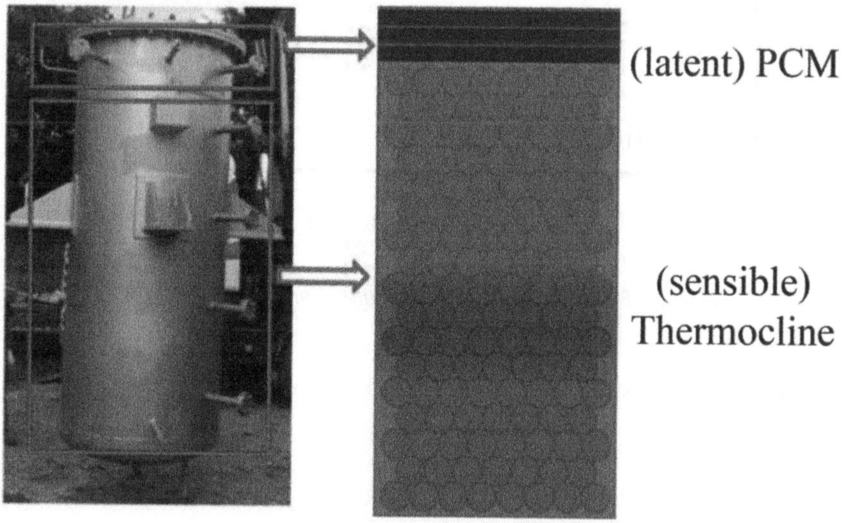

Figure 8.18. *Integration of an encapsulated PCM in a thermocline storage. For a color version of this figure, see www.iste.co.uk/flamant/energy.zip*

8.6.3. *Principle*

The use of PCM makes it possible to maintain a constant discharge temperature. The heat brought by the heat transfer fluid triggers a change in the state of the material, generally from the solid state to the liquid state. The change of state can be viewed in Figure 8.19. The temperature of the solid increases (1–2), the melting temperature of the material is reached (2). The energy released by the heat transfer fluid supports the melting at constant temperature (2–3). When the entire material is molten and is in a liquid form (3), the material continues to accumulate energy and the temperature of the liquid increases (3–4). Upon discharge, the reverse phenomenon occurs, and the heat generated by solidification is transferred to the heat transfer fluid of the power plant.

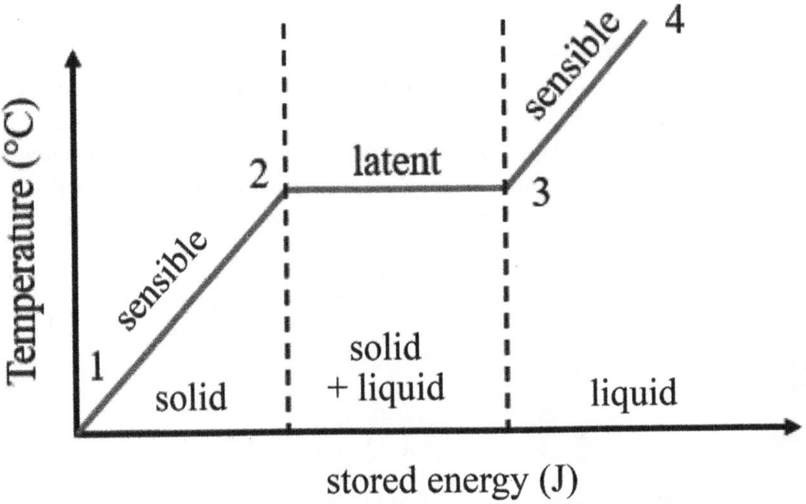

Figure 8.19. *Illustration of the state changes of a PCM*

The equation that is used to calculate the amount of latent heat for a solid/liquid state change is:

$$Q = m(Cp_{sol}(T_1 - T_{liq}) + \Delta h_{liq} + Cp_{liq}(T_{liq} - T_2)) \qquad [8.6]$$

where Δh_{liq} and T_{liq} are the liquefaction enthalpy and temperature, respectively.

The high-energy density of the storage by latent heat is due to the fact that state change enthalpies are high compared to the enthalpies related to sensible heat.

8.6.4. *Drawbacks of phase change materials*

PCMs have several drawbacks. They have low conductivity, which limits the propagation of heat inside the material, and therefore reduces the power of the process. Overmelting phenomena may appear, meaning that the liquid may go below its solidification point without solidifying, and thus perturb the system discharge. Furthermore, the thermal stability during charge/discharge sequence is not guaranteed for all PCMs.

Despite this, pilot research installations show that some of these difficulties can be overcome, and this may encourage a potential commercial deployment of solar power plants with direct steam generation.

8.7. Thermochemical storage

Heat can also be stored by chemical reaction. During the charge, heat is absorbed by the reagents and the products are formed during an endothermal chemical reaction. During the discharge, the process is reversed, releasing heat by exothermal reaction.

The stored thermochemical energy can be calculated as follows:

$$Q = a_r \Delta H_r \qquad [8.7]$$

where α_r is the reaction progress, and ΔH_r is the heat of reaction.

Unlike other types of storage in which heat dissipates, thermochemical storage has the advantage that there are no thermal losses, the chemical species being completely inert when at rest.

However, this type of storage is not quite developed, as the concept is relatively new. Thermochemical storage was tested at the laboratory scale and over short periods of discharge. The key to the development of this technology is the selection of the best chemical reactions, the improvement of reactors in order to maximize thermal transfers, and the selection of materials with high storage density. Concerning the potentially usable chemical reactions, the reader may refer to André et al. (2016) and Carrillo et al. (2019).

8.8. Comparison of the cost of stored solar power

While the development of batteries for electrochemical storage of energy is strongly encouraged and subsidized, asking what is the best option for the future is legitimate. What is the most economically viable form of energy to be stored in order to generate power: thermal energy or electrochemical energy? A trivial approach leads to a simple conclusion: heat must be stored when the conversion system involves a thermal conversion mode, as for the solar thermal power plants, and electrochemical energy must be stored when it is directly generated by the conversion system, as for the photovoltaic (PV) power plants. But other options are also possible, such as storing the electric power of solar thermal power plants with batteries or converting PV power into heat to store it, and then releasing energy in the electrical form using a thermodynamic converter. Finally, is it possible to maintain over the long term the advantage of bulk thermal storage associated with concentrated solar power compared to the battery storage associated with photovoltaics? These questions have been recently addressed. Obviously, there is no unique solution, and it depends on the evolution of the price of batteries and the storage duration.

Two studies conducted in different contexts (different solar resources), the western United States and southern Europe, lead to similar conclusions.

Feldman et al. (2016) compared the cost of electrochemical (battery) storage (BS) associated with photovoltaics (PV) and of the thermal storage (TS) associated with concentrated solar power (CSP) in the context of California. The parameters are storage duration and the possible evolution of the cost of batteries until 2030. Two hypotheses are retained for this last point: a decrease in cost according to the learning curve and a rapid decrease related to technological breakthroughs. For a three-hour duration and the hypothesis of a learning curve decrease trend, the cost of thermal storage (CSP-TS) and that of battery storage (PV-BS) are comparable in 2020 and 2030 for a service life of batteries of 15 years (around 0.11 USD/kWh). On the contrary, the ST storage is less expensive for a service life of batteries of 10 years. In the hypothesis of a drop in storage costs due to breakthrough, the PV-BS solution is less expensive starting in 2025, but the gaps are within the uncertainty margin of the estimations. On the contrary, for storage durations of 6 hours or 9 hours, the CSP-TS option is still less expensive, whatever the hypothesis of battery cost reduction.

In the European context, in Spain, Schöniger et al. (2020) compared three storage architectures shown in Figure 8.20: concentrated solar power coupled with TS, photovoltaics coupled with BS and photovoltaics coupled with a TS with electricity-heat conversion by an electric heater and heat-electricity conversion by a steam turbine. The power generated by the three devices is 100 MW.

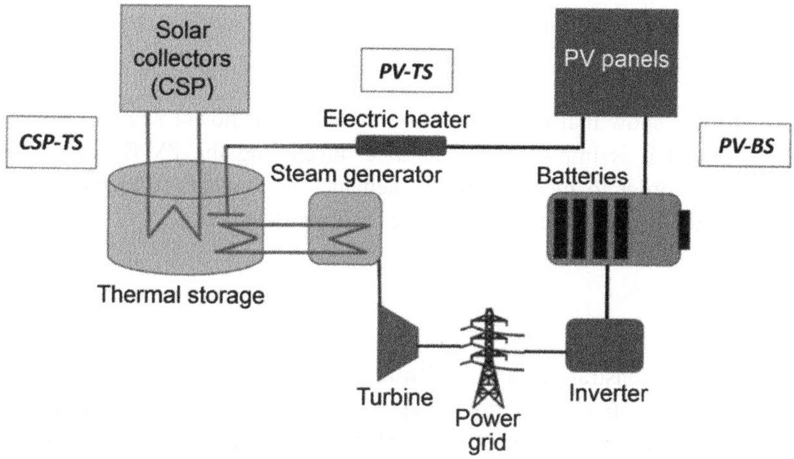

Figure 8.20. *Diagram of the three architectures for solar energy storage compared by Schöniger et al. (2020). For a color version of this figure, see www.iste.co.uk/flamant/energy.zip*

Before summarizing the main conclusions of this study, it is important to understand that the two technologies CSP-TS and PV-BS have very different cost structures. In the second case, the cost of the batteries governs the cost of kWh with battery storage. It represents about 80% of the cost of stored solar kWh irrespective of the duration. For the CSP-TS configuration, the storage cost is always below that of the collectors and of the power block. For example, for 12 h storage the two latter components account for approximately 75% of the cost of stored solar kWh. Consequently, the evolution of the global cost of electricity highly depends on the evolution of the storage cost for PV-BS technology and only marginally for the CSP-TS technology. On the contrary, the latter strongly depends on the storage duration.

Three cost hypotheses are studied, namely, actual, average and low cost. For battery storage (full system, Li-ion batteries, service life of 15 years), these hypotheses correspond to 500, 300 and 100 EUR/kWh$_e$ (electric kWh). For the thermal storage, the costs taken into account are 40, 25 and 10 EUR/kWh$_{th}$ (thermal kWh, the thermodynamic conversion efficiency is 42%). Taking into account the operational and maintenance costs, the results of this technical-economic calculation, applied to the sunshine conditions in Spain show that PV-BS storage is less expensive than CSP-TS storage for short durations of 2–3 h. The CSP-TS option is less expensive for longer storage durations except for the hypothesis of a cost reduction with BS at 100 EUR/kWh$_e$. In this case, the limit is pushed back to approximately 10 h. The third option, PV-TS, leads to a more expensive kWh than the CSP-TS option for all scenarios.

These results show that the comparative cost of solar power storage depends on its duration, and it is important to continue developing the PV-BS and CSP-TS technologies in order to meet the future demand for low-carbon electricity in the various contexts in which these technologies are implemented.

8.9. Conclusion

It has been demonstrated at the industrial scale that concentrated solar power has the capacity to store thermal energy at large scale (several GWh), this being an asset of this technology. Liquid solar salt (60% $NaNO_3$, 40% KNO_3 in weight) dominates the commercial development of thermal storage systems by sensible heat. When used with the two-tank technology in the tower power plants, this technology reaches specific storage capacities of 100 kWh/m^3 at a cost of 40 to 35 EUR/kWh$_{th}$. Thermocline storage is a credible alternative, but it must be demonstrated at large scale. The compared analysis of the cost of solar energy stored by electrochemical means (associated with PV) and by thermal means (associated with CSP) shows the economic advantage of the latter for storage durations above 3 h. The comparison of solar power solutions should integrate the results of the short-term analysis of life cycle, as the recycling problems will become increasingly constraining with the development of renewable energies and the associated storage technologies.

8.10. References

André, L., Abanades, S., Flamant, G. (2016). Screening of thermochemical systems based on solid-gas reversible reactions for high temperature solar thermal energy storage. *Renewable & Sustainable Energy Review*, 64, 703–715.

Angelini, G., Lucchini, A., Manzolini, G. (2013). Comparison of thermocline molten salt storage performances to commercial two-tank configuration. *Energy Procedia*, 49, 694–704. doi: 10.1016/j.egypro.2014.03.075.

Behar, O., Grange, B., Flamant, G. (2020). Design and performance evaluation of a modular solar power plant using a gas turbine combined cycle with fluidized particle-in-tube receiver and direct thermal storage. *Energy Conversion & Management*, 220, 113108.

Carrillo, A., González-Aguilar, J., Romero, M., Coronado, J. (2019). Solar energy on demand: A review on high temperature thermochemical heat storage systems and materials. *Chemical Reviews*, 119, 4777–4816.

Denholm, P. and Hummon, M. (2013). Simulating the value of concentrating solar power with thermal energy storage in a production cost model. Technical Report, U.S. Department of Energy [Online]. Available at: http://ieeexplore.ieee.org/lpdocs/epic03/wrapper.htm?arnumber=5484626.

Dowling, A.W., Zheng, T., Zavala, V.M. (2017). Economic assessment of concentrated solar power technologies: A review. *Renewable and Sustainable Energy Reviews*, 72, 1019–1032. doi: 10.1016/j.rser.2017.01.006.

Esence, T. (2017). Étude et modélisation des systèmes de stockage thermique de type régénératif solide/fluide. Thesis, Université Grenoble-Alpes, Grenoble.

Faas, S. (1986). 10 MWe solar thermal central receiver pilot plant – Test report. Sandia National Laboratory Test Reports, Albuquerque, NM.

Fasquelle, T. (2017). Modélisation et caractérisation expérimentale d'une boucle solaire cylindro-parabolique intégrant un stockage de type thermocline. Thesis, University of Perpignan, Perpignan.

Fasquelle, T., Falcoz, Q., Neveu, P. (2018). A temperature threshold evaluation for thermocline energy storage in concentrated solar power plants. *Applied Energy*, 212, 1153–1164. doi: 10.1016/j.apenergy.2017.12.105.

Feldman, D., Margolis, R., Denholm, P., Stekli, J. (2016). Exploring the potential competitiveness of utility-scale photovoltaics plus batteries with concentrating solar power, 2015–2030. NREL Technical Report, NREL/TP-6A20-66592.

Garcia, P., Olcese, M., Rougé, S. (2015). Experimental and numerical investigation of a pilot scale latent heat thermal energy storage for CSP power plant. *Energy Procedia*, 69, 842–849. doi: 10.1016/j.egypro.2015.03.102.

Good, P., Zanganeh, G., Ambrosetti, G., Barbato, M.C., Pedretti, A., Steinfield, A. (2014). Towards a commercial parabolic trough CSP system using air as heat transfer fluid. *Energy Procedia*, 49, 381–385. doi: 10.1016/j.egypro.2014.03.041.

Grange, B., Gal, A., Le Flamant, G. (2020). Simulation of the next-CSP solar loop including a hybrid gas turbine. Online event, SolarPaces Conference 2020, September 28–October 2.

Heath, G., Turchi, C., Decker, T., Burkhardt, J., Kutscher, C. (2009). Life cycle assessment of thermal energy storage: Two-tank indirect and thermocline. *ASME 2009 3rd International Conference on Energy Sustainability* [Online]. Available at: https://doi.org/10.1115/ES2009-90402.

IRENA (2016). The power to change: Solar and wind cost reduction potential to 2025. Report, International Renewable Energy Agency.

Jacob, R., Belusko, M., Cabesa, L., Saman, W., Bruno, F. (2016). Embodied energy and cost of high temperature thermal energy storage systems for use with concentrated solar power plants. *Applied Energy*, 180, 586–597. doi: 10.1016/j.apenergy.2016.08.027.

Kuravi, S., Trahan, J., Goswami, D.Y., Rahman, M., Stefanakos, K. (2013). Thermal energy storage technologies and systems for concentrating solar power plants. *Progress in Energy and Combustion Science*, 39(4), 285–319. doi: 10.1016/j.pecs.2013.02.001.

Niedermeier, K., Mueller-Trefzer, F., Daubner, M., Marocco, L., Weisenburger, A., Wetzel, T. (2020). Theoretical and experimental studies of dual-media thermal energy storage with liquid metal. Online event, SolarPaces Conference 2020, September 28–October 2.

Odru, P., Olivès, R., Nadau, L., Muguerra, P., Lazaro, A., Franquet, E., Doppiu, S., Palomo, E., Duquesne, M., Joma, W. (Forthcoming). *Heat and Cold Storage 1*. ISTE Ltd, London, and John Wiley & Sons, New York.

Pelay, U., Luo, L., Fan, Y., Stitou, D., Rood, M. (2017). Thermal energy storage systems for concentrated solar power plants. *Renewable and Sustainable Energy Reviews*, 82–100. doi: 10.1016/j.rser.2017.03.139.

Pihl, E., Kushnir, D., Sanden, B., Johnsson, F. (2012). Material constraints for concentrating solar thermal power. *Energy*, 44, 944–954. doi: 10.1016/j.energy.2012.04.057.

Schöniger, F., Thonig, R., Resch, G., Lilliestam, J. (2020). Making the sun shine at night: Comparing the cost of dispatchable concentrating solar power and photovoltaics with storage. *Energy Sources, Part B: Economics, Planning, and Policy*. doi.org/10.1080/15567249.2020.1843565.

Schumann, T.E.W. (1929). Heat transfer: A liquid flowing through a porous prism. *Journal of the Franklin Institute*, 208(3), 405–416. doi: 10.1016/S0016-0032(29)91186-8.

Steinmann, W.D. and Eck, M. (2006). Buffer storage for direct steam generation. *Solar Energy*, 80(10), 1277–1282. doi: 10.1016/j.solener.2005.05.013.

USEIA (2020). Levelized cost and levelized avoided cost of new generation resources in the annual energy outlook 2020. Report, United States Energy Information Administration.

9

Hybrid PV–CSP Systems

Alexis VOSSIER[1] **and Joya ZEITOUNY**[1,2]
[1] *PROMES-CNRS, Font-Romeu, France*
[2] *C2N, CNRS, Paris-Saclay University, Palaiseau, France*

9.1. Introduction

Despite the extremely low cost of photovoltaic energy in 2020 (recent announcements of PV power plant manufacturers relate to projects characterized by resale prices significantly below 0.03 EUR/kWh), the massive deployment of photovoltaic electricity is increasingly confronted with the lack of dispatchability of this technology: nowadays, it is obvious that not being able to match the photovoltaic power generation with power demand leads to many power grid management difficulties, as the share of PV technology in the energy mix becomes significant. This is notably reflected by difficulties in managing production (solar power being so-called "fatal", it is injected in the power grid when sunshine allows it, to the detriment of the "dispatchable" power sources nuclear, coal and gas). When photovoltaic power generation is a significant fraction of the power demand, operators may be in an overproduction situation (production and consumption balance must necessarily be maintained in order to ensure the stability of the power grid), which may lead to negative resale prices on the power markets. This regularly occurs in Germany, where the penetration rates of photovoltaic solar and wind power reached more than 30% of the total power consumption in the country.

The massive deployment of variable renewable energy sources also requires maintaining large dispatchable power sources in order to meet the need for electricity in the absence of sunshine (or wind, for wind power). There are several types of such sources: gas power plants, coal power plants, hydroelectric power plants, nuclear power plants (certain countries have load-following technologies, such as France). The required maintenance of CO_2-emitting technologies in order to compensate for the intermittence of the solar resource leads to this dilemma: the massive deployment of photovoltaic power plants will not necessarily be accompanied by substantial reduction of CO_2 emissions.

The development of energy storage solutions is quite obviously a promising way to address the challenge of solar intermittence, and many technologies are today the focus of intensive research and development work. Examples worth mentioning are batteries, capacitors, flywheels, compressed air storage systems, pumped energy transfer stations or hydrogen storage. To this day, none of these technologies seems to be able to meet the significant storage needs associated with a massive deployment of variable renewable energies. On the one hand, the high cost of most of these alternatives is presently preventing the latter from competing with conventional sources. On the other hand, many problems related to the availability of materials, recycling and pollution are obstacles for the wide deployment of these various storage technologies.

Even though nowadays the cost of concentrated solar power is significantly higher than the cost of photovoltaics, CSP offers a considerable specific advantage: the possibility of storing thermal energy.

Unlike many other energy storage solutions, thermal energy storage is an inexpensive and efficient solution, relying on abundant and low-polluting materials.

Figure 9.1 illustrates the positioning of the main solar technologies on a cost-dispatchability diagram (dispatchability designates the capacity of a power generation technology to meet the demand at a given moment) and schematically summarizes the interest of hybrid approaches. While photovoltaic power is definitely not very expensive at the turn of 2020, it is also poorly dispatchable. The integration of batteries (electrochemical storage) notably increases the dispatchability of photovoltaic power plants, but generates a significant additional cost given the current cost of batteries, for a relatively limited increase in dispatchability. Conversely, CSP power plants have significantly higher power costs than those of photovoltaic power plants, but they have the advantage of enhanced dispatchability. The recent expansion of hybrid PV–CSP technology is an expression of the willingness to reconcile two assets specific to each of these sectors: the simplicity and the low cost of photovoltaic conversion technology, together with the simplicity and the low cost of thermal storage technology inherent in CSP power plants.

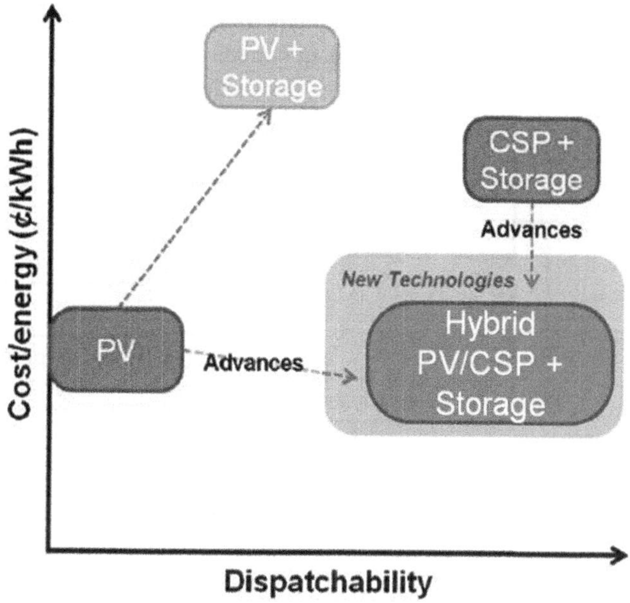

Figure 9.1. *Cost and dispatchability comparison between PV technologies (with and without storage), CSP systems and hybrid PV/CSP systems (source: FOCUS n.d.). For a color version of this figure, see www.iste.co.uk/flamant/energy.zip*

This chapter reviews the main hybrid PV–CSP power plant technologies, before going into further details on the main scientific and technological challenges to be addressed so that this strategy may be used to provide competitive and dispatchable solar power. It concludes with a brief presentation of several innovative hybrid technologies proposed in order to remedy the defects of conventional hybrid systems.

9.2. Hybrid strategies

Nowadays, there are numerous hybrid PV–CSP strategies that belong essentially to two large families (Figure 9.2):

1) Strategies referred to as "non-compact", for which PV and CSP power plants are physically separated and operated independently from one another.

2) Strategies referred to as "compact", which rely on the integration of the two technologies in one and the same power plant. Considering the diversity of strategies

implemented on compact hybrid systems, the numerous scientific and technological obstacles associated with them and the abundant literature focusing on this family of systems over the years, a deliberate choice was made to dedicate most of this chapter to this latter family of systems.

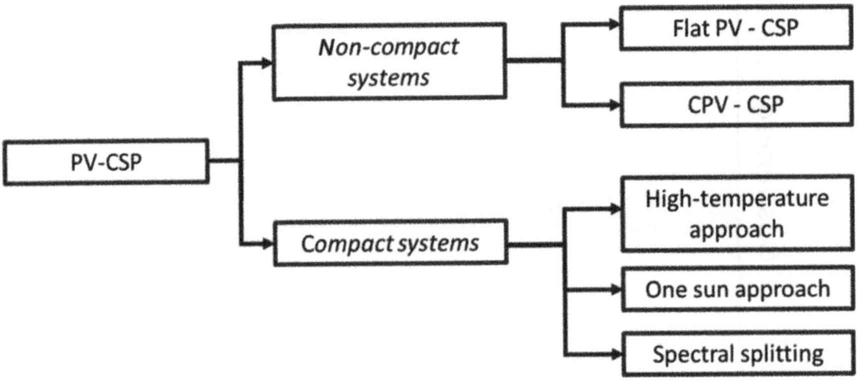

Figure 9.2. *Main families of hybrid PV–CSP systems (Ju et al. 2017)*

9.3. Non-compact hybrid strategies

"Non-compact" hybrid systems refer to solar power plants employing PV and CSP technologies built in close proximity, but operated independently. With such systems, the solar resource is optimally used, PV being generally used during the day, while CSP is preferentially used during cloudy spells or at night (depending on the preferred resale strategy). The use of non-compact hybrid systems significantly increases the capacity factor of solar power plants compared to conventional PV or CSP solar power plants (it is important to recall that the capacity factor of a power plant is defined as the ratio between the power generated by this power plant during the year and the power it would have generated if operated at rated power 100% of the time).

Non-compact hybridization strategies also offer high flexibility in terms of integration, given the very nature of the technology: the characteristics of the PV power plant do not influence the characteristics or performances of the CSP power plant, because they are built on two different (though neighboring) sites. On the contrary, the integration of these two technologies into one and the same power plant, in the case of compact hybrid approaches, is clearly more problematic and poses actual scientific and technological challenges. Finally, it is important to note that non-compact hybrid power plants rely on the use of mature, reliable and tested solar technologies, and their practical implementation is already considered, just like the hybrid PV/CSP power plant project of Copiapó in Chile (an artist's view of which is shown in Figure 9.3 (IEEE Spectrum n.d.).

Figure 9.3. *An artist's view of the non-compact hybrid PV–CSP power plant of Copiapó (Chile) (source: IEEE Spectrum n.d.). For a color version of this figure, see www.iste.co.uk/flamant/energy.zip*

9.4. Compact hybrid strategies

The "compact" hybrid strategies are essentially part of two large subfamilies relying on distinct physical principles:

– Hybrid systems involving a differentiated use of solar photons: such systems necessarily employ an optical component whose purpose is to direct photons to the most appropriate conversion system, according to their characteristic wavelength. Depending on the nature of the optical component employed (PV cell, filter, mirror), this family of systems can be subdivided into several subfamilies ("one-sun" approach, spectral division etc.).

– Hybrid systems involving the use of an integrated PV–CSP receiver. Further on in this chapter, it will be shown that despite its advantages, this approach (qualified as "high temperature" in what follows) presents many scientific and technological obstacles.

9.4.1. *High-temperature approach*

The hybrid approach referred to as high temperature (generally known as "PV-topping") relies on the use of an integrated PV–CSP receiver for which a concentrating photovoltaic module is used as thermal receiver (Figure 9.4). The heat absorbed by the module is then transferred to the thermal converter, and is stored for future use, or converted into electricity. This approach therefore involves a very high operating temperature of the PV module, which is definitely the main scientific and technological obstacle of this approach.

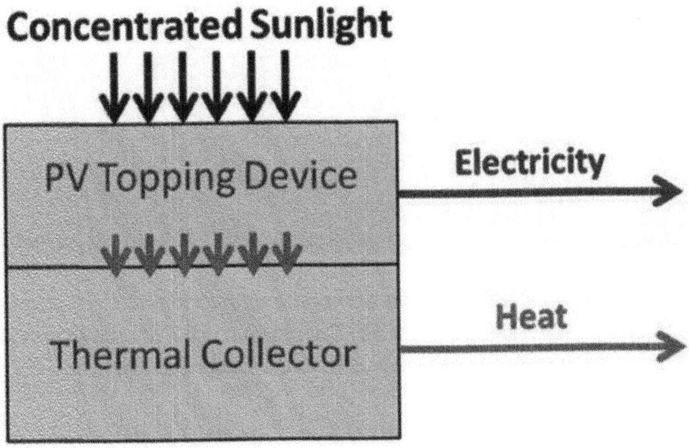

Figure 9.4. *Schematic representation of a "high-temperature" integrated PV–CSP receiver. The concentrated solar radiation is absorbed by a high-temperature PV module that is also a thermal collector. Part of the incident solar energy is directly converted into electricity while the rest is transferred in the form of heat to the thermal collector (Branz et al. 2015). For a color version of this figure, see www.iste.co.uk/flamant/energy.zip*

The conversion of the residual heat generated by the PV cells raised many hopes and was the subject of numerous discussions in the photovoltaic community during the past decades. The effect of temperature on the performances of PV cells is largely documented nowadays and essentially results from two distinct phenomena:

– The decrease in energy band gap of semiconductor materials used in PV cells yields a slight increase in the photogenerated current I_{sc}.

– The increase in intrinsic carrier density, which is the concentration of charge carriers – electrons or holes – ejected from the valence band towards the conduction band (electrons) or from the conduction band to the valence band (holes) under the sole effect of temperature, leads to a drop in the open circuit voltage V_{oc} of the cell.

The deleterious effect of temperature on the open circuit voltage is more significant, in terms of its amplitude, than the beneficial effect of temperature on the photogenerated current, and an increase in temperature leads overall to a decrease in the conversion efficiency of PV cells. Figure 9.5 illustrates this behavior: the electric power that can be extracted from a PV cell decreases significantly with the temperature of the latter. The amplitude of the phenomenon seems incompatible with the philosophy of the high-temperature hybrid approach, which requires characteristic temperatures of several hundred degrees for the PV module.

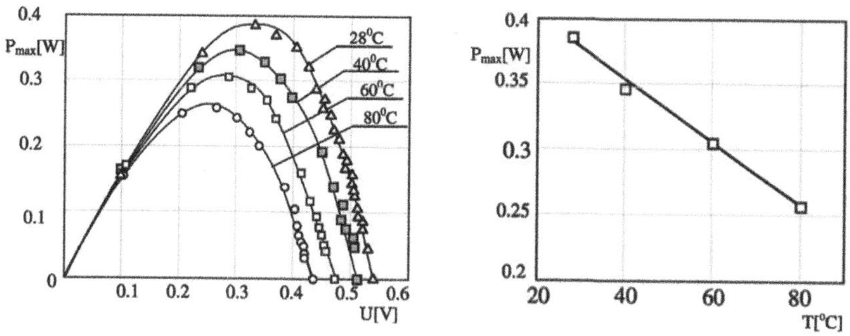

Figure 9.5. *Left: Power–voltage curves (left) illustrating the effect of temperature on the electric characteristics of a PV module: an increase in the module temperature is accompanied by a notable decrease of the open circuit voltage of the latter, and also of the maximal electric power. Right: The maximal electric power generated by a PV module depending on its temperature (Radziemska 2003)*

The study of the temperature dependence of concentrating photovoltaic cells was the object of a significant number of research works in the past decade, leading to a paradigm change: the temperature dependence of photovoltaic cells, and therefore, their capacity to efficiently convert sunlight at high temperature could be considerably improved with solar flux concentration. This important result is explained by the double-dependence of the open circuit voltage V_{oc} of PV cells, on the one hand, on temperature and, on the other hand, on the light power density to which the cells are subjected, as indicated by the following equation:

$$V_{oc}(X,T) \propto ln \frac{I_{ph}(X)}{I_0(T)} \quad [9.1]$$

where I_{ph} is the current photogenerated by the cell (the current associated with the absorption of incident solar photons) and I_0 is the diode current (the current associated with the recombination of charge carriers in the cell). T and X are, respectively, the cell temperature and the solar concentration factor, expressed in suns (where 1 "sun" is the equivalent of a solar power density of 1,000 W/m^2).

While the increase in the cell temperature leads to an increase in the recombination current I_0 (essentially because of the above mentioned increase in the concentration of intrinsic carriers), the exposure to concentrated solar flux is accompanied by an increase in the photogenerated current, which varies proportionally to the absorbed solar power density (in practice, and as indicated above, there is also a positive effect of temperature on the photogenerated current, but this can be considered negligible, in first approximation, compared to the solar concentration effect).

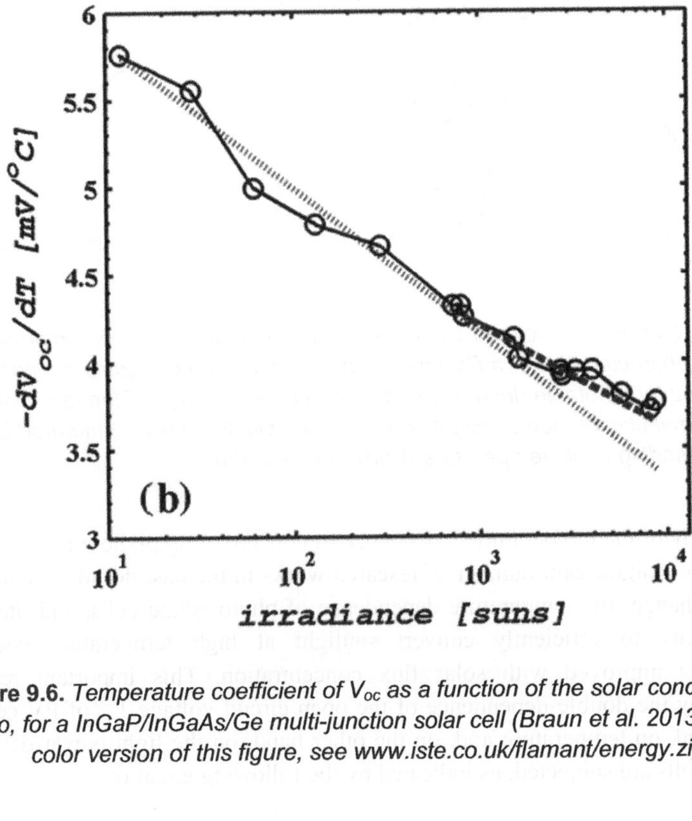

Figure 9.6. *Temperature coefficient of V_{OC} as a function of the solar concentration ratio, for a InGaP/InGaAs/Ge multi-junction solar cell (Braun et al. 2013). For a color version of this figure, see www.iste.co.uk/flamant/energy.zip*

The temperature dependence of a given electric parameter of a solar cell is generally quantified using the temperature coefficient associated with this very parameter, which describes how a 1°C temperature increase influences the numerical value of the respective parameter. The evolution of the temperature coefficient of V_{oc}, illustrated in Figure 9.6 as a function of the solar concentration factor, shows a clear tendency: temperature sensitivity notably decreases with solar concentration. In other terms, the decrease in the open circuit voltage associated with the cell heating is all the smaller as this very cell is subjected to higher solar powers.

In light of these recent results, the practical implementation of high-temperature hybrid systems can therefore be considered, since the PV cells integrated in the receiver are subjected to sufficiently high levels of solar concentration.

Figure 9.7 represents the various energy fluxes (in percentage of the incident solar power) involved in the case of high-temperature hybrid receivers, considering double-junction PV cells (composed of a stack of 2 *pn* junctions optimized to provide optimal conversion of solar energy for a given temperature and concentration ratio). The various energy fluxes associated with the high-temperature hybrid receiver are represented for concentration ratios of 100 suns (a) and 1,000 suns (b), as a function of the receiver temperature. Finally, the conversion efficiency of the turbine is taken equal to 2/3 of the Carnot efficiency.

The electric power generated by the hybrid system, which is interpreted as the sum of photovoltaic power (dark blue) and thermal power (light blue), varies significantly as a function of the temperature of the receiver and of the solar concentration ratio to which this receiver is subjected. As it can be expected, the increase in temperature of the hybrid receiver leads to a decrease of the power generated by the PV module, and to an increase in the power generated by the turbine. An increase in the concentration ratio of the solar flux proves to be beneficial on two levels: (1) it leads to a lower temperature sensitivity of the PV module and therefore to an increase in its conversion efficiency (and therefore of its power) for increasing temperatures and (2) it drastically reduces (radiative and convection) heat losses, and thus substantially increases the electric power generated by the turbine for increasing temperatures. It is important to stress that besides the beneficial effect on the maximal electric power, the increase in the solar concentration ratio significantly widens the range of temperatures for which the system offers maximal performances and enables a better control of the equilibrium of powers generated by each converter.

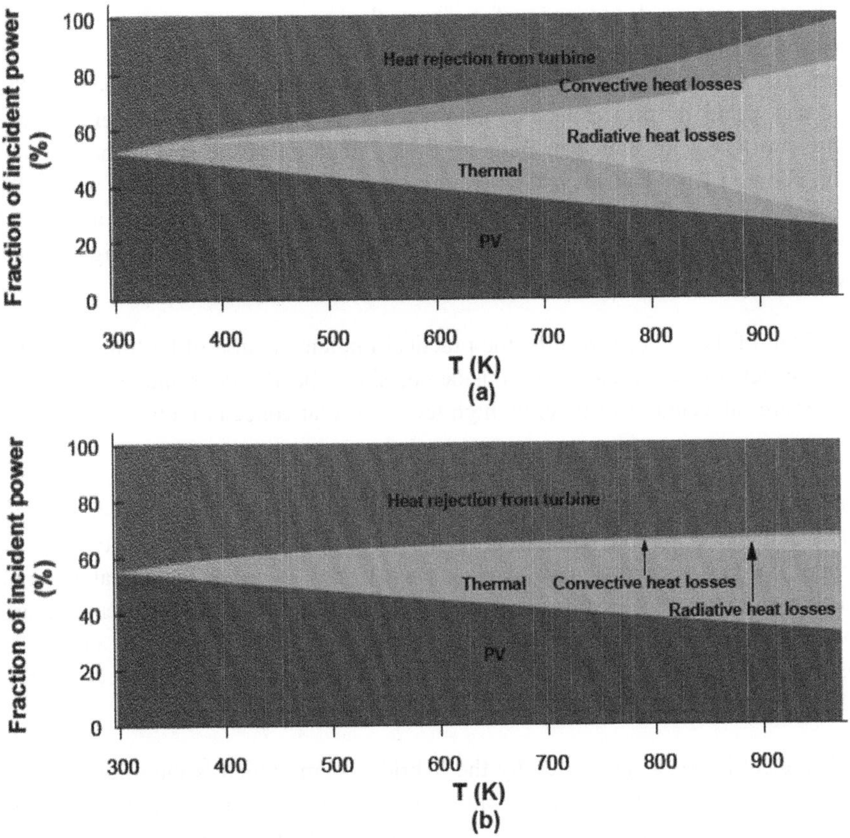

Figure 9.7. *PV, thermal and hybrid conversion efficiencies as a function of the operating temperature and the associated heat losses for cells operating at the radiative limit under a concentration of 100 suns (a) and at the radiative limit under 1,000 suns (b) (Zeitouny et al. 2018). For a color version of this figure, see www.iste.co.uk/flamant/energy.zip*

The main technological obstacle to this approach is quite obviously the development of photovoltaic cells that are efficient at temperature levels largely exceeding the typical operating temperatures of conventional PV modules.

Significant progress was nevertheless achieved during the last five years, mainly by NREL (Perl et al. 2016, 2018) and Arizona State University (Maros et al. 2015) that developed PV cells architectures with acceptable performances at 400°C and 600°C. Figure 9.8 illustrates the evolution of three important characteristics of high-temperature PV cells developed by NREL: open circuit voltage V_{oc}, fill factor

FF and conversion efficiency. Although a notable decrease of the value of these three parameters can be clearly observed for increasing temperatures, solar concentration allows a significant improvement of cell performances at a given temperature. These first results highlight the potential of solar concentration in the development of efficient high-temperature hybrid systems. It is important to note that the physics of PV cells at very high temperature is still widely unknown: the development of efficient high-temperature hybrid systems requires a better comprehension of the physical mechanisms involved and an adaptation of PV cell architecture to extreme conditions of both illumination and temperature.

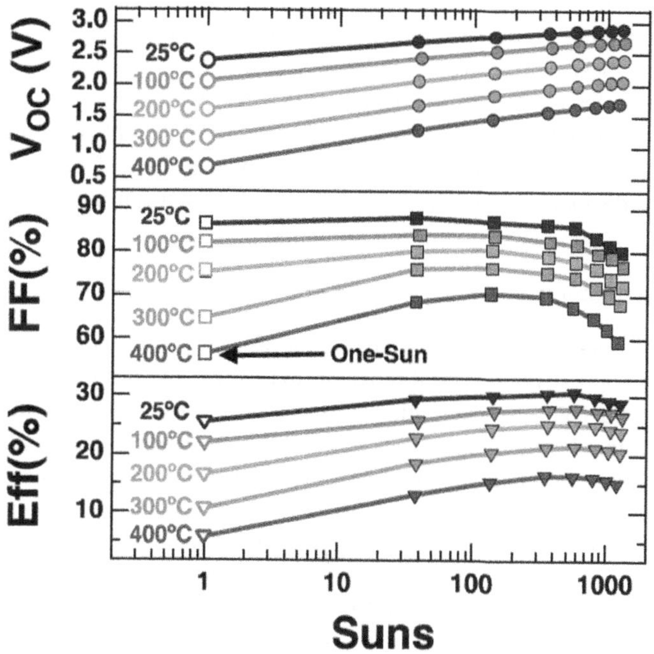

Figure 9.8. *Open circuit voltage (V_{OC}), fill factor (FF) and conversion efficiency of a GaInP/GaInAs tandem cell as a function of solar concentration and for temperatures ranging between 25°C and 400°C (Perl et al. 2018). For a color version of this figure, see www.iste.co.uk/flamant/energy.zip*

An alternative to high-temperature compact hybrid systems resides in the development of integrated receivers operating at significantly lower temperatures. The development of such systems was the subject of considerable research effort in the past years (notably CPV-ORC systems involving turbines relying on Rankine organic cycle and characterized by temperatures barely exceeding 100°C

(Ju et al. 2017). Such a temperature level significantly relaxes the constraints at the level of photovoltaic cells (which now operate at temperature levels much closer to the temperatures encountered on conventional PV or CPV systems). Nevertheless, the low efficiency of ORC turbines over this temperature range raises many questions on the relevance and economic viability of this strategy.

9.4.2. *Spectral splitting*

The last two hybrid approaches detailed in this section differ from the high-temperature approach in the sense that these two PV and CSP converters are physically separated from one another. The incident solar photons are sent to one or the other of these two converters depending on their wavelength (high-energy photons are preferentially absorbed by the PV cells, while low-energy photons are ideally absorbed by the thermal receiver). The nature of the optical component used for spectral separation, and the layout of these two conversion systems differentiates between two distinct sub-families, namely (1) the one-sun approach and (2) spectral separation.

9.4.2.1. *PV one-sun approach*

The "PV one-sun" approach involves the use of PV cells playing the role of photovoltaic converter and optical reflector. In practice, the surface of a concentrator is covered with PV cells integrating a mirror on the backside (or alternatively, a stack of selective layers on the frontside) enabling the highest-energy photons to be absorbed by PV cells. The photons whose energy is lower than the energy band gap of the PV cell are reflected towards the CSP receiver. This approach has many advantages in terms of PV cell operating conditions: these are indeed susceptible to convert a part of the diffuse solar radiation (where the systems involving solar concentration are limited to the use of the direct fraction of radiation). Moreover, the characteristic operating temperatures are significantly lower than in the case of high-temperature systems, contributing to reaching high photovoltaic efficiencies.

A schematic representation of a hybrid "PV one-sun" system is represented in Figure 9.9 for a parabolic trough geometry: the red and near infrared (IR) components of the solar spectrum are transmitted through a film covering the solar cells, while the UV, mid-IR components and a large fraction of visible radiation are reflected to the thermal receiver. Using a suitable optical engineering of the selective layer, it is thus possible to finely optimize the regions of the solar spectrum sent to each of the receivers, depending on the efficiency with which they will be converted. The inventors of the "PV-mirror" concept, affiliated with the University of Arizona, estimate that this system allows an approximately 15% increase in the efficiency of solar-to-electric power conversion, compared to a conventional PV system.

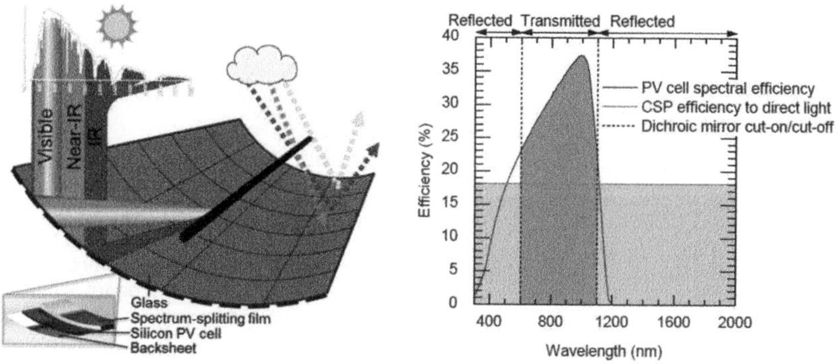

(a) (b)

Figure 9.9. *Schematic representation of the hybrid "PV-mirror" system developed at Arizona State University (a) and diagram representing spectral efficiencies of PV cells and CSP converter, and the ranges of wavelengths used by each converter (Fisher et al. 2017). For a color version of this figure, see www.iste.co.uk/flamant/energy.zip*

9.4.2.2. *Strategies based on the spectral separation of light*

A second form of the hybrid approach involving a spectral splitting of light relies on the use of an optical component integrated to the system and allowing a spectral separation of the incident light. The philosophy of this approach is very close to the previously described "PV one-sun" approach, with the exception that the optical component used for spectral separation is not simultaneously a solar converter. Similar to the "PV one-sun" approach, this strategy offers better control of the operating conditions on each of the converters (it is important to note that this strategy theoretically allows the control of the solar power density on the photovoltaic cells, unlike the "PV one-sun" approach). Figure 9.10 shows an example of hybrid solar power plant based on this approach. The solar flux is concentrated using a heliostat field, before being spectrally separated by a spectral splitter placed on the optical path of light. The photons whose energy is higher than the energy band gap of the cells integrated in the photovoltaic receiver are transmitted through the filter and continue their path to the latter, at the top of the tower. The low-energy photons are reflected to the thermal receiver, placed at the lower part. This configuration allows a decoupling between PV and CSP conversion while benefiting from the assets of solar concentration on the PV receiver.

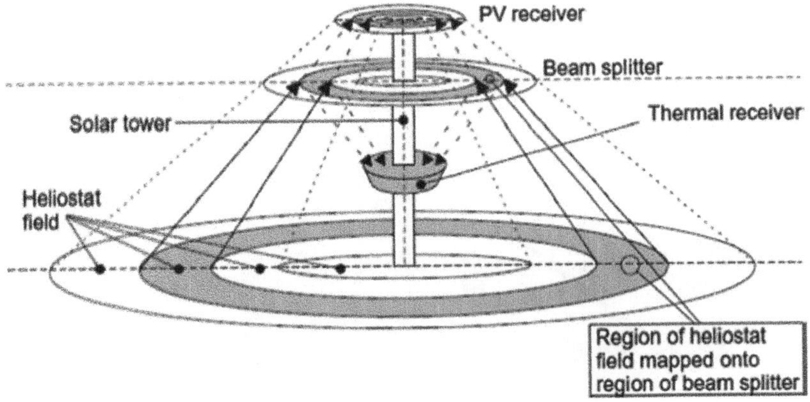

Figure 9.10. *Example of hybrid PV–CSP power plant based on the spectral separation of solar radiation (Imenes et al. 2006)*

The spectral separation of solar light can be achieved in various ways: band-pass filters can be used in order to absorb a part of the solar spectrum and reflect the other, using for this purpose the technology of corrugated filters or Bragg's filters. These filtering technologies are efficient, inexpensive, relatively easy to manufacture and particularly well suited for hybrid applications. Dichroic interference mirrors can be used to split the radiation into two contributions straight around a well-defined cut-off wavelength. Liquid absorption filters (liquids integrating certain types of nanoparticles, heat transfer fluids, etc.) can also be used in order to absorb a part of the solar radiation. Heat transfer fluids can also play the role of filters when they are associated with semiconductor-doped glasses preselected depending on the range of wavelengths to be transmitted or reflected. It is however important to note that the use of liquids has a certain number of drawbacks, given their tendency to rapidly degrade, particularly in extreme environments such as those that can be met under concentrated solar flux.

In summary, although the systems based on the spectral splitting of light are in all likelihood easier to implement than the above described high-temperature hybrid systems, a certain number of major scientific and technological obstacles have to be overcome in order to ensure an optimal efficiency, in particular:

– improve the chemical stability and the mechanical durability of the spectral separation technologies implemented;

– improve the transmission/reflection efficiency of the filters, which must ideally be close to a step function characterized by values of transmissivity and reflectivity varying abruptly between 0 and 100% around a defined cut-off wavelength;

– minimize the angular dependence of spectral separation systems. The angular response of filters has indeed the tendency to vary substantially with the incidence angle of photons, which may be particularly problematic for concentrating systems, generally characterized by a wide distribution of photons.

9.4.3. *Performance-based comparison of the main hybrid strategies*

The performance evaluation of a hybrid system is not as trivial as it may first seem. It is important to recall that such systems must simultaneously ensure a solar-to-electric conversion efficiency as high as possible, and a satisfactory equilibrium between photovoltaic electricity and thermal electricity. Figure 9.11 illustrates this difficulty: the graphic on the left represents the theoretical maximal conversion efficiency of "high-temperature" hybrid systems (red curve), of hybrid systems based on the spectral splitting of light (blue curve) and the "PV one-sun" approach (green curve), as a function of the energy band gap of the PV cell employed. The graphic on the right represents the equilibrium between PV generation and CSP generation, also as a function of the energy band gap of the PV cell (a PV fraction equal to 1 indicates that the entire electric power supplied by the system is of photovoltaic origin. Reciprocally, a 0 value indicates that the entire electricity results from the thermal converter).

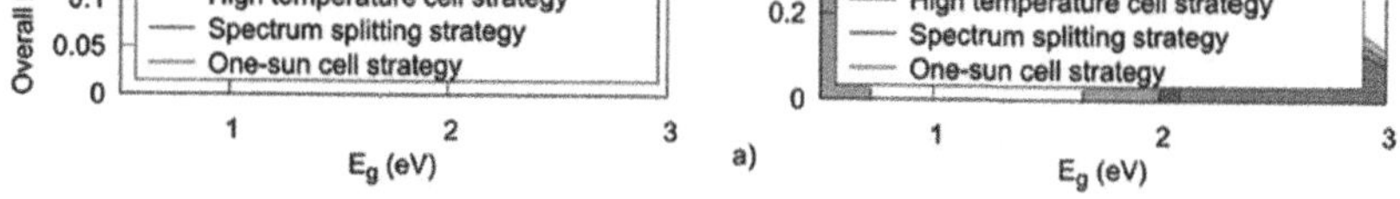

Figure 9.11. *Global conversion efficiency (left) and PV fraction (right) for three compact hybrid strategies, as a function of the energy band gap of the PV cell, and for PV cells operating at the radiative limit (Vossier et al. 2018). For a color version of this figure, see www.iste.co.uk/flamant/energy.zip*

Several important conclusions can be drawn from Figure 9.11. Among the three hybrid PV/CSP strategies represented, the "PV one-sun" and "high temperature" approaches have very similar global conversion efficiencies, although they are observed for significantly different energy band gaps (~1.1 eV for the "high-temperature" approach and ~1.5 eV for the "PV one-sun" approach). The last approach (using the spectral splitting of light) has, on the contrary, a significantly lower global conversion efficiency. A second important conclusion of this figure resides in the equilibrium between PV and CSP, which varies particularly as a function of the considered strategy: the "high-temperature" approach is characterized by a balanced distribution between PV and CSP over a wide range of band gaps (the PV fraction ranging between 0.4 and 0.6 for values of E_g between 0.4 and 2 eV). On the contrary, a strong imbalance in favor of PV can be noted for the two approaches relying on a spectral selection of incident photons, and this over a large range of band gaps. Moreover, although the "PV one-sun" approach has a high global conversion efficiency, the strong imbalance in favor of PV suggests that this approach will likely not be able to guarantee sufficient CSP contribution, limiting in fact the capacity of the system to generate electricity outside sunshine periods.

9.4.4. *Hybrid PV-TS systems*

It is important to recall that the first motivation of hybrid PV–CSP systems is the significant reduction of the cost of solar power due to a combination of two distinct low-cost technologies: photovoltaic solar power plants and thermal energy storage. In this sense, a radical evolution of this concept may be implemented by direct association of these two components, without using the CSP power plant generally associated with thermal energy storage (Gordon et al. 2021). Such a concept is schematically represented in Figure 9.12: a conventional PV power plant is connected to the grid, while the surplus of electric power is converted in the form of heat through resistive heating and transferred to thermal storage materials (generally molten salts). In case of cloudy spells, or during the night, the stored heat can then be converted into electricity using a turbine.

While at first glance this strategy may seem attractive due to its relative simplicity (as it does not require the construction of a conventional CSP solar power plant in addition to the photovoltaic power plant), a major defect hampers its practical implementation: its inefficiency (see also Chapter 8).

Figure 9.12. *Illustration of the PV + thermal storage + steam turbine concept (Gordon et al. 2021). For a color version of this figure, see www.iste.co.uk/flamant/energy.zip*

The efficiency of the heat into electricity conversion is in fact determined by the Carnot efficiency and is typically around 45%. Moreover, the global efficiency of solar power-to-electricity conversion of this approach is particularly low considering the number of conversion processes involved and their relative inefficiency (~20% efficiency for solar power-to-photovoltaic electricity conversion, and ~45% efficiency for thermal power-to-electricity conversion). Nevertheless, the very low cost of photovoltaic electricity generation at the turn of 2020, associated with the very low cost of heat thermal storage, renders this solution potentially attractive from an economic standpoint, compared to solutions of significantly higher performances, but requiring expensive components or technologies. Moreover, this strategy offers increased flexibility compared to the conventional hybrid systems (better use of the diffuse solar resource, which is not at all or poorly used by the hybrid systems, possibility of fully decoupling the photovoltaic field from the thermal energy storage and the turbine, possibility of implementing this approach under latitudes that are less favorable to the installation of concentrating solar power plants, etc.). This solution is still discussed and debated to date and its economic benefit remains to be clarified.

9.5. Innovative hybrid systems

In recent years, original concepts were developed, taking advantage of "exotic" properties and mechanisms of the materials, in order to optimize collection and conversion, while increasing dispatchability of the solar energy conversion systems. This section offers an inexhaustive review of several recently proposed solutions. They all rely on the use of a high-temperature storage fluid allowing power generation to be shifted in time.

9.5.1. *Mixed hybrid systems*

The various compact hybrid systems described in this chapter can essentially be grouped into two large families, which can be distinguished based on the nature of the physical mechanisms implemented in order to harvest, split and direct the solar resource towards each of the PV and CSP receivers. The systems based on the spectral splitting of light offer relative flexibility of use and a certain ease of implementation, each of the two converters being physically separated. Nevertheless, it is difficult to optimally exploit photovoltaic cells in such systems. Indeed, the large cell surfaces required, and the low power density to which these cells are exposed are actually banning the use of very high efficiency PV cells. High-temperature systems offer the advantage of receiver compactness and the possibility of exposing the photovoltaic cells to concentrated solar flux (thus benefiting from the advantages associated with photovoltaic concentration, notably improved conversion efficiency and significant reduction of the required cell surface). Nevertheless, the high temperatures to which these cells are necessarily subjected are definitely an obstacle for the development of this latter approach. One solution may be the development of mixed hybrid receivers, whose operating principle relies on an intersection of these two strategies, similar to the HEATS technology developed by MIT and schematically illustrated in Figure 9.13 (Weinstein et al. 2018).

The HEATS system represented in Figure 9.13 relies on the use of "light pipes" absorbing low-energy photons, while the highest-energy photons are transmitted to the photovoltaic cells on the back of the module. Unlike a conventional high temperature receiver, light pipes do not necessarily need to be in thermal contact with the PV cells, which may be operated at temperatures close to the ambient temperature. Moreover, the external aperture of the module is coated with a thermal insulator that is transparent in the solar spectrum, thus allowing effective transmission of light to the "light pipes", then to the PV cells, while maintaining the surface temperature of the module as low as possible. This provides optimal operating conditions for each converter, while significantly reducing thermal losses.

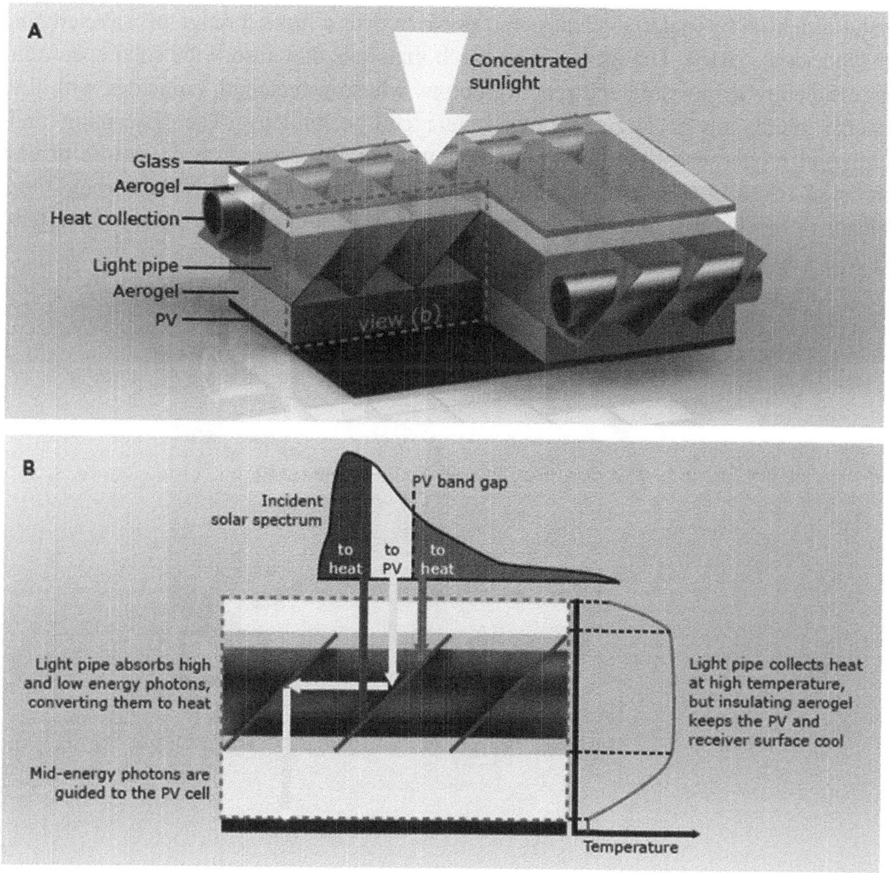

Figure 9.13. *Schematic representation of the HEATS concept. The module consists of a stack of several elements, notably a glass protection, an aerogel layer acting as thermal insulator, spectrally selective light pipes, another aerogel layer and, finally, PV cells. The light pipes contain parallel fins that are coated with a selective coating. These fins are attached to the light pipes, to which they transmit the harvested heat. The latter is absorbed by the heat transfer fluid and is then carried to the thermal storage system (Weinstein et al. 2018). For a color version of this figure, see www.iste.co.uk/flamant/energy.zip*

9.5.2. *Luminescent solar converters*[1]

An original concept recently suggested in order to adapt the spectral content of light to the absorption properties of solar converters resides in the use of a photo-luminescent absorber. This physical mechanism relies on the absorption of solar radiation by materials that reemit photons over a limited range of wavelengths (Manor et al. 2016). The idea is to research materials that absorb the solar spectrum as efficiently as possible and reemit photons whose wavelength coincides with the energy band gap of a given photovoltaic cell technology. The absorption and emission properties of the photoluminescent absorbers depend on the nature of the materials used in the absorber, and also on the temperature of this absorber. This physical mechanism inspired the development of luminescent solar converters (LSC) whose schematic representation is illustrated in Figure 9.14.

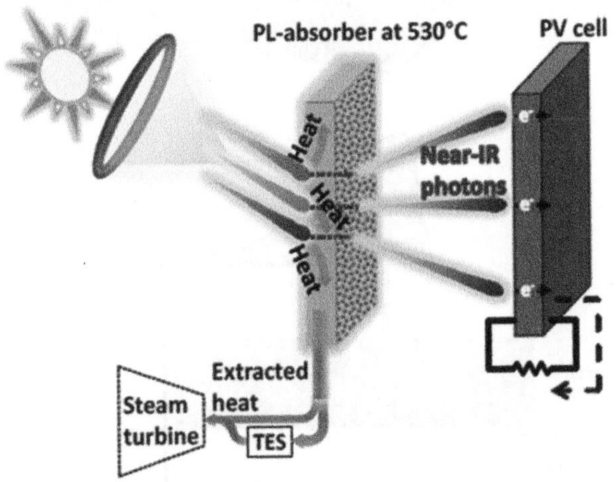

Figure 9.14. *Illustration of LSC concept. (a) The solar flux is concentrated on photo-luminescent absorber. The radiation emitted by photo-luminescence (whose spectrum is red shifted) is absorbed by a solar cell whose energy band gap is adapted, while the residual heat (at 530°C) is stored to be injected in a turbine (Haviv et al. 2020). For a color version of this figure, see www.iste.co.uk/flamant/energy.zip*

1 This concept should not be confused with "luminescent solar concentrators", which refer to the collection and conversion of light by luminescence.

In an LSC, solar radiation is concentrated on a photo-luminescent receiver absorbing high-energy photons, before reemitting mid-energy photons towards photovoltaic cells placed behind the absorber (Haviv et al. 2020). The spectral distribution of photons collected by the solar cells is adapted to the absorption properties of these cells, thus reducing the losses by thermalization, and guaranteeing maximal efficiency of the photovoltaic conversion. Low-energy photons, and the heat released during the conversion of photons in the PL absorber, are absorbed and then evacuated by a heat transfer fluid. The thermal energy thus harvested is then directly injected in a turbine or stored for differed use. It is important to note that the control of the transfer fluid flow rate in the PL absorber allows the temperature control of the latter and the fraction of photons sent to the photovoltaic cells or absorbed by the transfer fluid.

9.5.3. *Very high temperature thermal energy storage coupled with photovoltaic conversion*

Most "hybrid" solutions considered for the conversion of heat into electricity rely on the use of turbines. These offer indeed many advantages, in the sense that the technology is reliable and tested, and the conversion efficiencies are high. On the contrary, photovoltaic cells are relatively well adapted to efficiently convert high temperature heat sources (such as solar radiation, comparable in a first approximation with the radiation emitted by a black body at a temperature of 5,700 K) but their performances are much lower, since the characteristic temperature of the body emitting the radiation is low or moderate. Therefore, the interest in using one or the other of these technologies is largely dependent on the characteristic temperature of the source of energy to be converted. The large majority of the hybrid systems described in this chapter relies on the use of a thermal storage system using molten salts, characterized by a temperature typically ranging between 500 and 600°C. If the characteristic temperature of the storage medium is significantly higher, the use of PV cells as system for the conversion of heat into electricity can be considered. This is the principle on which rely the systems for photovoltaic conversion of thermal energy at very high temperatures (TEGS-MPV) (Amy et al. 2019). The operating principle of such a system is schematically represented in Figure 9.15.

The TEGS-MPV concept, proposed by an MIT team in 2019, involves the use of a silicon bath at very high temperature playing simultaneously the role of thermal storage and source of radiation. The liquid silicon is stored in a "cold" tank at a temperature of 1,900°C (discharged state). The system is charged by pumping liquid silicon with a graphite pump, through pipes irradiated by heating resistors of tungsten or graphite, making it possible to reach a "hot" temperature of 2,400°C. Pumping and heating are supplied by the electricity generated by a conventional (PV, wind, hydraulic, etc.) power plant. When the power plant operator wants to inject electricity in the grid, liquid silicon is pumped from the "hot" pump and injected into the MPV cycle. This cycle is composed of a set of graphite pipes inserted in a tungsten container, playing the role of protective barrier between the graphite pipes and the PV module inserted at the core of the assembly. The internal tungsten wall acts as a photon emitter (similar to an incandescent bulb), irradiating the photovoltaic cells, which convert a fraction of this radiation into electricity. While liquid silicon passes through the set of graphite pipes, it gradually cools and then returns to the "cold" tank, where it can be once again used. Although potentially efficient and inexpensive, this storage strategy is obviously confronted with many scientific and technological challenges, due to which its short-term application is probably unrealistic. Nevertheless, it illustrates the large variety of storage solutions that are studied nowadays in order to use solar electricity as an economical, efficient and dispatchable alternative.

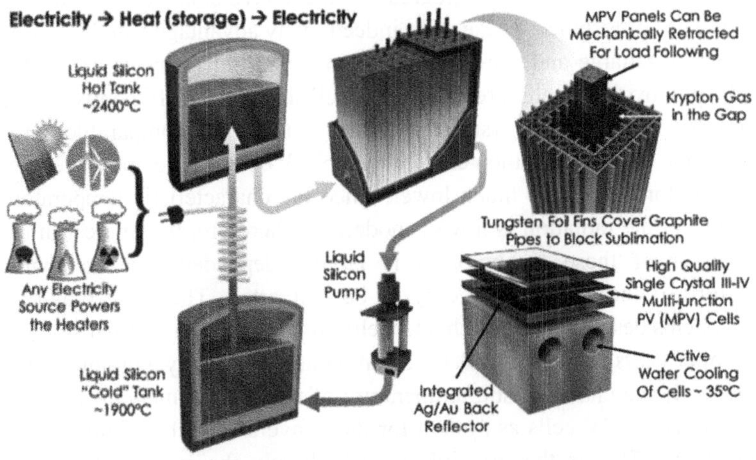

Figure 9.15. *Schematic representation of TEGS-MPV concept: electricity is converted in the form of heat through a heating resistor, before being transferred to a liquid storage in the form of latent heat (1,900–2,400°C). Low-cost metallurgical-grade silicon is used as storage material. When electricity must be generated, the liquid is pumped through a set of pipes, emitting radiation converted into electricity by multi-junction cells located at the center of the set (Amy et al. 2019). For a color version of this figure, see www.iste.co.uk/flamant/energy.zip*

9.6. Conclusion

The massive deployment of solar electricity is nowadays confronted with the lack of dispatchability of the photovoltaic technology. The essential failure of this technology to generate electricity that meets demand is indeed a significant obstacle to its integration in the power grid. Combining the simplicity and the low cost of photovoltaic technologies and thermal storage (inherent to CSP power plants) may lead to significant increase in the dispatchability of solar power plants, while benefiting from the low cost of exploitation of conventional photovoltaic power plants. This chapter reviewed a certain number of strategies suggested during the past decade in order to maximize the efficiency of these systems. With the exception of "non-compact" hybrid systems and PV-TS systems, which rely on the association of two mature and tested solar technologies, the majority of solutions mentioned here remain as of today largely speculative. Significant research and development efforts are required for overcoming the main obstacles to the practical implementation of these solutions. Two major aspects to be addressed relate to (1) the improvement of the temperature strength of materials and components for the strategies involving very high operating temperatures (such as the hybrid "high-temperature" approach or the TEGS-MPV concept), and (2) the development of optical components for the efficient selection and separation of solar photons depending on their wavelength (hybrid strategies relying on the spectral splitting of light, HEATS concept or luminescent solar converters).

9.7. References

Amy, C., Seyf, H.R., Steiner, M.A., Friedman, D.J., Henry, A. (2019). Thermal energy grid storage using multi-junction photovoltaics. *Energy & Environmental Science*, 12(1), 334–343.

Branz, H.M., Regan, W., Gerst, K.J., Borak, J.B., Santori, E.A. (2015). Hybrid solar converters for maximum exergy and inexpensive dispatchable electricity. *Energy & Environmental Science*, 8(11), 3083–3091.

Braun, A., Hirsch, B., Vossier, A., Katz, E.A., Gordon, J.M. (2013). Temperature dynamics of multijunction concentrator solar cells up to ultra-high irradiance. *Progress in Photovoltaics: Research and Applications*, 21(2), 202–208.

Fisher, K., Yu, Z.J., Striling, R., Holman, Z. (2017). PVMirrors: Hybrid PV/CSP collectors that enable lower LCOEs. *AIP Conference Proceedings*, 1850(1), 020004.

FOCUS (n.d.) FOCUS [Online]. Available at: https://arpa-e.energy.gov/technologies/programs/focus [Accessed 29 December 2020].

Gordon, J.M., Fasquelle, T., Nadal, E., Vossier, A. (2021). Providing large-scale electricity demand with photovoltaics and molten-salt storage. *Renewable and Sustainable Energy Reviews*, 135, 110261.

Haviv, S., Revivo, N., Kruger, N., Manor, A., Khachatryan, B., Shustov, M., Rotschild, C. (2020). Luminescent solar power–PV/thermal hybrid electricity generation for cost-effective dispatchable solar energy. *ACS Applied Materials & Interfaces*, 12(32), 36040–36045.

IEEE Spectrum (n.d.). Chile's hybrid PV-solar thermal power stations [Online]. Available at: https://spectrum.ieee.org/energywise/green-tech/solar/chiles-hybrid-pvsolar-thermal-power-stations [Accessed 28 December 2020].

Imenes, A.G., Buie, D., Mills, D.R., Schramek, P., Bosi, S.G. (2006). A new strategy for improved spectral performance in solar power plants. *Solar Energy*, 80(10), 1263–1269.

Ju, X., Xu, C., Hu, Y., Han, X., Wei, G., Du, X. (2017). A review on the development of photovoltaic/concentrated solar power (PV-CSP) hybrid systems. *Solar Energy Materials and Solar Cells*, 161, 305–327.

Manor, A., Kruger, N., Sabapathy, T., Rotschild, C. (2016). Thermally enhanced photoluminescence for heat harvesting in photovoltaics. *Nature Communications*, 7(1), 13167.

Maros, A., Gangam, S., Fang, Y., Smith, J., Vasileska, D., Goodnick, S., Bertoni, M.I., Honsberg, C.B. (2015). High temperature characterization of GaAs single junction solar cells. *2015 IEEE 42nd Photovoltaic Specialist Conference (PVSC)*, 1–5.

Perl, E.E., Simon, J., Geisz, J.F., Lee, M.L., Friedman, D.J., Steiner, M.A. (2016). Measurements and modeling of III-V solar cells at high temperatures up to 400°C. *IEEE Journal of Photovoltaics*, 6(5), 1345–1352.

Perl, E.E., Simon, J., Friedman, D.J., Jain, N., Sharps, P., McPheeters, C., Sun, Y., Lee, M.L., Steiner, M.A. (2018). (Al)GaInP/GaAs tandem solar cells for power conversion at elevated temperature and high concentration. *IEEE Journal of Photovoltaics*, 8(2), 640–645.

Radziemska, E. (2003). Thermal performance of Si and GaAs based solar cells and modules: A review. *Progress in Energy and Combustion Science*, 29(5), 407–424.

Vossier, A., Zeitouny, J., Katz, E.A., Dollet, A., Flamant, G., Gordon, J.M. (2018). Performance bounds and perspective for hybrid solar photovoltaic/thermal electricity-generation strategies. *Sustainable Energy & Fuels*, 2(9), 2060–2067.

Weinstein, L.A., McEnaney, K., Strobach, E., Yang, S., Bhatia, B., Zhao, L., Huang, Y., Loomis, J., Cao, F., Boriskina, S.V. et al. (2018). A hybrid electric and thermal solar receiver. *Joule*, 2(5), 962–975.

Zeitouny, J., Lalau, N., Gordon, J.M., Katz, E.A., Flamant, G., Dollet, A., Vossier, A. (2018). Assessing high-temperature photovoltaic performance for solar hybrid power plants. *Solar Energy Materials and Solar Cells*, 182, 61–67.

10

Synthetic Fuels from Hydrocarbon Resources

Sylvain RODAT and Stéphane ABANADES
PROMES-CNRS, Font-Romeu, France

10.1. Introduction to solar fuels

Concentrated solar energy can be used to convert liquid or gaseous carbonaceous materials (oils, natural gas, etc.) or solid carbonaceous materials (coal, biomass, organic waste, etc.) into gaseous or liquid solar fuels. This relates in particular to endothermal processes of hydrocarbon cracking and reforming, and also to the pyrolysis and gasification of carbonaceous materials (Figure 10.1). The use of concentrated solar power in providing heat for these processes offers various advantages:

– avoids the combustion of fossil resources to generate heat, which reduces the global consumption of hydrocarbons;

– limits the production of CO_2 and therefore the impact on climate change;

– saves raw materials, as they are no longer burnt;

– stores solar energy, having served for providing the enthalpy of reactions, in the products;

– produces cleaner fuels that can be transported and stored;

– delivers heat at very high temperatures with very significant radiative fluxes depending on the concentration ratios and therefore supporting certain reaction paths.

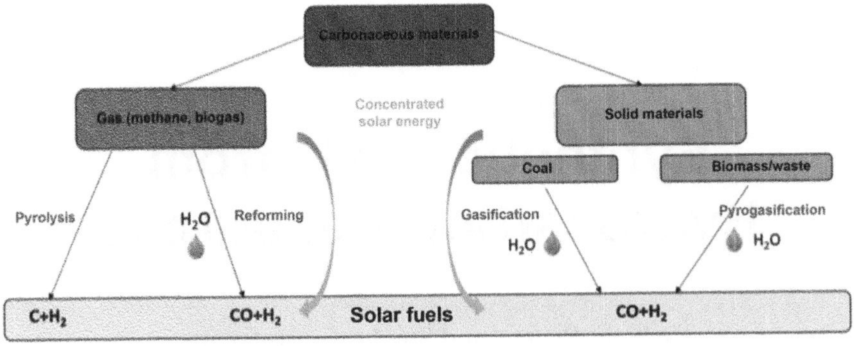

Figure 10.1. *Main processes for manufacturing synthetic fuels from hydrocarbon resources. For a color version of this figure, see www.iste.co.uk/flamant/energy.zip*

The performances of solar reactors mainly rely on two indicators. The energy upgrade factor, often denoted by U, is the ratio between the energy (higher calorific value, HCV) contained in the products and the energy contained in the reactants. The objective is to have U>1, which means that a part of the solar energy is stored in the products:

$$U = \frac{HCV\ products}{HCV\ reactants} \quad [10.1]$$

The efficiency of solar energy conversion into fuel compares the energy of the products with the energy of the reactants plus the solar energy used. This ratio, always below 1, accounts for the use of solar energy; if it is 1, this means that all the solar energy used was stored in the products:

$$\eta = \frac{HCV\ products}{HCV\ reactants + Solar\ energy\ used} \quad [10.2]$$

10.2. Conversion of carbonaceous materials using solar energy

10.2.1. *Solar cracking and reforming of hydrocarbons*

10.2.1.1. *Solar cracking of hydrocarbons*

10.2.1.1.1. Process

Solar cracking, also known as the thermal dissociation, involves the decomposition of a hydrocarbon under the effect of temperature. The general reaction can be written as:

$$C_xH_y \rightarrow y/2\ H_2 + x\ C(s) \quad [10.3]$$

For methane cracking, the reaction becomes:

$$CH_4 \rightarrow 2\ H_2 + C(s)\ \Delta H° = 75\ \text{kJ mol}^{-1} \quad [10.4]$$

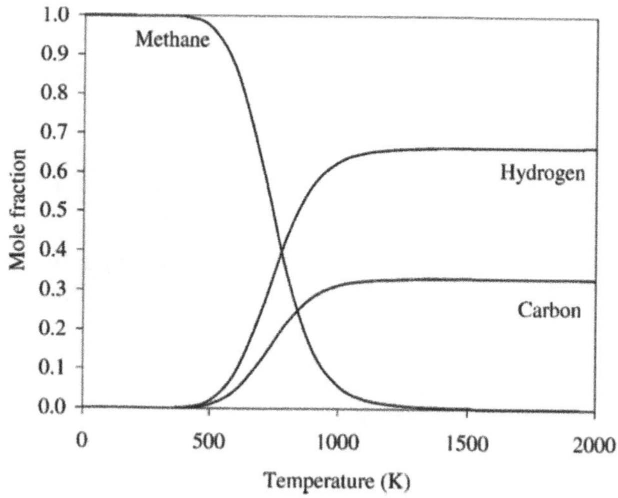

Figure 10.2. *Simplified equilibrium diagram of the dissociation of CH_4 depending on the temperature (Hirsch 2001)*

The dissociation reaction is endothermal; therefore, it requires an energy input that can be supplied by solar energy concentration. The resulting products are a gaseous fuel (H_2) and solid carbon in the form of nanoparticles (carbon black) that are essentially used industrially for the manufacturing of tires. As indicated in Figure 10.2, the reaction is total at thermodynamic equilibrium for temperatures above 1,300 K. Kinetic aspects should also be considered. Depending on the operating conditions (temperature, residence time), sub-products can be formed, particularly acetylene, which is one of the main reaction intermediates.

The energy balance of the reaction is illustrated in Figure 10.3. The energy contained in one mole of CH_4 is 889 kJ (HCV). Following the solar dissociation of the molecule, the energy value of the products increases by the enthalpy of reaction supplied as the solar energy. The interest of having this reaction using the solar energy is to save fossil energy (and therefore avoid CO_2 emissions) and at the same time store the solar energy: 8% of the energy contained in the products is of solar origin (U=1.08). Moreover, 59% of the energy of products is contained in hydrogen and 41% in carbon.

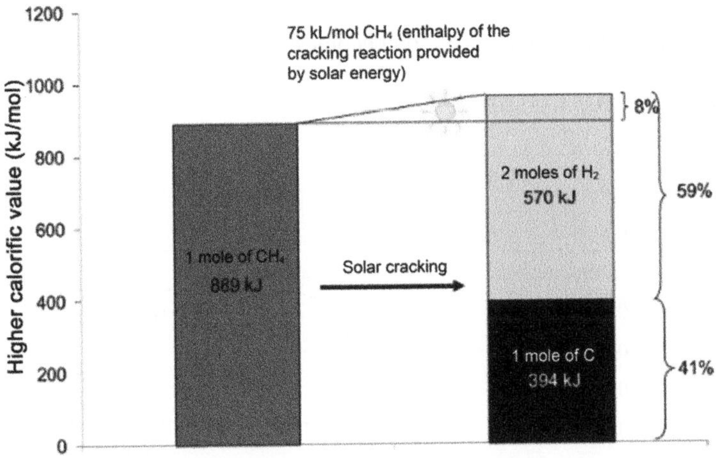

Figure 10.3. *Energy balance of the solar cracking of methane (adapted from Rodat et al. 2010). For a color version of this figure, see www.iste.co.uk/flamant/energy.zip*

It is important to note that this process can be used to convert a fossil resource (methane) to a carbon-free gas (hydrogen). This process is thus considered a transition stage of an economy relying on fossil energies to a sustainable economy based on the use of renewable energies. Indeed, provided that the energy required for the conversion process is renewable, thermal dissociation would be a means to continue using the widely deployed gas infrastructures while avoiding CO_2 emissions from methane combustion. This is the perspective from which solar cracking has been researched for many years.

10.2.1.1.2. State of the art

The first works date back to 1978 (Lédé et al. 1978). A review of the most recent works was proposed by Ozalp et al. (2009). Various reactors were studied, which can be classified into two categories:

– Direct heating reactors:

In direct heating reactors, the reactant gas directly receives the solar radiation, a seeding of particles playing the role of radiative absorbers is required and a large transfer surface is thus obtained.

Hirsch and Steinfeld (Hirsch 2004) worked on a direct heating solar reactor (Figure 10.4). The reactor is composed of a cylindrical cavity separated from the external environment by a window. The methane is injected tangentially. It is seeded with active coal particles, allowing the absorption of concentrated solar light in order to convert it into heat for the reaction. Due to the injection of 10 μm particles, very significant transfer surfaces are obtained. The maximal conversion of methane into hydrogen was of 67% at 1,600 K for a 5 kW power.

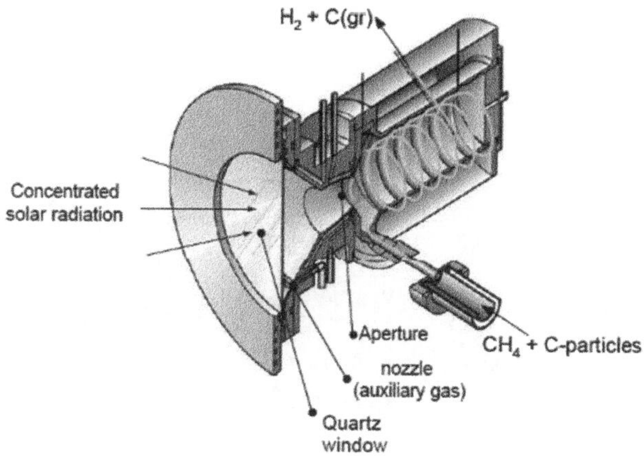

Figure 10.4. *Direct heating solar reactor for methane cracking (Hirsch 2004)*

– Indirect heating reactors:

In indirect heating reactors, the reactant gas does not directly receive the solar radiation, as it is absorbed by the walls of an absorber, then released by radiation, conduction and convection to the reactant gas. The heat transfer surface is smaller than that for direct heating reactors; however, it has two advantages, namely the particle seeding is not necessary and there is no problem of particle deposit on the window.

Two prototypes of indirect heating solar reactors at the scale of 10–50 kW thermal were developed at PROMES-CNRS (Rodat et al. 2011). They use tubes inserted in a cavity, playing the role of the black body (Figure 10.5). The material used is graphite, which has good optical, thermal and mechanical properties at high temperatures. In order to avoid the oxidation of the material, the receiving cavity is insulated by a window and swept by an inert gas. The use of oxidation-resistant materials may allow the operation without a window, which is a weak point from the perspective of technology extrapolation. The heat transfers to gas are also a limiting point. For 900 g/h of CH_4 injected (mole fraction of 50%, the rest being argon) at 1,800 K, a 50 kW reactor generates 200 g/h of H_2 (88% efficiency in H_2), 330 g/h of carbon black (49% of efficiency in carbon) and 340 g/h of C_2H_2 with a thermal efficiency of 15%. Acetylene is the most important sub-product. It can be used as such, or its content can be reduced by increasing the residence time. In order to limit the reaction temperatures, metallic or carbon-based catalysts can be used (Abanades et al. 2015).

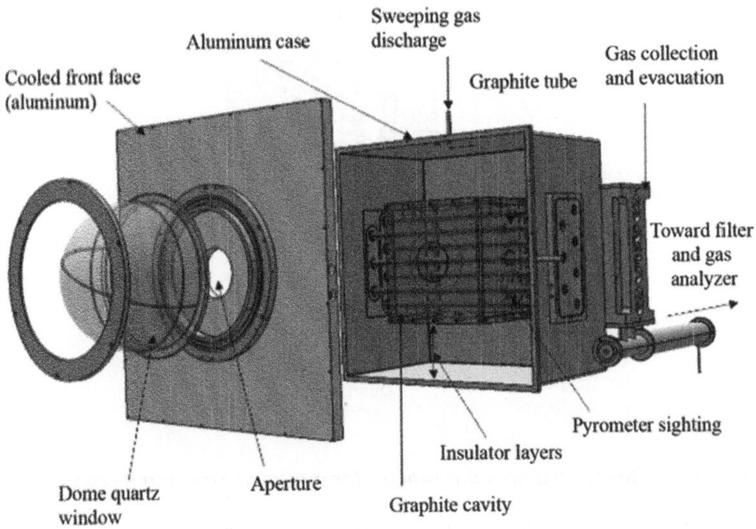

Figure 10.5. *Indirect heating solar reactor for methane cracking. For a color version of this figure, see www.iste.co.uk/flamant/energy.zip*

Other reactors were developed and could be implemented using the solar energy. For example, Shpilrain et al. (1999) worked on a packed-bed reactor. Methane is dissociated in a vector gas (H_2 or N_2) heated through particles. Steinberg (1999) proposed a liquid metal reactor (tin or copper) in which heat is transferred to methane by gas bubbling and carbon separation by density gradient (similar to recovering slags on steel baths). Indeed, carbon (and pyrolytic deposits on the walls) control remains a difficult task for the operation of gas-phase cracking solar reactors. The introduction of water vapor in the system leads to the reforming reaction presented in the following section.

10.2.1.2. *Solar reforming of natural gas*

10.2.1.2.1. Process

Reforming involves the reaction of methane with water or CO_2 in order to produce a mixture of hydrogen and carbon monoxide, commonly known as synthetic gas. This is currently the most widespread process for hydrogen generation. The reaction of methane with carbon dioxide is referred to as dry reforming, in contrast with steam reforming. The reactions at play are the following:

$$CH_4 + H_2O_{(l)} \rightarrow 3H_2 + CO \quad \Delta H° = 249 \text{ kJ mol}^{-1} \qquad [10.5]$$

$$CH_4 + CO_2 \rightarrow 2H_2 + 2CO \quad \Delta H° = 247 \text{ kJ mol}^{-1} \qquad [10.6]$$

These very endothermal reactions are operated at temperatures of the order of 800–900°C on metal catalysts (nickel type). Heat is conventionally added by the combustion of a part of methane, which leads to significant CO_2 emissions. The input of solar thermal energy at high temperatures makes it possible to decrease the methane consumption of the process by about 40%. The temperature effect on the composition at the thermodynamic equilibrium of the equimolar mixture CH_4/H_2O is presented in Figure 10.6. Above a temperature of the order of 900°C, the resulting products are exclusively H_2 and CO with a H_2/CO ratio of 3. The reaction is also facilitated by low pressures (Le Chatelier's principle).

In order to convert carbon monoxide into hydrogen, reforming is generally followed by a stage of conversion through the exothermal reaction of the water–gas shift:

$$CO + H_2O \rightarrow CO_2 + H_2 \quad \Delta H° = -42 \text{ kJ mol}^{-1} \qquad [10.7]$$

From an energy perspective, the steam reforming reaction is equivalent to the succession of the methane cracking reaction and then of the carbon gasification obtained according to:

$$C + H_2O \rightarrow CO + H_2 \quad \Delta H° = 131 \text{ kJ mol}^{-1} \qquad [10.8]$$

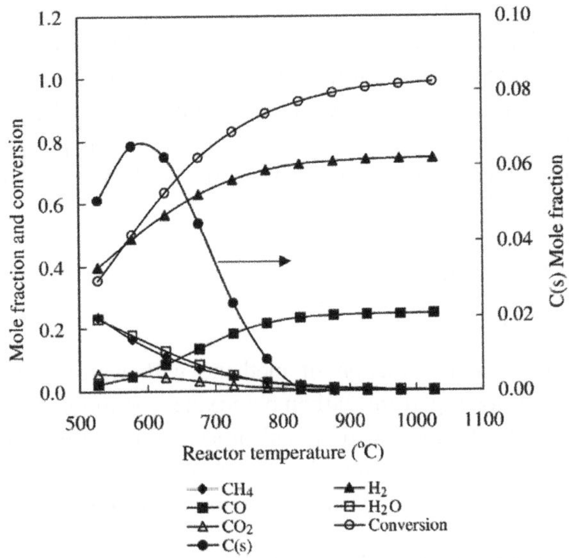

Figure 10.6. *Diagram of equilibrium and conversion of methane during the steam reforming reaction (pressure 1 bar, stoichiometric ratio $CH_4/H_20=1$) (Seo et al. 2002)*

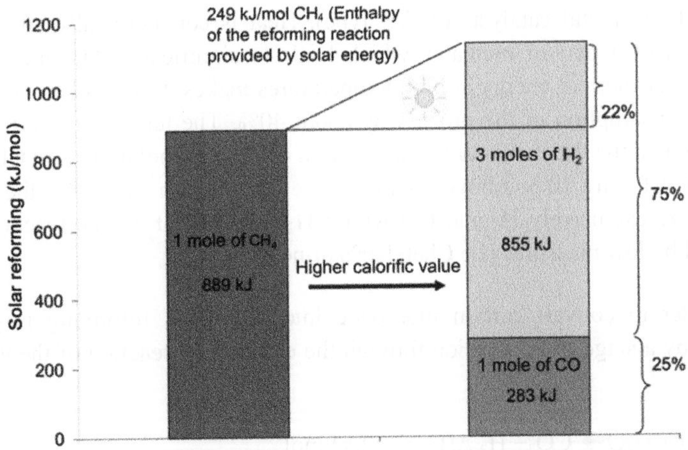

Figure 10.7. *Energy balance of the solar steam reforming of methane. For a color version of this figure, see www.iste.co.uk/flamant/energy.zip*

The energy balance of the reaction is shown in Figure 10.7. It can be noted that the reaction is definitely more endothermal than the pyrolysis reaction, with 22%

(U=1.22) of the final energy of products of solar origin (compared to 8% for pyrolysis). The three-fourths of the final energy are obtained in the form of H_2 and the remaining one-fourth in the form of CO.

10.2.1.2.2. State of the art

Literature reviews proposed by Agrafiotis et al. (2014) or Said et al. (2016) have provided an overview of various works in the field of solar reforming. Three types of reactors are identified: reactors heated by an intermediate fluid, tubular reactors (indirect solar heating) and volumetric reactors (direct solar heating).

Various reformers heated by an intermediate fluid (sodium, air) itself heated by the solar energy have been studied. For example, the ASTERIX (Advanced Steam Reforming of Methane) project, conducted at the end of the 1980s, demonstrated the operation of a reformer whose heat was supplied by the air heated in a solar tower power plant (Böhmer et al. 1991). A maximum of 0.36 kg/s of air at 1,000°C and 9 bars was generated. It supplied a 170 kW tubular reformer with a packed bed of catalyst that allowed the generation of a synthetic gas of industrial quality with a chemical conversion within the range of 68–93%. The decoupled (receiver–reactor) operation can be used for intermediate heat storage means to ensure the continuous operation of the process.

An indirect solar heating tubular reactor of 480 kW was also tested (Epstein 2011) on top of the tower of Weizmann Institute of Science (Israel). It is a cavity in which tubes are installed, which is very similar to a conventional reformer (Figure 10.8). Concentrated solar energy is absorbed through a circular aperture in order to provide the energy required by the reaction. The cavity contains eight tubes (6.35 cm in diameter) that are 4.5 m long. The working pressure is 18 bars for a production of synthetic gas at 800°C by steam or dry reforming. Since this type of reactor represents a significant mass for the receiver, the concept of the "on ground" reformer was considered. It uses an optical system on top of the tower, which redirects solar radiation to the ground (*beam down optics*), which may theoretically allow solar hybridization of a conventional reformer for continuous operation. Indeed, the device may use the solar energy during the day and operate with burners during the periods of low sunshine.

Finally, direct heating reactors were designed, manufactured and tested, such as the direct heating reformer based on foamed ceramics (Figure 10.9) (Möller et al. 2006a). The principle is the direct heating of a substrate offering a large specific surface maximizing both the absorption of solar radiation and catalytic reactions. The expected temperature is 950°C for a pressure of 15 bars.

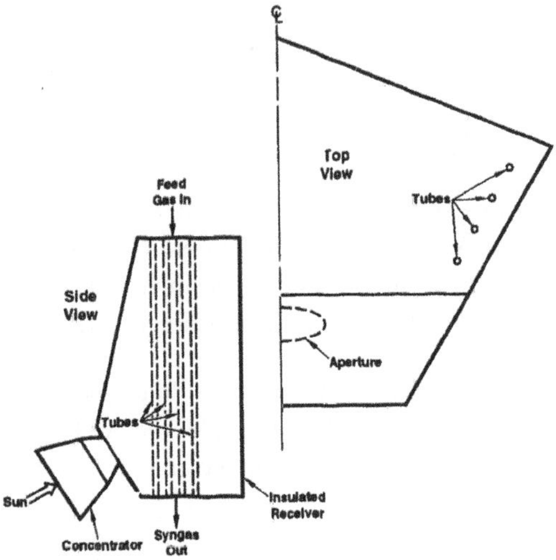

Figure 10.8. *Tubular solar reformer (Epstein 2011)*

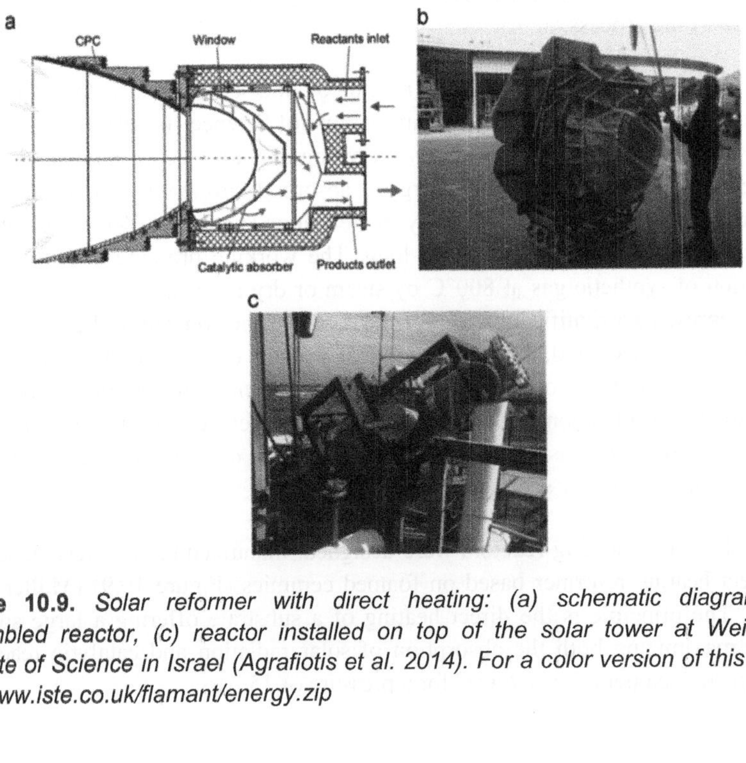

Figure 10.9. *Solar reformer with direct heating: (a) schematic diagram, (b) assembled reactor, (c) reactor installed on top of the solar tower at Weizmann Institute of Science in Israel (Agrafiotis et al. 2014). For a color version of this figure, see www.iste.co.uk/flamant/energy.zip*

10.2.1.3. *Economic aspects*

The cost of hydrogen generation associated with solar-based cracking/reforming technologies is obviously highly dependent on the cost of natural gas. Moreover, concerning methane cracking, the cost of hydrogen depends on the use of the co-produced carbon black. It is also difficult to estimate the costs associated with the solar installation, as until now there is no industrial deployment, and the costs are estimated based on the data available for commercial power plants. For example, Figure 10.10 shows the evolution of the hydrogen generation cost for conventional reforming, solar reforming and cracking of methane depending on the sale price of carbon black. The cost of natural gas considered in the study is 0.24 USD/Nm^3. It can be noted that solar cracking technologies may compete with conventional technologies for a sale price of carbon black of about 1 USD/kg. On the contrary, according to the works of Möller et al., solar reforming may be competitive, starting with a price of natural gas of about 0.35 USD/Nm^3 (Möller et al. 2006b). These results may be influenced in the future by the carbon tax.

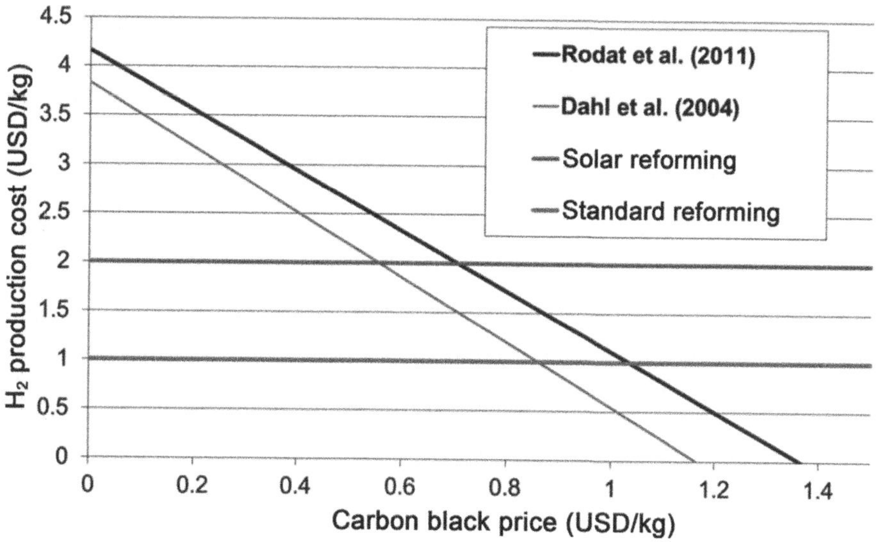

Figure 10.10. *Hydrogen production cost depending on the carbon black price compared to solar reforming and conventional reforming (Rodat et al. 2011). For a color version of this figure, see www.iste.co.uk/flamant/energy.zip*

The first part presented the pyrolysis and reforming of gaseous hydrocarbons; reforming can be assimilated to pyrolysis followed by gasification. The stages of pyrolysis and then gasification are also implemented in the conversion of solid carbonaceous resources.

10.2.2. *Solar pyrolysis and gasification of solid carbonaceous materials*

10.2.2.1. *Process*

As suggested by its name, gasification involves the reaction of a solid carbonaceous material (coal, biomass, etc.) to transform it into synthetic gas according to the following general equation:

$$C_xH_yO_z + (x-z)\ H_2O \rightarrow (y/2+x-z)\ H_2 + x\ CO \qquad [10.9]$$

For a typical biomass whose formula is $C_6H_9O_4$, the equation becomes:

$$C_6H_9O_4 + 2\ H_2O \rightarrow 6.5\ H_2 + 6CO \quad \Delta H^\circ = 800\ kJ\ mol^{-1} \qquad [10.10]$$

Depending on the composition of the initial carbonaceous charge, the reaction mechanisms can be very complex. Nevertheless, three main stages can be identified (Figure 10.11). First, the pyrolysis (or devolatilization) reaction decomposes the solid into char, tars and non-condensable gases in the temperature range of 300–1,000°C. Ashes can also be recovered (mineral matter).

$$\text{Biomass} \rightarrow \text{char} + CO + CO_2 + H_2 + CH_4 + \text{tars} \qquad [10.11]$$

This stage is particularly sensitive to the heating rate that conditions the distribution into solid and gaseous products. The high fluxes and temperatures attainable by solar energy facilitate gas production. The reaction can be slightly endothermal or exothermal depending on the raw material employed (Roberts 1971).

Second, the resulting char is gasified by an oxidizing agent (water vapor or CO_2) according to the following main reactions:

$$C + H_2O \rightarrow CO + H_2 \quad \Delta H^\circ = 131.3\ kJ\ mol^{-1} \qquad [10.12]$$

$$C + 2H_2 \rightarrow CH_4 \quad \Delta H^\circ = -74.6\ kJ\ mol^{-1} \qquad [10.13]$$

$$C + CO_2 \rightarrow 2CO \quad \Delta H^\circ = 172.4\ kJ\ mol^{-1}\ \text{(Boudouard equilibrium)} \qquad [10.14]$$

Finally, the gas phase is governed by two main reactions, namely the reforming reaction (equation [10.5]) and the reaction of the water–gas shift (equation [10.7]) (Dupont et al. 2007).

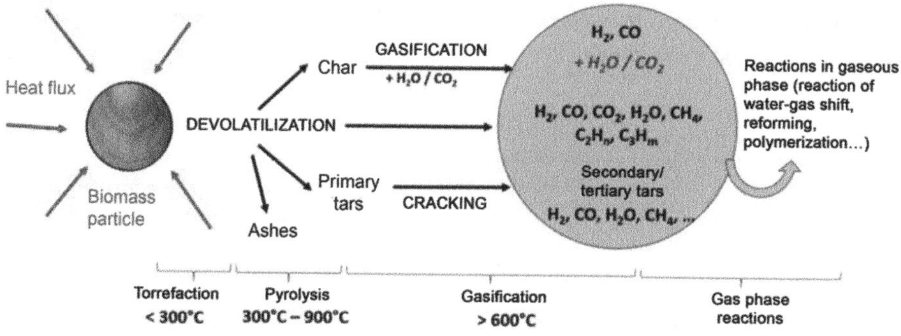

Figure 10.11. *Thermochemical conversion of biomass. For a color version of this figure, see www.iste.co.uk/flamant/energy.zip*

As indicated in Figure 10.12, above a temperature of 1,000°C, the thermodynamic equilibrium of the ($C_6H_9O_4$+$2H_2O$) mixture leads essentially to the formation of H_2 and CO.

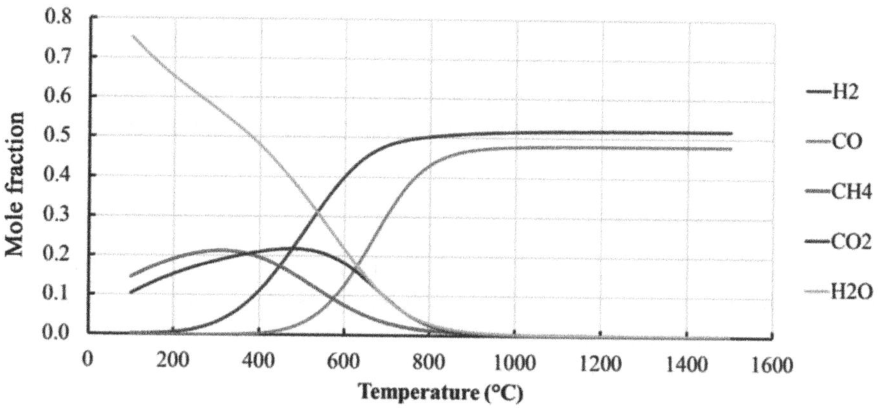

Figure 10.12. *Evolution of the composition at thermodynamic equilibrium of synthetic gas during wood gasification with water depending on the temperature ($C_6H_9O_4$+$2H_2O$) (adapted from Bellouard et al. 2020). For a color version of this figure, see www.iste.co.uk/flamant/energy.zip*

The energy balance of the gasification reaction is illustrated in Figure 10.13 for a typical biomass (beech wood, formula $CH_{1.66}O_{0.69}$). The resulting synthetic gas contains 26% solar energy (U=1.26), which makes solar gasification a good means to store the solar energy in gaseous form (comparable to solar reforming). Its composition is balanced between CO and H_2, with 53% of the energy in the form of H_2 and 47% in the form of CO.

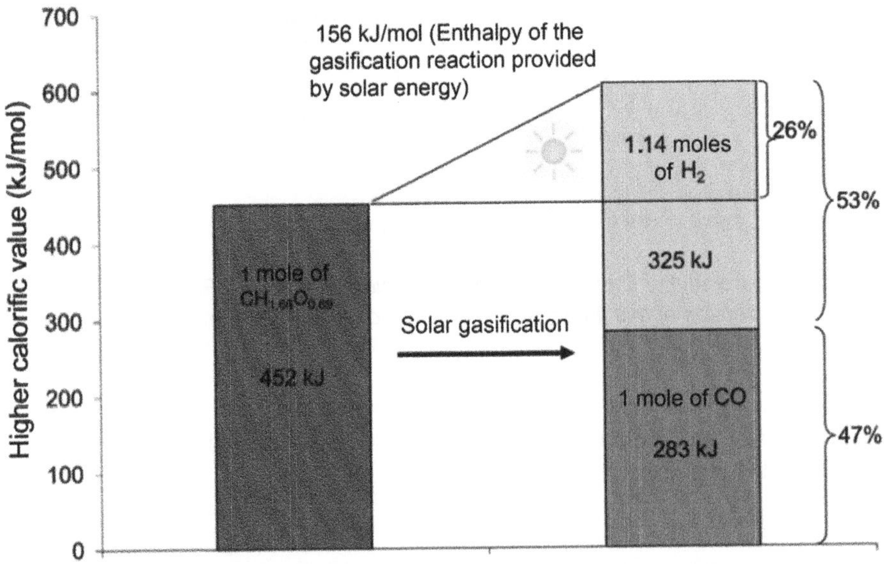

Figure 10.13. *Energy balance of solar gasification of biomass (beech wood, formula $CH_{1.66}O_{0.69}$). For a color version of this figure, see www.iste.co.uk/flamant/energy.zip*

10.2.2.2. *State of the art*

Various reactors were tested for the solar gasification of biomass. The first works date back before the 1980s (Lédé 2016). Various types of reactors were developed (Nzihou et al. 2012); these are in particular packed-bed reactors, fluidized/entrained bed reactors and spouted bed reactors.

A 150 kW packed-bed solar reactor was developed by Wieckert et al. (2013) and tested for the gasification of various types of carbonaceous resources (wood, waste, tires, etc.) at temperatures up to 1,100°C. It simply has two overlapping cavities: the first cavity is filled with the matter to be gasified, which is surmounted by a second one that serves as a solar absorber (Figure 10.14). This type of reactor can be used to reach significant conversion rates due to a long residence time. It may accept a wide range of raw materials, both in terms of composition and particle size. On the contrary, the thickness of the packed bed drives significant temperature gradients in its height, which limits the control of the reaction products and facilitates the formation of tars. Furthermore, it operates in batch (discontinuous) mode. A maximal energy upgrade factor of 1.3 and solar energy conversion efficiencies between 22 and 35% are reached.

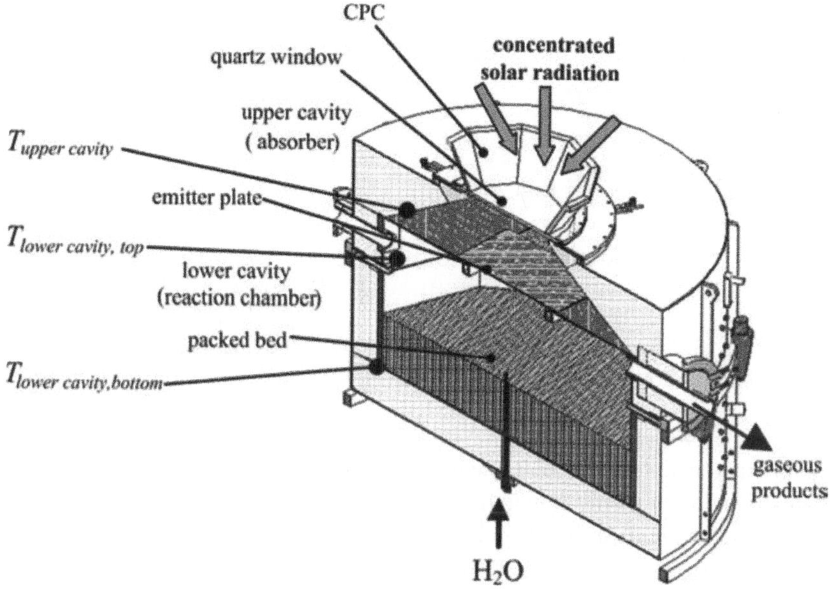

Figure 10.14. *A 150 kW packed-bed reactor for solar gasification (Piatkowski and Steinfeld 2011)*

In order to limit the temperature gradients, various concepts of circulating beds (fluidized beds, entrained beds, cyclones) were developed in direct or indirect heating modes. The drawback of this type of device is that it generally requires small (<1 mm) particles. The option of a spouted bed reactor is also considered (Bellouard et al. 2017). This type of reactor (Figure 10.15) can be used for processing various biomasses with particle size over millimeter scale and at temperatures up to 1,400°C. It has a vertical cone cavity that is able to directly or indirectly receive solar radiation. Biomass is supplied through a hopper that carries the particles in the upper part of the cavity. The particles fall due to gravity at the bottom of the cone where the oxidizing agent (CO_2, H_2O) is injected. The effect of this spout is that it drives these particles into a cyclic motion until the complete reaction. The residence time of the solid is therefore maximized and the stirring facilitates heat and mass transfers. An upgrade factor of the initial energy of the biomass of 1.24 is obtained. The injection of oxygen, in addition to water or carbon dioxide, facilitates the burning of a part of the biomass during the low sunshine periods, in order to keep the system operating.

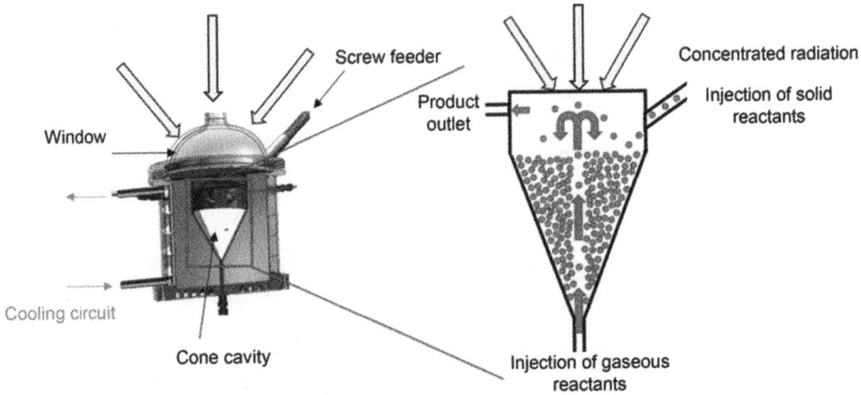

Figure 10.15. *3D view and schematic representation of the spouted bed reactor (1.5 kW). For a color version of this figure, see www.iste.co.uk/flamant/energy.zip*

In order to maintain the solar reactor operating even during the night, research is being carried out on gasification in molten salts (Li_2CO_3, Na_2CO_3, K_2CO_3) at 1,200 K. An intermediate fluid can be used to enhance heat transfers, in order to consider a heat storage means, and it also offers catalytic properties. Salt stability at high temperatures is a point to be optimized. A 3 kW prototype is presented in Figure 10.16. It is composed of a receiving cavity with a double wall through which the molten salt flows, in which cellulose and water vapor are injected. An energy upgrade factor of 0.9 is obtained for a 78% carbon conversion.

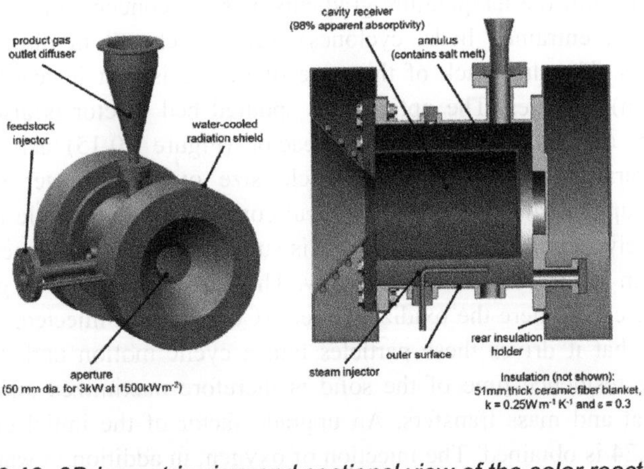

Figure 10.16. *3D isometric view and sectional view of the solar reactor for the gasification of cellulose into molten salts (Hathaway and Davidson 2020)*

10.2.2.3. *Economic aspects*

Concerning the solar gasification of carbonaceous materials, the cost-effectiveness of the process compared to conventional gasifiers is not obvious. Indeed, the autothermal process of biomass gasification using a part of the charge to provide the energy required for the reaction does not emit fossil CO_2; therefore, the conventional process is not penalized by the carbon tax. The solar process is cost-effective provided that the cost of the part of matter that is conventionally burned compensates the cost of the solar installation, which is not the case under the current economic circumstances (Nickerson et al. 2015). On the contrary, the case of the solar gasification of coal cuts the CO_2 emissions by 26%, according to the works of Saw et al. (2015). According to the authors' estimates, the price of the liquid fuel produced ranges between 5.5 and 5.8 USD/kg for a hybrid solar process. This process uses the solar energy when it is available and continues to operate by coal combustion in addition to or instead of the solar energy as needed (notably during the night). By comparison, the conventional process of coal gasification delivers liquid fuel at 4.8 USD/kg. Other carbonaceous materials can be used for solar gasification, particularly organic waste and other plastic materials. Solar energy could thus provide a sustainable solution to the societal problem of waste management.

10.3. Conclusion and perspectives

This chapter focused on the main pathways for producing synthetic fuels from concentrated solar energy and carbonaceous materials. On the one hand, it is possible to convert hydrocarbons (gases, liquids) by thermolysis or reforming. On the other hand, carbonaceous solids can be gasified to produce a synthetic gas. The solar energy input in these endothermal reactions allows the supply of the processes with renewable energy and the storage of a part of the solar energy in the form of gaseous fuel. Gaseous fuels can be transformed into other fuels by the Fischer–Tropsch process, which allows the conversion of a synthetic gas into liquid hydrocarbon. There are other pathways for the solar-based use of carbonaceous matter, such as the carboreduction of metal oxides (metallurgy) or the chemical looping reforming and gasification (Chuayboon and Abanades 2020). All these processes can contribute to a transition from an economy based on fossil energies to a sustainable economy by progressively integrating concentrated solar energy as a source of high-temperature heat.

10.4. References

Abanades, S., Kimura, H., Otsuka, H. (2015). A drop-tube particle-entrained flow solar reactor applied to thermal methane splitting for hydrogen production. *Fuel*, 153, 56–66.

Agrafiotis, C., von Storch, H., Roeb, M., Sattler, C. (2014). Solar thermal reforming of methane feedstocks for hydrogen and syngas production: A review. *Renewable and Sustainable Energy Reviews*, 29, 656–682.

Bellouard, Q., Abanades, S., Rodat, S. (2017). Biomass gasification in an innovative spouted-bed solar reactor: Experimental proof of concept and parametric study. *Energy & Fuels*, 31(10), 10933–10945.

Bellouard, Q., Rodat, S., Grateau, M., Abanades, S. (2020). Solar biomass gasification combined with iron oxide reduction for syngas production and green iron metallurgy. *Frontiers in Energy Research*, 8, 66.

Böhmer, M., Langnickel, U., Sanchez, M. (1991). Solar steam reforming of methane. *Solar Energy Materials*, 24(1–4), 441–448.

Chuayboon, S. and Abanades, S. (2020). An overview of solar decarbonization processes, reacting oxide materials, and thermochemical reactors for hydrogen and syngas production. *International Journal of Hydrogen Energy*, May, S0360319920314543.

Dupont, C., Boissonnet, G., Seiler, J.-M., Gauthier, P., Schweich, D. (2007). Study about the kinetic processes of biomass steam gasification. *Fuel*, 86(1–2), 32–40.

Epstein, M. (2011). Solar thermal reforming of methane. SFERA Winter School, Zürich.

Hathaway, B.J. and Davidson, J.H. (2020). Solar steam gasification of cellulose in a molten alkali salt: Salt chemistry and reactor performance. *Energy & Fuels*, 34(2), 1811–1821.

Hirsch, D. (2001). The solar thermal decarbonization of natural gas. *International Journal of Hydrogen Energy*, 26(10), 1023–1033.

Hirsch, D. (2004). Solar hydrogen production by thermal decomposition of natural gas using a vortex-flow reactor. *International Journal of Hydrogen Energy*, 29(1), 47–55.

Lédé, J. (2016). Pyrolysis and gasification of biomass in solar and simulated solar environments: The pioneering works of Michael J. Antal in the period of 1976–1989. *Energy & Fuels*, 30(10), 7896–7903.

Lédé, J., Weber, C., Villermaux, J. (1978). Evaluation of the efficiency of a device which uses solar energy to apply thermal shocks to gas. *Academie des Sciences (Paris), Comptes Rendus, Serie B – Sciences Physiques*, 286(22), 299–302.

Möller, S., Friedmann, S., Walter, M., ten Dam, J. (2006a). SOLREF: Development of an advanced solar high-temperature reformer. *Solar Energy. ASME 2006 International Solar Energy Conference. ASMEDC*, Denver, Colorado, 269–276 [Online]. Available at: https://asmedigitalcollection.asme.org/ISEC/proceedings/ISEC2006/47454/269/319156 [Accessed 23 June 2020].

Möller, S., Kaucic, D., Sattler, C. (2006b). Hydrogen production by solar reforming of natural gas: A comparison study of two possible process configurations. *Journal of Solar Energy Engineering*, 128(1), 16–23.

Nickerson, T.A., Hathaway, B.J., Smith, T.M., Davidson, J.H. (2015). Economic assessment of solar and conventional biomass gasification technologies: Financial and policy implications under feedstock and product gas price uncertainty. *Biomass and Bioenergy*, 74, 47–57.

Nzihou, A., Flamant, G., Stanmore, B. (2012). Synthetic fuels from biomass using concentrated solar energy: A review. *Proceedings of the ICE – Energy*, 42(1), 121–131.

Ozalp, N., Kogan, A., Epstein, M. (2009). Solar decomposition of fossil fuels as an option for sustainability. *International Journal of Hydrogen Energy*, 34(2), 710–720.

Piatkowski, N. and Steinfeld, A. (2011). Solar gasification of carbonaceous waste feedstocks in a packed-bed reactor – Dynamic modeling and experimental validation. *AIChE Journal*, 57(12), 3522–3533.

Roberts, A.F. (1971). The heat of reaction during the pyrolysis of wood. *Combustion and Flame*, 17(1), 79–86.

Rodat, S., Abanades, S., Sans, J.-L., Flamant, G. (2010). A pilot-scale solar reactor for the production of hydrogen and carbon black from methane splitting. *International Journal of Hydrogen Energy*, 35(15), 7748–7758.

Rodat, S., Abanades, S., Flamant, G. (2011). Methane decarbonization in indirect heating solar reactors of 20 and 50 kW for a CO2-free production of hydrogen and carbon black. *Journal of Solar Energy Engineering*, 133(3), 031001.

Said, S.A.M., Waseeuddin, M., Simakov, D.S.A. (2016). A review on solar reforming systems. *Renewable and Sustainable Energy Reviews*, 59, 149–159.

Saw, W., Kaniyal, A., van Eyk, P., Nathan, G., Ashman, P. (2015). Solar hybridized coal-to-liquids via gasification in Australia: Techno-economic assessment. *Energy Procedia*, 69, 1819–1827.

Seo, Y.-S., Shirley, A., Kolaczkowski, S.T. (2002). Evaluation of thermodynamically favourable operating conditions for production of hydrogen in three different reforming technologies. *Journal of Power Sources*, 108(1–2), 213–225.

Shpilrain, E., Shterenberg, V., Zaichenko, V. (1999). Comparative analysis of different natural gas pyrolysis methods. *International Journal of Hydrogen Energy*, 24, 613–624.

Steinberg, M. (1999). Fossil fuel decarbonization technology for mitigating global warming. *International Journal of Hydrogen Energy*, 24(8), 771–777.

Wieckert, C., Obrist, A., Zedtwitz, P., von Maag, G., Steinfeld, A. (2013). Syngas production by thermochemical gasification of carbonaceous waste materials in a 150 kW_{th} packed-bed solar reactor. *Energy & Fuels*, 27(8), 4770–4776.

11

Solar Fuel Production by Thermochemical Dissociation of Water and Carbon Dioxide

Stéphane ABANADES and Sylvain RODAT
PROMES-CNRS, Font-Romeu, France

11.1. Introduction

The solar thermochemical production of synthetic fuels without greenhouse gas emission can be achieved using water and carbon dioxide, and concentrated solar power as a source of heat at high temperature. The various solar fuels considered are hydrogen, synthetic gas (with H_2 and CO as main components) and the derived fuels (methanol, synthetic liquid fuels). Hydrogen is an energy vector allowing long-term storage and transportation of solar energy, for its use in direct combustion or in a fuel cell. The use of CO_2 emissions resulting, for example, from industrial processes is also a major challenge, notably for the conversion of CO_2 into synthetic fuel using solar energy. Hydrogen production and CO_2 recycling into liquid fuel can be expected to develop into an economic sector in the medium term. This sector would include three main industrial components:

– ***Resources and energy harvesting***. Although capturing atmospheric CO_2 is not yet at an industrial stage, many industrial applications already have the potential to provide the CO_2 required for the startup of this sector, this subject being rapidly expanding particularly in terms of CO_2 capture and sequestration. H_2O can be obtained by desalination of seawater. Finally, among the available energy sources,

Concentrating Solar Thermal Energy,
coordinated by Gilles FLAMANT. © ISTE Ltd 2022.

concentrated solar energy has the potential to cover the energy needs of thermochemical processes.

– ***H_2O and CO_2 dissociation.*** Among the studied paths, photo-electrolysis seems the most difficult to conduct at large scale, given the many constraints weighing on the development of materials that compose the cells and the low efficiencies associated. Electrolysis is the closest to a commercial stage thanks to its proximity to fuel cells, but more adapted and less expensive materials are still being explored. This process remains largely dependent on an inexpensive source of electricity and its implementation with a renewable (solar, wind) source is considered within the framework of storage of these intermittent energies. Finally, thermochemical cycles based on metallic oxides are in an intermediate stage of development. Their interest resides in a higher efficiency potential, heat being directly converted into chemical energy without intermediate conversion into electrical energy.

– ***Liquid fuel production*** is undoubtedly the best-managed point of this potential sector through the Fischer–Tropsch industrial catalytic process.

This chapter describes the various thermochemical paths for H_2O and CO_2 dissociation (electrochemical, photochemical and photobiological pathways are not considered). Among these, the dissociation of H_2O and CO_2 by thermochemical cycles using metallic oxides seems to be the most efficient, solar energy being directly used as heat, without losses related to the transformation into electricity (as it would be the case for electrolysis), and with an energy efficiency that is potentially higher than that of thermolysis due to a reduced temperature, and therefore less radiative losses. Various categories of oxide-based cycles and the solar reactors for their implementation are presented.

11.2. Direct H_2O and CO_2 thermolysis

Thermolysis consists of breaking a bond with an oxygen atom at high temperature, according to the reaction:

$$CO_2 \rightarrow CO + \frac{1}{2}O_2 \qquad [11.1]$$

$$H_2O \rightarrow H_2 + \frac{1}{2}O_2 \qquad [11.2]$$

The variations of enthalpy and free energy are given in Figure 11.1 as a function of temperature, at atmospheric pressure. The temperature ranges corresponding to the various methods considered are also mentioned.

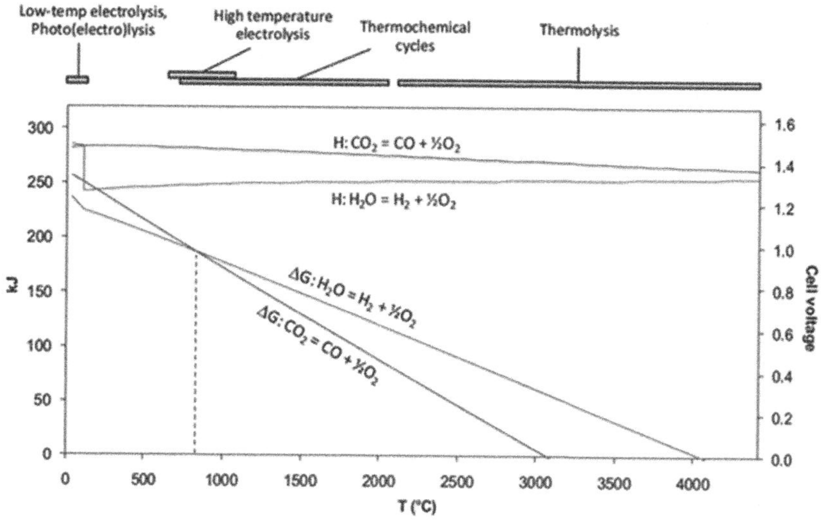

Figure 11.1. *Evolution of enthalpy and free energy associated with CO_2 and H_2O thermolysis as a function of temperature. For a color version of this figure, see www.iste.co.uk/flamant/energy.zip*

CO_2 thermolysis is very endothermal, and it cannot occur spontaneously below 3,000°C. This temperature threshold is even higher for H_2O thermolysis. Several simple observations tend to show that working at these high temperatures is not relevant:

– Designing a reactor capable of supporting extended thermal cycles is a costly challenge, since the materials that can support such temperatures are rare and expensive.

– The reaction can be prevented from getting reversed (i.e. recombination of products), either by a means to separate H_2 (or CO) and O_2 (a potentially explosive mixture) at very high temperature, or by conducting a rapid quenching of this gas mixture. Currently, no technology is able to solve these issues.

– Since radiative losses depend on the fourth power of temperature, it is unrealistic to expect high-energy efficiency for this type of process.

– The only renewable source that could be used in order to reach such high temperature levels is concentrated solar power. It is difficult to directly heat a gas (H_2O, CO_2) by solar radiation, though these molecules absorb a part of the solar spectrum, and the complexity of the process needs to be increased by adding, for example, particles for radiation absorption and diffusion.

Despite these difficulties, a patent was filed by Jensen et al. (2000). This process relies on the evolution of the CO_2 absorption spectrum as a function of temperature. Indeed, according to their observations, by preheating the gas to 1,800 K, solar energy is correctly absorbed and generates a photolysis of CO_2 into CO, O and O_2. The mixture is cooled by bulk addition of CO_2 at ambient temperature, which reacts with O to form more CO. The announced efficiency of transformation of CO_2 into CO is of about 6%. This process is at the limit of thermolysis, as even though high temperature is at the origin of the reaction, concentrated photons are the ones triggering it, and not the temperature on its own.

11.3. Thermochemical cycles

11.3.1. *Principle*

One method to overcome the difficulties related to direct thermolysis is to decompose this reaction in several stages, using thermochemical cycles. Advanced for the first time by Funk et al. (1966), this technique drew a lot of attention from laboratories during the first oil crisis in 1973, in the intent to produce hydrogen from water using the heat released by a nuclear reactor (up to 950°C). The interest then faded away and, beginning in the 2000s, was strongly revived, especially due to the potential of concentrated solar energy to provide heat at very high temperatures (Abanades et al. 2006).

The most widely studied cycles for a high-temperature application (>1,000°C) are the cycles based on oxidation–reduction reactions of metal oxides. The two stages are (1) the reduction of the metal oxide (MO_x) at high temperature, producing a reduced oxide and oxygen (equation [11.3]), and (2) the reoxidation of this product in the presence of CO_2 (or H_2O), regenerating the metal oxide and producing CO (or H_2) (equation [11.4]):

$$MO_x \rightarrow MO_{x-y} + \frac{y}{2}O_2 \quad T>1{,}000°C,\ \Delta H>>0 \qquad [11.3]$$

$$MO_{x-y} + yCO_2(H_2O) \rightarrow MO_x + yCO(H_2) \quad T<1{,}000°C,\ \Delta H<0 \qquad [11.4]$$

The estimated cost of the hydrogen produced with this cycle would range between 46 and 66 USD/GJ (5.5–7.9 USD/kg) (which is higher than electrolysis using nuclear electricity) (Charvin et al. 2008a; Graf et al. 2008), but with better prospects of being reduced, thanks to the rapid drop in infrastructure costs.

The separated or simultaneous production of H_2 and CO allows the production of a syngas with a controlled H_2/CO ratio, which can be used for the synthesis of liquid fuels (synthetic fuels). This stage is essential for the development of this pathway, as it ensures a transition towards a future economy of hydrogen and fuel cells, and a storage of hydrogen in the form of a fuel with high-energy density by weight. Indeed, CO and H_2 have good energy density by weight, but low energy density by volume (Figure 11.2; note that diesel is close to gasoline and DMF, that is, dimethylfuran or third-generation biofuel).

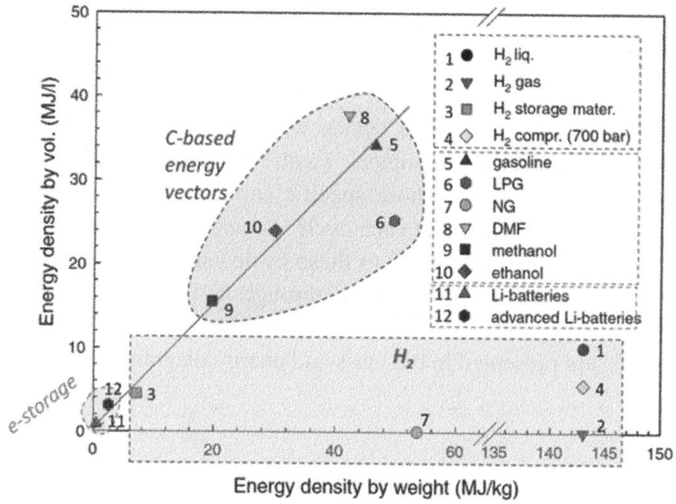

Figure 11.2. *Comparison of energy densities by weight and by volume for various energy vectors (Centi and Perathoner 2011). For a color version of this figure, see www.iste.co.uk/flamant/energy.zip*

Once the syngas mixture obtained, the most "direct" way to obtain liquid fuels is the Fischer–Tropsch catalytic synthesis, this method being commercially used since the 1920s for the generation of liquid fuel from syngas obtained by the gasification of carbon or hydrogen produced by steam reforming of natural gas. The process can be represented by the following reaction (Dry 2002):

$$2H_2(g) + CO(g) \rightarrow -CH_2- (l) + H_2O(g), \Delta H = -165 \text{ kJ/mol} \quad [11.5]$$

where $-CH_2-$ is a link in the carbon chain. The length of the carbon chain, which determines the fuel quality, depends on temperature, pressure, composition of the syngas used and on the catalyst (Espinoza et al. 1999; Steynberg et al. 1999). Moreover, the separated production of the two components of syngas facilitates the fine adjustment of its composition and influences the selectivity to one of the Fischer–Tropsch process products. Finally, catalyst poisoning by sulfur is prevented.

Several types of cycles can be differentiated, notably based on the technology to be implemented for using them (Abanades et al. 2006; Abanades 2019). Cycles with more than two steps have the advantage of requiring lower temperatures, compatible with a nuclear heat source, but involve a more complicated management of the flow of materials, separation and corrosion problems, which will not be addressed here. Concerning the two-step cycles, a distinction can be made between the cycles for which the reduction step leads to a partial reduction of the oxide (non-stoichiometric cycles) or a different species (stoichiometric cycle). This species can then be obtained in gaseous form (cycles with volatile oxides) or not (cycle with non-volatile oxides). Non-stoichiometric cycles have a lower productivity in H_2/CO per unit mass of oxide. As for the cycles with volatile oxides, they generate a gaseous product that easily recombines (with O_2) and must be recovered by condensation and filtration, but whose specific surface area is significant (several dozens of m^2/g), and regenerated at each cycle. Besides these technological criteria, the thermodynamic properties related to these cycles are highly important, as they will determine the maximal efficiencies through a theoretical exergy analysis (Steinfeld and Palumbo 2001; Abanades et al. 2006; Galvez et al. 2008). The various sources and wells are presented in the mass and energy diagrams (Figure 11.3).

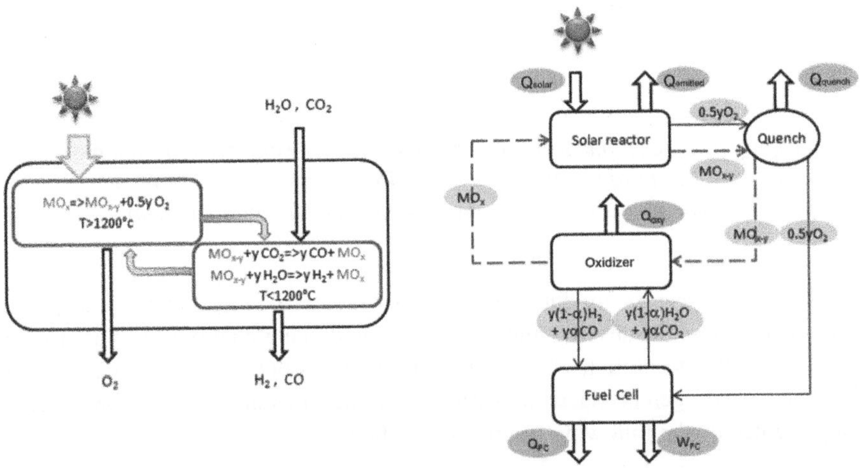

Figure 11.3. *Mass and energy diagram for a thermochemical cycle. For a color version of this figure, see www.iste.co.uk/flamant/energy.zip*

11.3.2. *Cycles with volatile oxides*

ZnO/Zn. The ZnO/Zn cycle was widely studied for its significant theoretical energy efficiency (efficiency of the conversion of solar energy into chemical energy of 35% and up to 50% depending on the heat recovery degree (Steinfeld 2002; Loutzenhiser et al. 2010).

The kinetics of the reduction of zinc oxide was studied by means of a solar-driven thermogravimetry setup (Schunk and Steinfeld 2009; Levêque and Abanades 2015), giving for a temperature of 1,800–2,100 K and a zero-order reaction, an activation energy of 361±53 kJ/mol. Other solutions were considered (Levêque and Abanades 2013), notably the reduction in the form of aerosol, which improves by several orders of magnitude the pre-exponential factor (Perkins et al. 2007; Perkins 2008). The global efficiency of this stage largely depends on the heat transfer to reactant (direct irradiation is preferred) and on the separation of $Zn_{(g)}$ and O_2 to prevent recombination (Chambon et al. 2010a). A rapid quenching is generally retained as the preferable solution (zinc vaporization temperature being of about 1,200 K and its recrystallization following a heterogeneous process from nucleation sites), although the energy thus extracted from the products cannot be reused in the process.

Zn reoxidation takes place in two stages: a first rapid stage corresponding to the oxidation of the available zinc surface, followed by a second longer stage that is limited by the diffusion inside zinc particles (passivation of particle surface). The size of particles has a direct impact on the reaction rate, finer particles limiting the diffusion mechanism (Ernst et al. 2009; Abanades 2012a). In the case of an oxidation involving only H_2O (Chambon et al. 2009; Lv et al. 2010) or CO_2 (Abanades and Chambon 2010; Loutzenhiser et al. 2011), the oxidizer concentration has a small influence on the kinetics, unlike temperature. The main difference between the reactions with H_2O and CO_2 is in terms of kinetics: the activation energies of the second step were estimated respectively at 87.7±7 kJ/mol (Ernst et al. 2009) and 162±25.3 kJ/mol (Loutzenhiser et al. 2009) for similar particles. Despite this strong disparity, the co-reduction of H_2O and CO_2 was studied (Stamatiou et al. 2010). A strong connection between the input gas composition and the quality of the generated syngas was observed. The oxidation of gaseous Zn by CO_2 or H_2O was studied (Melchior et al. 2009), showing 10 times higher kinetics than for solid Zn. Despite this performance, the generalization of this technique is unlikely, because being difficult to implement. Although it is generated in gaseous phase, Zn must be condensed to prevent recombination and allow its separation from O_2 and the energy released during its condensation is lost through quenching.

The main characteristics of this cycle resulting from the previous studies are:

– the temperature of reduction step is high (above 2,000 K);

– the quenching of gases from >2,000 K to >1,200 K is a key stage: its quality is decisive for the next step (size of zinc particles and purity), and energy efficiency of the process;

– the reoxidation step is limited by a passivation phenomenon due to the formation of a zinc oxide layer.

The addition of carbon to zinc oxide allows the decrease in the temperature of the reduction step. Indeed, carbon is a reducer that facilitates the reaction, generating CO from the first step (Wieckert et al. 2007). The first step of the cycle becomes:

$$ZnO + C \rightarrow Zn(g) + CO, \Delta G<0 \text{ for } T>1{,}230 \text{ K} \quad [11.6]$$

The carbon sources studied were methane (Kräupl and Steinfeld 2003) and coals (Osinga et al. 2004a) (graphite, charcoal, petroleum coke, activated carbon). The type of coal is essential for the reduction, charcoal and activated carbon being the best reducers. The retained reaction mechanism is composed of two solid-gas reactions (Osinga et al. 2004b):

$$ZnO + CO \rightarrow Zn(g) + CO_2 \quad [11.7]$$

$$CO_2 + C(s) \leftrightarrow 2CO \quad [11.8]$$

SnO₂/SnO. Since tin can be oxidized to several degrees, +II or +IV, several redox pairs can be considered: SnO_2/SnO, SnO_2/Sn and SnO/Sn. A first cycle with three steps was patented in 1971 by D. Souriau aiming to use the heat released by a nuclear reactor as a source of energy. The process involves the high temperature reduction of SnO_2 into SnO (equation [11.9]), which disproportionates at 600°C into liquid metallic tin and SnO_2 (equation [11.10]). Then, liquid tin reacts with water to produce hydrogen and regenerate the oxide (equation [11.11]):

$$SnO_2 \rightarrow SnO + \frac{1}{2}O_2 \quad [11.9]$$

$$2SnO \rightarrow Sn(l) + SnO_2 \quad [11.10]$$

$$Sn(l) + 2H_2O \rightarrow SnO_2 + 2H_2 \quad [11.11]$$

In a study conducted by Charvin et al. (2008b), although dismutation occurs, the formation of two phases was not observed, which impedes the recovery of Sn(l) and the rest of the process becomes obsolete. Since then, the SnO_2/SnO cycle was highlighted (Abanades et al. 2008; Abanades 2012b; Levêque et al. 2014): the first step corresponds to that of Souriau cycle, while the second one ignores the dismutation stage and uses directly SnO as reducer:

$$SnO_2 \rightarrow SnO(g) + \frac{1}{2}O_2 \qquad [11.12]$$

$$SnO + \alpha CO_2 + (1-\alpha)H_2O \rightarrow SnO_2 + \alpha CO + (1-\alpha)H_2, \ \ 0 < \alpha < 1 \qquad [11.13]$$

It was indeed shown that metallic tin is not necessarily more efficient than SnO for water reduction (Charvin et al. 2008b). This cycle was patented for the production of hydrogen by PROMES (WO 2008113944).

Similar to zinc, product recombination is a limiting factor for the reduction step (Chambon et al. 2010a). Reducing the pressure in the reactor chamber facilitates the production of gaseous species according to Le Chatelier's principle, and allows a decrease in the temperature required for the reaction (1,900 K at 20 kPa compared to a theoretical temperature of 2,342 K at atmospheric pressure). The dilution by a neutral gas allows the decrease in the partial pressure of O_2, and therefore limits recombination. Similarly, the temperature required for the reaction is also reduced. Finally, a rapid quenching of gases allows the condensation of SnO and the decrease of temperature below the reaction temperature. Condensation follows a heterogeneous process from the nucleation sites.

During the reoxidation step, several factors should be taken into account. The composition of the reduced oxide powder may vary depending on the operating conditions (SnO or Sn/SnO_2). Indeed, partial SnO dismutation may occur during the rise in temperature, or during the cooling in the first step (Levêque et al. 2014). Similarly, powder morphology plays a predominant role, the reoxidation reaction being a solid/gas reaction. From this point of view, the recovery of SnO by condensation makes it possible to obtain powders composed of grains of the order of several dozen nanometers, which yields significant specific surface area (Abanades 2012b). Finally, the reduction of CO_2 or H_2O is not always responsive to the same kinetics (Abanades 2012b). H_2O is reduced from 798 K with an activation energy of 51 +/− 7 kJ/mol, while CO_2 is only reduced from 973 K, with an activation energy of 88 +/− 7 kJ/mol.

Several advantages with respect to the zinc cycle are listed (Abanades et al. 2008) as follows:

– The temperature of reduction is below that of the zinc cycle, and may be further lowered by the pressure decrease inside the chamber. The decrease in the operating temperature directly implies higher energy efficiency, and therefore a better economic viability of the process.

– The oxide dissociation rate is significant, and is less dependent on the gas quenching at the reactor outlet than for zinc (a 72% pure product is obtained, compared to 48% for ZnO; see Chambon et al. 2011). Indeed, the boiling point of zinc is 1,180 K, while that of SnO is 1,800 K, condensation of the latter being therefore facilitated.

– Finally, SnO is a good water reducer.

11.3.3. *Non-volatile oxide cycles*

Non-volatile oxide cycles have the advantage that they do not form gaseous species during the reduction step. This prevents any manipulation of the material between the steps of the cycle, and the two reactions (reduction and hydrolysis) can take place in the same reactor with only one temperature alternation if the phase change (melting) does not occur. The first non-volatile oxide cycle proposed was the cycle based on iron oxides (magnetite/wüstite). The magnetite (Fe_3O_4) is reduced at a temperature of 2,100 K and is then in a liquid form (melting at 1,811 K). The hydrolysis step occurs at a minimal temperature of 450 K (Nakamura 1977). Experimental works have shown a total reduction under inert gas, while the hydrolysis step has a good reactivity, which is limited by the formation of a passivating oxide layer (Charvin et al. 2007; Abanades and Villafán-Vidales 2011). Moreover, the experiments revealed the necessity of quenching after magnetite reduction to prevent the dismutation of wüstite into iron and magnetite at a temperature below 570°C. The theoretical energy efficiency of this cycle is 35% (based on the HHV of H_2):

$$Fe_3O_4 \rightarrow 3FeO + \frac{1}{2} O_2 \qquad [11.14]$$

$$3FeO + H_2O \rightarrow Fe_3O_4 + H_2 \qquad [11.15]$$

In order to reduce the temperature of the cycle, mixed compounds were developed, based on the formation of a solid solution of iron oxide with a M_3O_4/MO redox pair having better reducibility at lower temperature and a better thermal stability of the material remaining in solid phase (Tamaura et al. 1999). The $M_xFe_{3-x}O_4$/ $M_xFe_{1-x}O$ redox pair thus obtained can be reduced at a lower temperature and remains in solid phase during the high-temperature step. The dopant is inserted in the spinel structure of the magnetite thus forming a ferrite. The reduction is further facilitated as the doping induces a structural deformation of the spinel structure, which supports oxygen mobility. Various dopants were tested, such as Ni, Mn, Zn, Co, Cu and Mg.

Another cycle relies on the CeO_2/Ce_2O_3 system (Abanades and Flamant 2006). The two reactions of the cycle are the thermal reduction of cerium +IV into cerium +III followed by hydrolysis:

$$2\ CeO_2 \rightarrow Ce_2O_3 + \tfrac{1}{2}\ O_2 \qquad [11.16]$$

$$Ce_2O_3 + H_2O \rightarrow 2\ CeO_2 + H_2 \qquad [11.17]$$

A complete reduction in liquid phase can be obtained at 1,950°C under reduced pressure, Ce_2O_3 has very good reactivity (100% conversion) and a rapid kinetics during the hydrolysis from 400°C (complete reoxidation in 5 minutes compared to 30% in 120 minutes for FeO), and the size of particles has no or small influence thanks to the diffusion of oxygen in the crystal. Nevertheless, the temperature of reduction remains high, and a partial sublimation of CeO_2 was observed. The energy efficiency of this cycle is 40%. Ceria doping with another metal can be considered in order to enhance the ceria reduction rate and achieve this step at lower temperatures (Le Gal and Abanades 2011).

11.3.4. *Non-stoichiometric oxide cycles*

Ferrite-based cycles

Ferrites were focused by the first studies on the subject (Tamaura et al. 1999), and are among the only ones to have reached the stage of a 100 kW demonstrator (Roeb et al. 2011). The Fe_3O_4/FeO cycle is difficult to use as such. On the contrary, besides the fact that the temperature required for Fe_3O_4 reduction is extremely high, it is above its melting point, therefore the reaction occurs in liquid phase. The product will then need to be ground in order to have a certain specific surface area. Moreover, the resulting FeO dismutes into Fe_3O_4 and Fe above 570°C; the product must therefore be quenched. A method to overcome these limitations is to substitute some iron atoms in the initial crystal with various metals, the idea being to add a more readily reducible metal, which is however less reactive to water (case of cobalt). The resulting doped ferrites (of $(Fe_{1-x}M_x)_3O_4$ type, where M = Mn, Co, Mg, Ni, Zn, etc.), can be partially reduced to the $(Fe_{1-x}M_x)_3O_{4-\delta}$ form at lower temperatures, and remain reactive to water.

After the various studies conducted on the subject, nickel ferrites appear to have the highest performances in terms of hydrogen production and thermal stability for a reduction temperature below 1,500°C (Fresno et al. 2009). However, no ferrite has sufficient cycling capacity due to inevitable sintering. To avoid this phenomenon, the coating of a ceramic substrate was widely studied. The results indicate that the substrates with the highest performances are yttrium-stabilized zirconia (YSZ, Gokon et al. 2008) and monoclinic zirconia (m-ZrO_2, Gokon et al. 2009), the reduction rates remaining below 30%. Since the substrate may account for 90% of the weight of the reactant material, the hydrogen mass productivity of this type of solution is low.

Ceria-based cycles

Cerium oxide (ceria, CeO_2) is widely used in the car industry as a catalyst for the oxidation of exhaust gases. This use relies on the significant diffusion of oxygen ion in the crystal, which may also be beneficial for the hydrolysis step. The first ceria-based cycle was proposed in 2006 (Abanades and Flamant 2006). By lowering the temperature of reduction, a partial reduction of ceria occurs (or non-stoichiometric reduction), which involves the $CeO_2/CeO_{2-\delta}$ system (Chueh et al. 2010). If ceria is doped with a trivalent or tetravalent cationic element, the reduction capacity of the material can bc increased. Various works focused on the formulation of doped ceria, their synthesis and implementation, and have, for example, proven the interest of doping with zirconia (Le Gal et Abanades 2012; Le Gal et al. 2013). By comparison, the studied nickel-doped ferrites allowed a maximal production of 250 $\mu mol/g_{oxide}$, compared to 333 $\mu mol/g_{oxide}$ for the ceria-zirconia and better results in terms of cycling and kinetics of hydrogen production. The structure of the material is essential, and so is its stability at high temperature (Haeussler et al. 2019; Pullar et al. 2019). These two properties may result from the method used for the powder synthesis or its a posteriori structuring. An interesting concept has recently been evaluated based on the use of ceria in the form of porous structures, that is, reticulated foams with dual-scale porosity (Furler et al. 2014). Macroscopic porosity allows the efficient heating of the entire reactant, itself microstructured to increase the specific surface area (Haeussler et al. 2020). The elaboration of ordered macroporous structures using biomimetics has also been studied (Oliveira et al. 2018).

Despite the recent progress, several points must still be developed in order to fully take advantage of the strengths of ceria-based cycles:

– The formulation of ceria can be further improved (addition of several dopants, use of composite materials, etc.).

– In case the powder is used as such, synthesis methods ensuring high-temperature resistance are needed.

– If the reactant is structured, resistant substrates and efficient coating methods or architectured porous structures elaboration methods need to be developed.

Perovskite structure-based cycles

The perovskite structure is generally denoted by ABO_3, where A is a cation whose size is greater than that of B. Moreover, A is bonded to 12 oxygen atoms and B to 6 (Figure 11.4a). Its main advantage is the possibility of substituting certain A and B sites by other atoms and alter its stability. It is therefore possible to obtain compounds of the type $A_{1-x}A'_xB_{1-y}B'_yO_3$ (0<x,y<1) with finely adjusted properties, as needed.

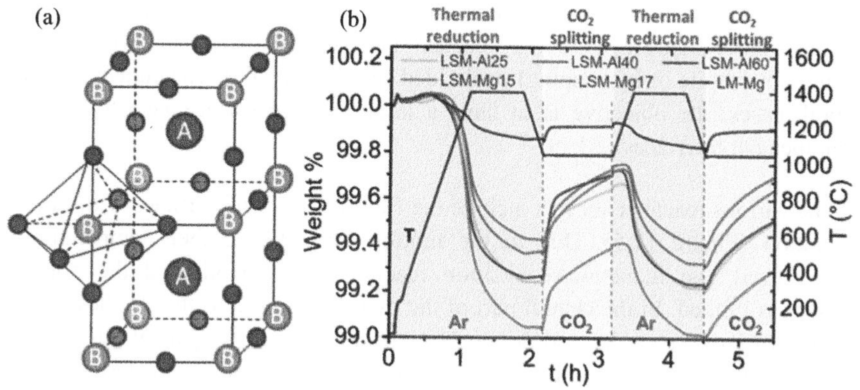

Figure 11.4. *(a) ABO_3 perovskite structure; (b) reactivity of LSM perovskites doped with Al or Mg at B site during two successive cycles (reduction under Ar at 1,400°C and oxidation under CO_2 at 1,050°C). For a color version of this figure, see www.iste.co.uk/flamant/energy.zip*

By carefully choosing the atoms filling the main sites and the doping cations, it is possible to obtain deformations of the structure and of the electronic configuration of the crystal that may facilitate the displacement and the escape of the oxygen from it. The combination of this characteristic with the electronic stability of the structure allows the consideration of a significant oxygen non-stoichiometry. Materials with adjustable reduction and oxygen diffusion capacities that are higher than those of ferrites and ceria can therefore be considered (Demont et al. 2014; Haeussler et al. 2018).

Perovskite formulations such as $La_xSr_{1-x}MnO_3$ (LSM) were studied (production of H_2 or CO above 200 µmol/g, for a temperature of reduction of 1,400°C, Figure 11.4b), showing a hydrogen production potential higher than that of CeO_2 (Demont and Abanades 2015; Nair and Abanades 2016, 2018). These results remain comparable to those obtained for doped ceria, but they show a strong development potential.

11.3.5. *Solar reactor concepts for cycle implementation*

Two main families of reactors can be considered: reactors allowing the two steps of the cycle to proceed in a single chamber and those whose reduction and reoxidation functions are decoupled. Integrating solar heat is essential for obtaining an optimal efficiency of the reactor.

Single-chamber reactors

This type of reactor groups the majority of technologies for the implementation of non-volatile oxide cycles. The reactant material can be either structured (by coating a substrate or by shaping) or divided (pure material or coated as thin film). In both cases, the objective is to have a large specific surface area to facilitate phenomena at interfaces.

The various reactor concepts include the fluidized beds of particles with internal circulation (Figure 11.5). Due to the motion of particles, their temperature is homogenized, the reduction/reoxidation reactions are smooth and the potential sintering is limited. In the central part of the cylindrical reactor, the gas flow allows the fluidization of particles and drives them towards the higher part of the bed. The gas flow rate is adapted to the size and density of the particles (determining the conditions required for their fluidization), so that the upper part of the bed is at the focal point of the concentration system, allowing the reduction. Once arrived at the irradiation zone, the particles go back down to the periphery at the bottom of the bed, transmitting a part of the heat from the upper part to the lower part. A better vertical homogenization of the temperature can be obtained compared to the classical beds, and the flow rate of neutral gas required for fluidization is reduced to the central section. Once reduction is completed, irradiation stops, and a part of the neutral gas flow is replaced by water vapor. This type of reactor has, for example, been implemented with ZrO_2 particles covered with a nickel ferrite ($NiFe_2O_4$) (Gokon et al. 2008).

Another concept of monolithic reactor has been developed, focusing on solar-based production of hydrogen, with a 100 kW pilot project (Roeb et al. 2011). The reactant material (doped ferrite) is deposited on a honeycomb monolithic ceramic structure, which acts both as a volumetric radiation absorber and hydrolysis reactor (Figure 11.6). To ensure continuous operation, two similar reactors are positioned in parallel. While one of them is subjected to concentrated solar flux allowing reduction, the other one hosts the ferrite hydrolysis and reoxidation and vice versa. The reactors are therefore fixed, the steps being alternated under the control of heliostats. Other monolithic reactors integrating massive porous ceria structures (reticulated foams) have more recently been developed (Haeussler et al. 2020).

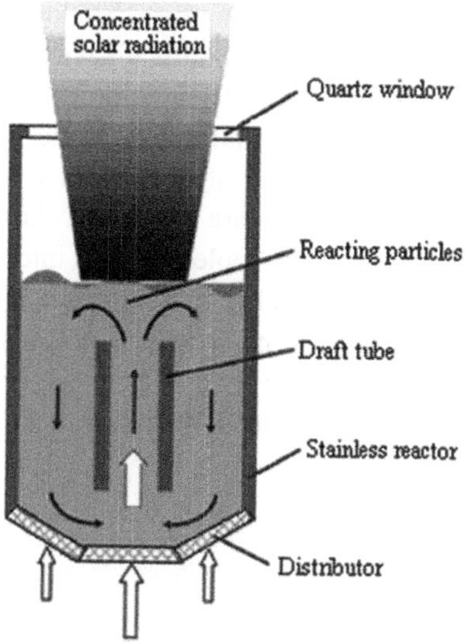

Figure 11.5. *Diagram of a fluidized bed with internal circulation*

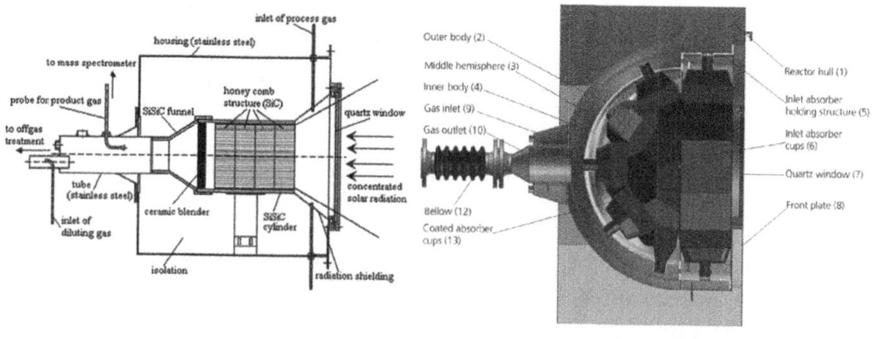

Figure 11.6. *Diagrams of monolithic reactors integrating a porous structure*

Decoupled reactors

This category covers all the reactors for which the solar receiver is used in the reduction step, H_2/CO being produced in a second reactor at lower temperature. The heating/cooling cycle of the cavity is avoided, this being specifically optimized for this reaction, and the assembly can potentially operate continuously. The most common example concerns the reactors for the reduction of volatile oxides.

A concept of rotating cavity reactor was developed (Koepf et al. 2016, Figure 11.7). Due to the rotation of the cylindrical cavity, the ZnO particles, fed in by a screw feeder in the axial part, are distributed on the walls, which prevents the formation of hot points due to the non-uniformity of the incident flux. A constant flow of gas is injected in the back face of the quartz window to protect it from particle deposition. The resulting gases are evacuated through an annular section around the screw feeder, where they are cooled by bulk injection of argon at ambient temperature (quench). The zinc particles (50% purity) are finally recovered by filtration. Another concept of rotating reactor was also developed allowing continuous operation with injection of ZnO particles and recovery of Zn products in the form of a fine powder at the outlet (Abanades et al. 2007; Chambon et al. 2010b).

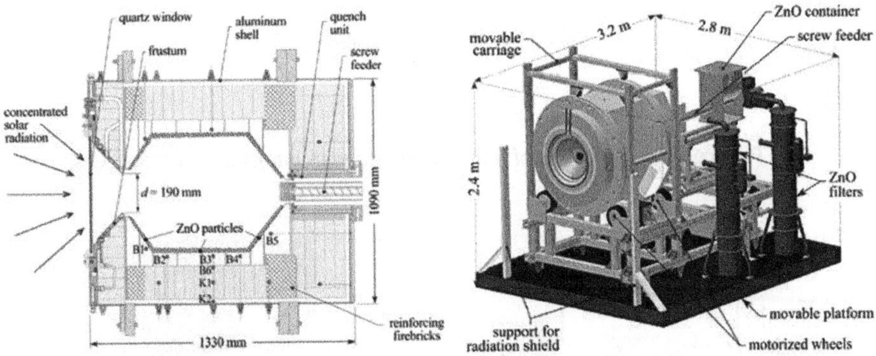

Figure 11.7. *100 kW reactor for ZnO dissociation*

The highest power prototype for the reduction of metal oxide was built during the Solzinc European project (Wieckert et al. 2007) (Figure 11.8). This reactor implements the carboreduction of zinc oxide, thus reducing notably the thermal constraints and the technological difficulties of the gas quench to the detriment of carbon balance.

The reactor is composed of two cavities. The upper part of the reactor, composed of SiC plates, receives and absorbs radiation and re-emits it to the lower part containing a packed bed of particles, heated by the radiation of the separation plate. This allows a significant production of reduced particles without risking their deposition on the quartz window of the reactor. A 50 kg/h production of 95% pure zinc was obtained, for an operating temperature of 1,300–1,500 K and a 30% efficiency of the process.

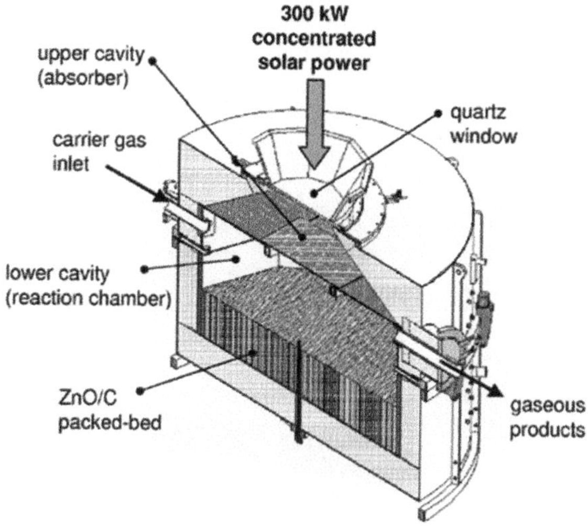

Figure 11.8. *300 kW reactor for the carboreduction of ZnO*

A different concept of reactor with vertical irradiation axis was proposed for the thermal dissociation of ZnO (gravity-fed solar thermochemical reactor; Koepf et al. 2012, Figure 11.9). To homogenize the distribution of particles in the reactor, particles are fed through 15 distributors located on the circumference of the reactor. By gravity, the particles must form a curtain falling on the plates of refractory material composing the cavity, whose slope and surface were chosen for this purpose. The particles must be reduced and vaporized during their progression, the gases being driven by a revolving flow of neutral gas towards the outlet at the center of the reactor. The particles that did not react are recovered in the annular part between this gas outlet and the lower part of the reactor.

The PROMES laboratory has also developed technological solutions adapted to the continuous reduction of volatile oxides. A first set-up with a rotating cavity was assembled (Chambon et al. 2010b). One of the objectives and the originality of the system was the possibility of having reactions under reduced pressure in order to increase the kinetics (Levêque and Abanades 2015). A second different prototype with vertical irradiation axis was developed, operating also at reduced pressure (Chambon et al. 2011). The reactant, in compressed form (pellets), is continuously fed into the insulated cavity through the lower part and is simultaneously heated by the direct concentrated radiation and the radiation from the cavity walls (Figure 11.10). The main advantage of this reactor is reduced distance between the reduction zone and the evacuation connected to the particle filtration system, which limits the recombination. The powders recovered at the outlet contain approximately 40%

reduced species for ZnO and 70% for SnO_2, for an operating pressure of about 20 kPa and an efficiency of solar energy into chemical energy conversion estimated at 2–3%.

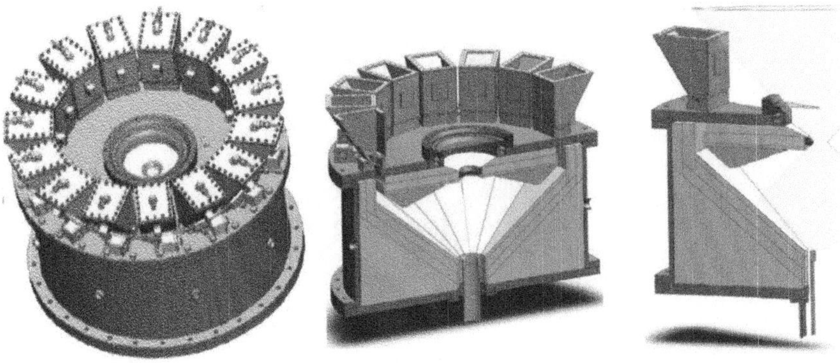

Figure 11.9. *Diagram of the reactor with gravity-driven injection of particles. For a color version of this figure, see www.iste.co.uk/flamant/energy.zip*

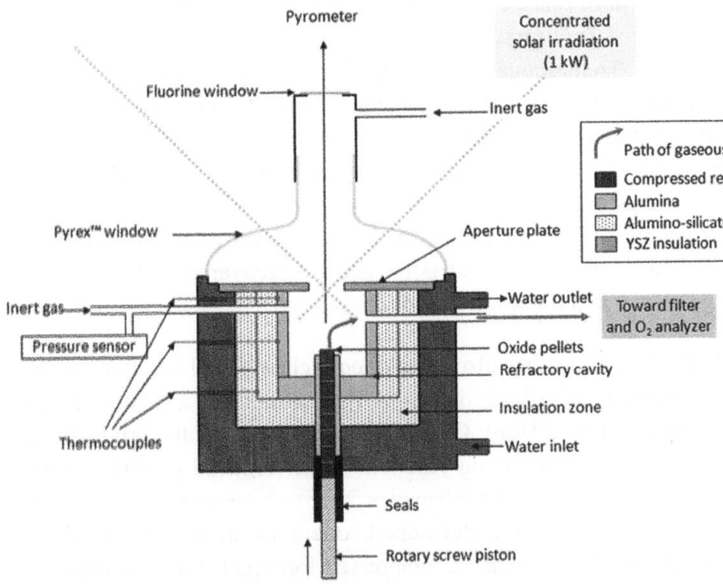

Figure 11.10. *Reactor for the dissociation of volatile oxides (ZnO, SnO_2) with continuous injection of reactant in compressed form. For a color version of this figure, see www.iste.co.uk/flamant/energy.zip*

For the purpose of integrating heat between the two steps and of continuous operation, two reactors with solid oxides were developed, with integrated displacement of the reactant from one chamber to another. The CR5 concept (Counter-Rotating-Ring Receiver/Reactor/Recuperator, Figure 11.11a) uses two chambers separated by a stack of rotating rings (Diver et al. 2008). Rotation drives the ring surface covered with metal oxide from the reduction zone to the oxidation zone; the hot part exiting the reduction zone of one reheats the colder parts getting in of its two neighbors, allowing a 30% heat recovery. The challenge of this concept resides in this assembly of rings, which must also separate the two chambers one from the other. Although the concept was experimentally validated with a prototype using cobalt ferrites, there are not sufficient data available to assess its actual efficiency.

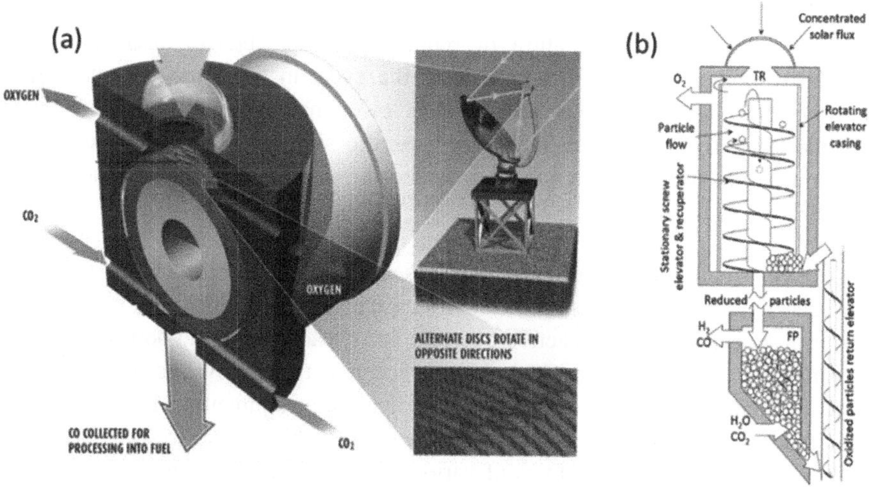

Figure 11.11. *(a) Diagram of CR5 reactor and (b) prototype of circulating dense bed reactor. For a color version of this figure, see www.iste.co.uk/flamant/energy.zip*

In the other concept (Figure 11.11b, Ermanoski et al. 2013), the particles are driven by a stationary screw towards the upper part of the reactor where they are subjected to concentrated radiation. Due to gravity, they go back down through the hollow axis of the screw, which acts as a heat exchanger between upward and downward particles, to the reoxidation chamber. The authors highlight the possibility of continuously operating, unlike the fluidized bed, which allows only batch operation. This concept was however not proven, and achieving tightness due to stacked particles seems difficult, similar to the transportation of particles to the high temperature zone.

11.4. Conclusion

The production of hydrogen and the recycling of CO_2 by the thermochemical cycles coupled to a source of concentrated solar energy is a relevant approach. Radiation is converted into heat with a reasonable efficiency, and then it is directly used to supply the thermochemical cycle. This involves a better global efficiency of the conversion of the solar resource into fuel than the paths using electricity as intermediate energy.

The volatile oxide cycles have a double advantage of relatively high specific productivity, and of their specific surface area being regenerated at each cycle thanks to the vaporization/condensation phenomenon. However, recombination is not easily preventable during cooling. ZnO/Zn and SnO_2/SnO cycles are the most attractive. Non-volatile oxide cycles mainly relate to non-stoichiometric oxides. They allow a lower temperature of reduction, phase change absence and the possibility of having the two reactions of the cycle in the same reactor without transfer of reactant. However, they have a lower mass productivity than for the stoichiometric oxides, a possible loss of specific surface area during the cycling (sintering), and the necessity to optimize the formulation and the method for the synthesis of materials.

The first materials studied were iron-based, but given the limitations related to their deactivation or low mass productivity, research on cerium oxides, either doped or not, has been particularly active over the last decade due to a good reactivity of the material with H_2O or CO_2. More recently, the perovskite materials are particularly drawing attention due to a significant development potential related to their redox properties (large substoichiometry accessible thanks to the mobility of oxygen) and to many possibilities of various formulations (high flexibility offering the possibility of incorporating many dopants). The transport properties of oxygen in cerium- and perovskite-based materials are particularly used in the H_2O and CO_2 dissociation cycles. Research is being carried on in various fields related to the formulation (choice of chemical/crystalline characteristics), synthesis and forming of materials (elaboration of cycling-stable macroscopic structures) for their integration in solar reactors. Extreme operating conditions (high temperatures and radiative fluxes) and the importance of the management of thermal losses should be taken into account in the design of receivers/reactors in order to optimize their energy and thermochemical efficiency. Finally, the system scale up and integration in the concentrated solar power plants must be taken into consideration for a production of clean fuels relying exclusively on the dissociation of water and carbon dioxide.

11.5. References

Abanades, S. (2012a). Thermogravimetry analysis of CO_2 and H_2O reduction from solar nanosized Zn powder for thermochemical fuel production. *Industrial & Engineering Chemistry Research*, 51(2), 741–750.

Abanades, S. (2012b). CO_2 and H_2O reduction by solar thermochemical looping using SnO_2/SnO redox reactions: Thermogravimetric analysis. *International Journal of Hydrogen Energy*, 37(10), 8223–8231.

Abanades, S. (2019). Metal oxides applied to thermochemical water-splitting for hydrogen production using concentrated solar energy. *ChemEngineering*, 3(3), 63.

Abanades, S. and Chambon, M. (2010). CO_2 dissociation and upgrading from 2-step solar thermochemical processes based on ZnO/Zn and SnO_2/SnO redox pairs. *Energy & Fuels*, 24(12), 6667–6674.

Abanades, S. and Flamant, G. (2006). Thermochemical hydrogen production from a two-step solar-driven water-splitting cycle based on cerium oxides. *Solar Energy*, 80(12), 1611–1623.

Abanades, S. and Villafán-Vidales, H.I. (2011). CO_2 and H_2O conversion to solar fuels via two-step solar thermochemical looping using iron oxide redox pair. *Chemical Engineering Journal*, 175, 368–375.

Abanades, S., Charvin, P., Flamant, G., Neveu, P. (2006). Screening of water-splitting thermochemical cycles potentially attractive for hydrogen production by concentrated solar energy. *Energy*, 31(14), 2805–2822.

Abanades, S., Charvin, P., Flamant, G. (2007). Design and simulation of a solar chemical reactor for the thermal reduction of metal oxides: Case study of zinc oxide dissociation. *Chemical Engineering Science*, 62(22), 6323–6333.

Abanades, S., Charvin, P., Lemont, F., Flamant, G. (2008). Novel two-step SnO_2/SnO water-splitting cycle for solar thermochemical production of hydrogen. *International Journal of Hydrogen Energy*, 33(21), 6021–6030.

Centi, G. and Perathoner, S. (2011) CO_2-based energy vectors for the storage of solar energy. *Greenhouse Gases: Science and Technology*, 1(1), 21–35.

Chambon, M., Abanades, S., Flamant, G. (2009). Kinetic investigation of hydrogen generation from hydrolysis of SnO and Zn solar nanopowders. *International Journal of Hydrogen Energy*, 34(13), 5326–5336.

Chambon, M., Abanades, S., Flamant, G. (2010a). Solar thermal reduction of ZnO and SnO_2: Characterization of the recombination reaction with O_2. *Chemical Engineering Science*, 65(11), 3671–3680.

Chambon, M., Abanades, S., Flamant, G. (2010b). Design of a lab-scale rotary cavity-type solar reactor for continuous thermal reduction of volatile oxides under reduced pressure. *ASME Journal of Solar Energy Engineering*, 132(2), 021006.

Chambon, M., Abanades, S., Flamant, G. (2011). Thermal dissociation of compressed ZnO and SnO_2 powders in a moving front solar thermochemical reactor. *AIChE Journal*, 57(8), 2264–2273.

Charvin, P., Abanades, S., Flamant, G., Lemort, F. (2007). Two-step water-splitting thermochemical cycle based on iron oxide redox pair for solar hydrogen production. *Energy*, 32(7), 1124–1133.

Charvin, P., Abanades, S., Lemort, F., Flamant, G. (2008a). Analysis of solar processes for hydrogen production from water-splitting thermochemical cycles. *Energy Conversion and Management*, 49(6), 1547–1556.

Charvin, P., Abanades, S., Lemont, F., Flamant, G. (2008b). Experimental study of SnO_2/SnO/Sn thermochemical systems for solar production of hydrogen. *AIChE Journal*, 54(10), 2759–2767.

Chueh, W.C., Falter, C., Abbott, M., Scipio, D., Furler, P., Haile, S.M., Steinfeld, A. (2010). High-flux solar-driven thermochemical dissociation of CO_2 and H_2O using nonstoichiometric ceria. *Science*, 330(6012), 1797–1801.

Demont, A. and Abanades, S. (2015). Solar thermochemical conversion of CO_2 into fuel via two-step redox cycling of non-stoichiometric Mn-containing perovskite oxides. *Journal of Materials Chemistry A*, 3, 3536–3546.

Demont, A., Abanades, S., Beche, E. (2014). Investigation of perovskite structures as oxygen-exchange redox materials for hydrogen production from thermochemical two-step water-splitting cycles. *The Journal of Physical Chemistry C*, 118(24), 12682–12692.

Diver, R.B., Miller, J.E., Allendorf, M.D., Siegel, N.P., Hogan, R.E. (2008). Solar thermochemical water-splitting ferrite-cycle heat engines. *Journal of Solar Energy Engineering*, 130(4), 041001.

Dry, M.E. (2002). The Fischer–Tropsch process: 1950–2000. *Catalysis Today*, 71(3–4), 227–241.

Ermanoski, I., Siegel, N.P., Stechel, E.B. (2013). A new reactor concept for efficient solar-thermochemical fuel production. *Journal of Solar Energy Engineering*, 135(3), 031002.

Ernst, F., Steinfeld, A., Pratsinis, S. (2009). Hydrolysis rate of submicron Zn particles for solar H_2 synthesis. *International Journal of Hydrogen Energy*, 34(3), 1166–1175.

Espinoza, R.L., Steynberg, A.P., Jager, B., Vosloo, A.C. (1999). Low temperature Fischer–Tropsch synthesis from a Sasol perspective. *Applied Catalysis A: General*, 186(1), 13–26.

Fresno, F., Fernández-Saavedra, R., Belén Gómez-Mancebo, M., Vidal, A., Sánchez, M., Rucandio, M., Quejido, A.J., Romero, M. (2009). Solar hydrogen production by two-step thermochemical cycles: Evaluation of the activity of commercial ferrites. *International Journal of Hydrogen Energy*, 34(7), 2918–2924.

Funk J.E. and Reinstrom R.M. (1966). Energy requirements in production of hydrogen from water. *Industrial & Engineering Chemistry Process Design and Development*, 5(3), 336–342.

Furler, P., Scheffe, J., Marxer, D., Gorbar, M., Bonk, A., Vogt, U., Steinfeld, A. (2014). Thermochemical CO_2 splitting via redox cycling of ceria reticulated foam structures with dual-scale porosities. *Physical Chemistry Chemical Physics*, 16(22), 10503.

Gálvez, M.E., Loutzenhiser, P.G., Hischier, I., Steinfeld, A. (2008). CO_2 splitting via two-step solar thermochemical cycles with Zn/ZnO and FeO/Fe_3O_4 redox reactions: Thermodynamic analysis. *Energy & Fuels*, 22(5), 3544–3550.

Gokon, N., Hasegawa, T., Takahashi, S., Kodama, T. (2008a). Thermochemical two-step water-splitting for hydrogen production using Fe-YSZ particles and a ceramic foam device. *Energy*, 33(9), 1407–1416.

Gokon, N., Takahashi, S., Yamamoto, H., Kodama, T. (2008b). Thermochemical two-step water-splitting reactor with internally circulating fluidized bed for thermal reduction of ferrite particles. *International Journal of Hydrogen Energy*, 33(9), 2189–2199.

Gokon, N., Murayama, H., Nagasaki, A., Kodama, T. (2009). Thermochemical two-step water splitting cycles by monoclinic ZrO_2-supported $NiFe_2O_4$ and Fe_3O_4 powders and ceramic foam devices. *Solar Energy*, 83(4), 527–537.

Graf, D., Monnerie, N., Roeb, M., Schmitz, M., Sattler, C. (2008). Economic comparison of solar hydrogen generation by means of thermochemical cycles and electrolysis. *International Journal of Hydrogen Energy*, 33(17), 4511–4519.

Haeussler, A., Abanades, S., Jouannaux, J., Julbe, A. (2018). Non-stoichiometric redox active perovskite materials for solar thermochemical fuel production: A review. *Catalysts*, 8(12), 611.

Haeussler, A., Abanades, S., Jouannaux, J., Drobek, M., Ayral, A., Julbe, A. (2019). Recent progress on ceria doping and shaping strategies for solar thermochemical water and CO_2 splitting cycles. *AIMS Materials Science*, 6(5), 657–684.

Haeussler, A., Abanades, S., Julbe, A., Jouannaux, J., Cartoixa, B. (2020). Solar thermochemical fuel production from H_2O and CO_2 splitting via two-step redox cycling of reticulated porous ceria structures integrated in a monolithic cavity-type reactor. *Energy*, 201, 117649.

Jensen, R., Lyman, J., King, J., Guettler, R. (2000). Solar reduction of CO_2. United States Patent, Patent Number: 6.066.187, Date of Patent: May 23.

Koepf, E., Advani, S.G., Steinfeld, A., Prasad, A.K. (2012). A novel beam-down, gravity-fed, solar thermochemical receiver/reactor for direct solid particle decomposition: Design, modeling, and experimentation. *International Journal of Hydrogen Energy*, 37(22), 16871–16887.

Koepf, E., Villasmil, W., Meier, A. (2016). Pilot-scale solar reactor operation and characterization for fuel production via the Zn/ZnO thermochemical cycle. *Applied Energy*, 165, 1004–1023.

Kräupl, S. and Steinfeld, A. (2003). Operational performance of a 5-kW solar chemical reactor for the co-production of zinc and syngas. *Journal of Solar Energy Engineering*, 125(1), 124.

Le Gal, A. and Abanades, S. (2011). Catalytic investigation of ceria–zirconia solid solutions for solar hydrogen production. *International Journal of Hydrogen Energy*, 36(8), 4739–4748.

Le Gal, A. and Abanades, S. (2012). Dopant incorporation in ceria for enhanced water-splitting activity during solar thermochemical hydrogen generation. *The Journal of Physical Chemistry C*, 116(25), 13516–13523.

Le Gal, A., Abanades, S., Bion, N., Le Mercier, T., Harle, V. (2013). Reactivity of doped ceria-based mixed oxides for solar thermochemical hydrogen generation via two-step water-splitting cycles. *Energy & Fuels*, 27(10), 6068–6078.

Levêque, G. and Abanades, S. (2013). Kinetic analysis of high-temperature solid-gas reactions by an inverse method applied to ZnO and SnO_2 solar thermal dissociation. *Chemical Engineering Journal*, 217, 139–149.

Levêque, G. and Abanades, S. (2015). Investigation of thermal and carbothermal reduction of volatile oxides (ZnO, SnO_2, GeO_2, and MgO) via solar-driven vacuum thermogravimetry for thermochemical production of solar fuels. *Thermochimica Acta*, 605, 86–94.

Levêque, G., Abanades, S., Jumas, J.-C., Olivier-Fourcade, J. (2014). Characterization of two-step tin-based redox system for thermochemical fuel production from solar-driven CO_2 and H_2O splitting cycle. *Industrial & Engineering Chemistry Research*, 53(14), 5668–5677.

Loutzenhiser, P.G., Gálvez, M.E., Hischier, I., Stamatiou, A., Frei, A., Steinfeld, A. (2009). CO_2 splitting via two-step solar thermochemical cycles with Zn/ZnO and FeO/Fe_3O_4 redox reactions II: Kinetic analysis. *Energy & Fuels*, 23(5), 2832–2839.

Loutzenhiser, P.G., Meier, A., Steinfeld, A. (2010). Review of the two-step H_2O/CO_2-splitting solar thermochemical cycle based on Zn/ZnO redox reactions. *Materials*, 3(11), 4922–4938.

Loutzenhiser, P.G., Barthel, F., Stamatiou, A., Steinfeld, A. (2011). CO_2 reduction with Zn particles in a packed-bed reactor. *AIChE Journal*, 57(9), 2529–2534.

Lv M., Zhou J., Yang W., Cen K. (2010). Thermogravimetric analysis of the hydrolysis of zinc particles. *International Journal of Hydrogen Energy*, 35(7), 2617–2621.

Melchior, T., Piatkowski, N., Steinfeld, A. (2009). H_2 production by steam-quenching of Zn vapor in a hot-wall aerosol flow reactor. *Chemical Engineering Science*, 64, 1095–1101.

Nair, M.M. and Abanades, S. (2016). Insights into the redox performance of non-stoichiometric lanthanum manganite perovskites for solar thermochemical CO_2 splitting. *ChemistrySelect*, 1(15), 4449–4457.

Nair, M.M. and Abanades, S. (2018). Experimental screening of perovskite oxides as efficient redox materials for solar thermochemical CO_2 conversion. *Sustainable Energy & Fuels*, 2, 843–854.

Nakamura, T. (1977). Hydrogen production from water utilizing solar heat at high temperatures. *Solar Energy*, 19(5), 467–475.

Oliveira, F.A.C., Barreiros, M.A., Abanades, S., Caetano, A.P.F., Novais, R.M., Pullar, R.C. (2018). Solar thermochemical CO_2 splitting using cork-templated ceria ecoceramics. *Journal of CO_2 Utilization*, 26, 552–563.

Osinga, T., Frommherz, U., Steinfeld, A., Wieckert, C. (2004a). Experimental investigation of the solar carbothermic reduction of ZnO using a two-cavity solar reactor. *Journal of Solar Energy Engineering*, 126(1), 633.

Osinga, T., Olalde, G., Steinfeld, A. (2004b). Solar carbothermal reduction of ZnO: Shrinking packed-bed reactor modeling and experimental validation. *Industrial & Engineering Chemistry Research*, 43(25), 7981–7988.

Perkins, C. (2008). Thermal ZnO dissociation in a rapid aerosol reactor as part of a solar hydrogen production cycle. *International Journal of Hydrogen Energy*, 33(2), 499–510.

Perkins, C., Lichty, P., Weimer, A.W. (2007). Determination of aerosol kinetics of thermal ZnO dissociation by thermogravimetry. *Chemical Engineering Science*, 62(21), 5952–5962.

Pullar, R.C., Novais, R.M., Caetano, A.P.F., Barreiros, M.A., Abanades, S., Costa Oliveira, F.A. (2019). A review of solar thermochemical CO_2 splitting using ceria-based ceramics with designed morphologies and microstructures. *Frontiers in Chemistry*, 7, 601.

Roeb, M., Säck, J.-P., Rietbrock, P., Prahl, C., Schreiber, H., Neises, M., de Oliveira, L., Graf, D., Ebert, M., Reinalter, W. et al. (2011). Test operation of a 100kW pilot plant for solar hydrogen production from water on a solar tower. *Solar Energy*, 85(4), 634–644.

Schunk, L.O. and Steinfeld, A. (2009). Kinetics of the thermal dissociation of ZnO exposed to concentrated solar irradiation using a solar-driven thermogravimeter in the 1800–2100 K range. *AIChE Journal*, 55(6), 1497–1504.

Stamatiou, A., Loutzenhiser, P.G., Steinfeld, A. (2010). Solar syngas production via H_2O/CO_2-splitting thermochemical cycles with Zn/ZnO and FeO/Fe_3O_4 redox reactions. *Chemistry of Materials*, 22(3), 851–859.

Steinfeld, A. (2002). Solar hydrogen production via a two-step water-splitting thermochemical cycle based on Zn/ZnO redox reactions. *International Journal of Hydrogen Energy*, 27(6), 611–619.

Steinfeld, A. and Palumbo, R. (2001). Solar thermochemical process technology. *Encyclopedia of Physical Science and Technology*, 15(1), 237–256.

Steynberg, A.P., Espinoza, R.L., Jager, B., Vosloo, A.C. (1999). High temperature Fischer–Tropsch synthesis in commercial practice. *Applied Catalysis A: General*, 186(1), 41–54.

Tamaura, Y., Ueda, Y., Matsunami, J., Hasegawa, N., Nesuka, M., Sano, T., Tsuji, M. (1999). Solar hydrogen production by using ferrites. *Solar Energy*, 65, 55–57.

Wieckert, C., Frommherz, U., Kräupl, S., Guillot, E., Olalde, G., Epstein, M., Santén, S., Osinga, T., Steinfeld, A. (2007). A 300 kW solar chemical pilot plant for the carbothermic production of zinc. *Journal of Solar Energy Engineering*, 129(2), 190.

List of Authors

Stéphane ABANADES
PROMES-CNRS
Font-Romeu
France

Françoise BATAILLE
PROMES-CNRS
UPVD
Perpignan
France

Dorian DUPUY
PROMES-CNRS
UPVD
Perpignan
and
CERFACS
Toulouse
France

Quentin FALCOZ
PROMES-CNRS
UPVD
Perpignan
France

Gilles FLAMANT
PROMES-CNRS
Fort Romeu
France

Benjamin GRANGE
PROMES-CNRS
Font-Romeu-Odeillo-Via
France

Stéphane GRIEU
PROMES-CNRS
UPVD
Perpignan
France

Antoine GROSJEAN
PROMES-CNRS
UPVD
Perpignan
France

François HÉNAULT
IPAG-CNRS
Grenoble
France

Sylvain RODAT
PROMES-CNRS
Font-Romeu
France

Audrey SOUM-GLAUDE
PROMES-CNRS
Fort Romeu
France

Stéphane THIL
PROMES-CNRS
UPVD
Perpignan
France

Adrien TOUTANT
PROMES-CNRS
UPVD
Perpignan
France

Aubin TOUZO
PROMES-CNRS
UPVD
Perpignan
France

Alexis VOSSIER
PROMES-CNRS
Font-Romeu
France

Joya ZEITOUNY
PROMES-CNRS
Font-Romeu
and
C2N, CNRS
Paris-Saclay University
Palaiseau
France

Index

A, C

D, E

F, G

H, I

L, M

N, O

P, R

S, T

V, W

Printed and bound by CPI Group (UK) Ltd, Croydon, CR0 4YY
13/09/2026
14960996-0006